The
Vector Calculus Companion, 2e

with over 325 worked-out examples

The Vector Calculus Companion, 2e
With over 325 worked-out examples

by Scott Surgent

Copyright © 2016, 2018, 2022 Scott Surgent. Printed and bound in the United States of America. All Rights Reserved. No part of this book may be reproduced or transmitted in any form or by any means, electronic or mechanical, including photocopying, recording, or by an information storage or retrieval system without permission in writing from the publisher. Short passages used for the purposes of reviews are permitted.

ISBN-13: 9798839547957

First edition published 2016.

Second (current) edition published July 2022.

www.surgent.net/vcbook

This is an original work, researched and written by the author.

Front cover art by the author.

Back cover image: Sundad Rock Art Site, Arizona (photo by the author)

Manuscript development assisted by www.pdfmerge.com.

Introduction

This book is intended to be a guide for students taking Multivariable and Vector Calculus. It came about as a result of my own experiences teaching this class, both in the traditional and on-line modalities. By this point, most students who have achieved the pre-requisites for this course have the necessary skills to be successful with this material. But exposure and practice are key components to eventual long-term success. This is how I see this book possibly being useful to students in this course.

Most textbooks are limited by space in how many examples can be included, and often, the examples that are included may skip steps, again in the interest of space. For my students, I would create extra worked-out examples, and it is these examples that I have collected and augmented into this book.

This book is **not** a textbook. It does not include proofs. By itself, this book is not sufficient to act in place of a traditional textbook. It could be viewed as an addendum to most textbooks that cover this material. In many sections, I go right to the main theorem or formula, assuming that the student's textbook will supply the proofs or exposition leading up to the main points.

Learning takes on many forms and everyone approaches learning from slightly different angles. It's fair to say that learning encompasses two broad themes: conceptual and mechanical. In the "concept", the main idea is presented, and how it can be applied to real-world problems. In the "mechanics", the actual grunt-work of working out the problems is encountered.

Some may claim that the concept should always come first, then the mechanics. Some may say it should be the other way. I believe that both can be developed simultaneously. The concept can be presented, then the student can try a couple starter problems, which may help cement the main concepts more solidly. In a sense, both the conceptual and mechanical development help the other. In an extreme case, it is possible that a student may be able to successfully work out problems of a certain topic, all the while being a little unclear on the concept. But is this necessarily a bad thing? Through repeated practice, the mechanics become more solid, and along the way, the student may suddenly "see" the concept.

This book is a "mechanics" book. I show hundreds of examples, concentrating on the steps, often showing the tedious algebra steps that many textbooks skip over or take for granted. The anecdotal feedback from students is that this is appreciated. I am convinced that the majority of students at this level understand the main concepts almost immediately, but then get bogged down by the algebra, the littlest of errors being the most frustrating roadblocks. Often, the errors are very minor: forgetting a negative sign, forgetting what $\sin \pi$ evaluates to, forgetting the rules about symmetry of certain functions or issues of quadrants as relates to angular measurement, and so on. It boils down to repeated practice. There is no shortcut to mastering this material.

I don't pretend to assume that this book will magically make Multivariable and Vector Calculus suddenly "plain as day". My hope is that a student can follow through an example and possibly see an important step, or being reminded of an important concept, in order to gain that incremental increase in mastery. I strongly suggest to all students to work out each example alongside my work. Go slow and deliberate and watch for the smallest steps, those "minor" steps that often are the ones that cause the most consternation.

There are many books on the market, and websites as well, that attempt to augment the learning experience in Multivariable and Vector Calculus. Many are quite well done, and each one has its own approach and philosophy. All require active participation from the student. Often, though, a slightly different viewpoint, or a different way of working through the steps, may be enough to help vault a student over the hump of a frustrating problem or concept.

I see this book as an on-going project. While the examples here have been seen by my students, and some have pointed out errors, I would not be surprised if some errors remain. Thus, I encourage anyone using this book to provide feedback in ways the book can be improved. Is one example particularly confusing? Please let me know. Do you see an error? I want to know. Do you have an idea for an example? Pass it along.

I maintain a website for this book at

www.surgent.net/vcbook

There, you can look over any updates that may have already been suggested, while my email listed on the website allows you to contact me directly.

Scott Surgent
August 2022

Table of Contents

1. **The *xyz* Coordinate Axis System** (6 examples) 4
2. **Distance & Midpoint** (3 examples) 8
3. **Triangles & Collinearity** (2 examples). 11
4. **Spheres & Ellipsoids** (9 examples). 13
5. **Multivariable Functions** (19 examples) 18
6. **Vectors** (14 examples) 32
7. **The Dot Product** (6 examples) 43
8. **Projections** (3 examples) 47
9. **The Cross Product** (4 examples) 50
10. **The Scalar Triple Product** (3 examples) 53

11. **Work & Torque** (4 examples) 55
12. **Lines in R^3** (8 examples) 59
13. **Planes in R^3** (15 examples) 64
14. **Distances in R^3** (4 examples) 73
15. **Vector Valued Functions** (10 examples) 78
16. **Vector Valued Functions: Limits & Continuity** (2 examples) 83
17. **Vector Valued Functions: Differentiation** (6 examples) . . 84
18. **Vector Valued Functions: Integration** (4 examples). . . 87
19. **Arc Length** (4 examples) 90
20. **Unit Tangent & Unit Normal Vectors** (5 examples). . . 93
21. **Curvature** (4 examples) 97
22. **Projectile Motion** (6 examples) 101
23. **Level Curves & Contour Maps** (2 examples) . . . 110
24. **Partial Differentiation** (11 examples) 119
25. **The Chain Rule** (5 examples). 130
26. **Directional Derivatives & The Gradient** (9 examples) . . 136
27. **Tangent Planes & Approximations** (7 examples) . . 143
28. **Differentials** (4 examples). 148
29. **Unconstrained Optimization** (6 examples). . . . 151
30. **Constrained Optimization** (5 examples) 159
31. **The Extreme Value Theorem** (2 examples). . . . 166
32. **Method of Lagrange Multipliers** (6 examples) . . . 172
33. **Riemann Summation over Rectangular Regions** (2 ex's) . 185
34. **Double Integration over Rectangular Regions** (6 ex's) . 189
35. **... over Non-Rectangular Regions of Type I** (5 examples) . 195
36. **... over Non-Rectangular Regions of Type II** (3 examples) . 201
37. **Double Integration in Polar Coordinates** (8 examples) . 205
38. **Triple Integration over Rectangular Regions** (4 examples) . 217
39. **... over Non-Rectangular Regions of Type I** (6 examples) . 221
40. **Spherical Coordinate System** (8 examples). . . . 236
41. **Integration with Spherical Coordinates** (4 examples) . . 244
42. **Change of Variables: The Jacobian** (3 examples) . . 251
43. **Vector Fields** (10 examples) 256
44. **Scalar Line Integrals** (6 examples) 264
45. **Vector Line Integrals: Work & Circulation** (5 examples) . 269
46. **Vector Line Integrals: Flux** (7 examples) 277
47. **Conservative Vector Fields** (6 examples) 283
48. **Fundamental Theorem of Line Integrals** (5 examples) . 287
49. **Green's Theorem** (6 examples) 292
50. **Surface Area Integrals** (8 examples) 300
51. **General Surface Integrals** (4 examples) 313
52. **The Del Operator: Curl & Divergence** (8 examples) . 321
53. **Flux Integrals** (7 examples) 331
54. **The Divergence Theorem** (5 examples) 345
55. **Stokes Theorem** (2 examples). 348
    Test Yourself 354
    Appendix 365

1. The *xyz* Coordinate Axis System

The ***xyz* coordinate axis system**, denoted R^3, is represented by three real number lines meeting at a common point, called the **origin**. The three number lines are called the ***x*-axis**, the ***y*-axis**, and the ***z*-axis**. Together, the three axes are called the **coordinate axes**.

Perspective (and other forms of artistic license) is used to represent three physical dimensions on a two-dimensional sheet of paper. Below is a common way to represent the three coordinate axes of R^3. At left are the entire three axes with their labels. To the right is a "cleaner" version where only the positive *x*, *y* and *z* axes are drawn. The three axes meet at right angles to one another.

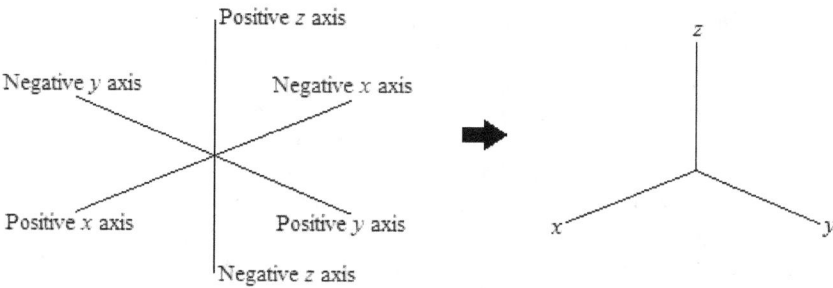

The three axes divide R^3 into eight regions, called **octants**. The region in which *x*, *y* and *z* are positive is called the **first octant** or the **positive octant**. The other octants are not numbered in any conventional way. Negative axes are drawn in only if the problem requires it.

A point is represented by an **ordered triple** (x, y, z), in which from the origin (whose ordered triple is (0,0,0)), one moves *x* units along the *x*-axis, then *y* units parallel to the *y*-axis, and then *z* units parallel to the *z*-axis, to arrive at the point. The values *x*, *y* and *z* are the **coordinates** of the point.

♦ • ♦ • ♦ • ♦ • ♦

Example 1.1: Represent the point (2,3,5) on an *xyz*-coordinate axis system.

Solution: Using perspective, draw in guidelines to form a "box" in which one corner is the origin, and the opposite corner is the desired point:

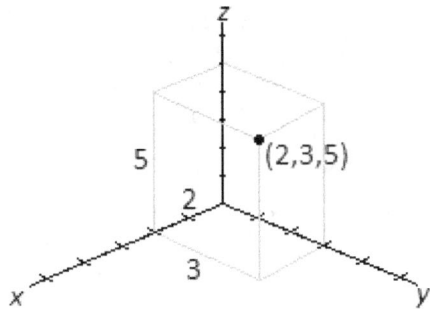

Other points are identified to show their relative positions in R^3:

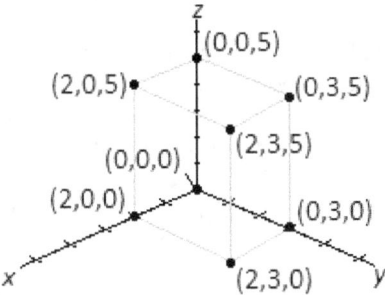

The point (2,3,0) is called a **projection** of (2,3,5) onto the *xy*-plane, found by setting $z = 0$. Other projections can be found similarly.

The three coordinate axes, taken two at a time, form three **coordinate planes**.

- The *x*-axis and the *y*-axis form the ***xy*-coordinate plane** and contains points whose ordered triples are of the form $(x, y, 0)$. The equation $z = 0$ represents the *xy*-plane.

- The *x*-axis and the *z*-axis form the ***xz*-coordinate plane** and contains points whose ordered triples are of the form $(x, 0, z)$. The equation $y = 0$ represents the *xz*-plane.

- The y-axis and the z-axis form the **yz-coordinate plane** and contains points whose ordered triples are of the form $(0, y, z)$. The equation $x = 0$ represents the yz-plane.

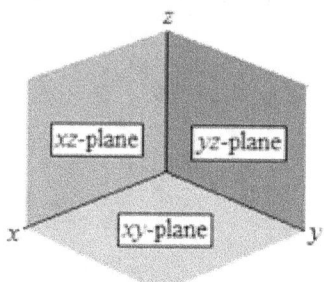

The three coordinate planes.

♦ • ♦ • ♦ • ♦ • ♦

Example 1.2: The point (100,6,4) is closest to which coordinate plane?

Solution: Since the z-value of 4 is the smallest of the three coordinates, the point (100,6,4) is closest to the xy coordinate plane.

♦ • ♦ • ♦ • ♦ • ♦

Example 1.3: Given the point $(4, -1, 2)$, find its projections onto the xy-plane, the xz-plane and the yz-plane.

Solution: The xy-plane is described by the equation $z = 0$, so the projection of $(4, -1, 2)$ onto the xy-plane is $(4, -1, 0)$. Similarly, the projection of $(4, -1, 2)$ onto the xz-plane is $(4, 0, 2)$, and $(4, -1, 2)$ onto the yz-plane is $(0, -1, 2)$.

♦ • ♦ • ♦ • ♦ • ♦

Example 1.4: Given the point $(4, -1, 2)$, find its reflections across the xy-plane, the xz-plane, the yz-plane, and the origin.

Solution: Points reflected across the xy-plane are found by negating the z coordinate. Thus, the reflection of $(4, -1, 2)$ across the xy-plane is $(4, -1, -2)$.

In a similar way, the reflection of $(4, -1, 2)$ across the xz-plane is $(4, 1, 2)$, and the reflection of $(4, -1, 2)$ across the yz-plane is $(-4, -1, 2)$.

To reflect across the origin, we negate all three coordinates. This is equivalent to reflecting a point across the xy-plane, then the xz-plane, then the yz-plane (in any order). Thus, the reflection of $(4, -1, 2)$ across the origin is $(-4, 1, -2)$,

Example 1.5 Describe the intersection of the planes $x = 0$ and $y = 0$.

Solution: The equation $x = 0$ is the yz-plane, and the equation $y = 0$ is the xz-plane, and they intersect at the z-axis. Points on the z-axis are described using set notation:

$$\{(x, y, z) \mid x = 0, y = 0, z \in R\}.$$

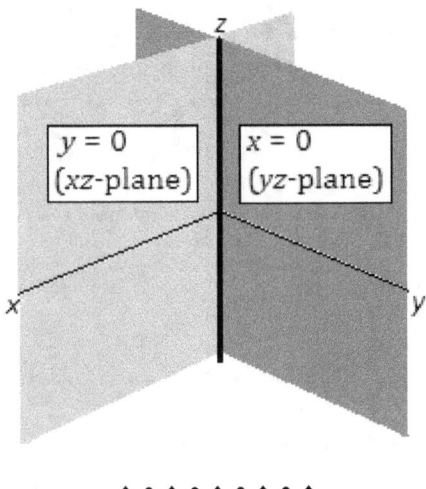

♦ • ♦ • ♦ • ♦ • ♦

Example 1.6: Describe the equation $x = 2$ as it appears in R^3.

Solution: The equation $x = 2$ includes all points of the form $(2, y, z)$. More generally, it can be described using set notation:

$$\{(x, y, z) \mid x = 2, y \in R, z \in R\}.$$

It is a plane that is parallel to the yz-plane; equivalently, it is the yz-plane shifted two units in the positive x direction. Note that the equation $x = 2$ does not imply any restriction on the variables y and z. They can assume any real number value. It is important to remember the "space" in which $x = 2$ is defined. In R^3, it is a plane. In R^2, it would be a vertical line passing through (2,0). In R^1 (or R), it is a point on the real number line.

The graph is on the next page.

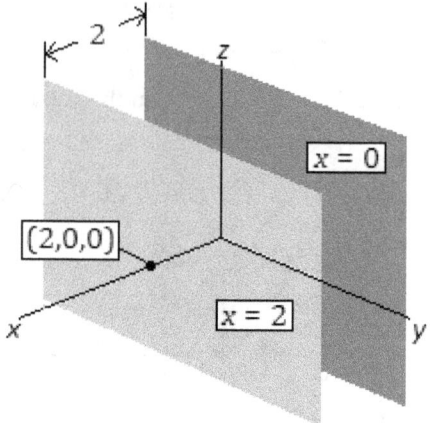

The plane $x = 2$ is parallel to the plane $x = 0$, shifted 2 units in the positive x direction.

♦ • ♦ • ♦ • ♦ • ♦

2. Distance & Midpoint

Given two points $A = (x_0, y_0, z_0)$ and $B = (x_1, y_1, z_1)$ in R^3, the **distance** between A and B is given by

$$D_{A,B} = \sqrt{(x_1 - x_0)^2 + (y_1 - y_0)^2 + (z_1 - z_0)^2},$$

and the **midpoint** between A and B is given by

$$M_{A,B} = \left(\frac{x_0 + x_1}{2}, \frac{y_0 + y_1}{2}, \frac{z_0 + z_1}{2}\right).$$

Note that the distance formula is the Pythagorean formula, and that the midpoint formula calculates the arithmetic mean (one at a time) of the x-coordinates, the y-coordinates and the z-coordinates.

♦ • ♦ • ♦ • ♦ • ♦

Example 2.1: Find the distance from the origin to the point $(3, -1, 5)$.

Solution: The origin is $(0,0,0)$, so the distance is

$$D = \sqrt{(3 - 0)^2 + (-1 - 0)^2 + (5 - 0)^2} = \sqrt{3^2 + (-1)^2 + 5^2} = \sqrt{35}.$$

Example 2.2: Given $A = (-2,1,4)$ and $B = (5,0,-7)$. Find the distance between A and B, and the midpoint of A and B.

Solution: The distance between A and B is

$$D_{A,B} = \sqrt{(5-(-2))^2 + (0-1)^2 + (-7-4)^2}$$
$$= \sqrt{7^2 + (-1)^2 + (-11)^2}$$
$$= \sqrt{171}$$
$$\approx 13.077 \text{ units.}$$

The midpoint between A and B is

$$M_{A,B} = \left(\frac{-2+5}{2}, \frac{1+0}{2}, \frac{4+(-7)}{2}\right) = \left(\frac{3}{2}, \frac{1}{2}, -\frac{3}{2}\right).$$

◆ • ◆ • ◆ • ◆ • ◆

Example 2.3: Given $A = (-2,1,4)$ and $B = (5,0,-7)$. Find all points in R^3 that are equidistant from A and B.

Solution: Let $P = (x,y,z)$ represent a point (represented as an ordered triple) equidistant from A and from B. Using the distance formulas,

$$D_{P,A} = \sqrt{(x-(-2))^2 + (y-1)^2 + (z-4)^2} = \sqrt{(x+2)^2 + (y-1)^2 + (z-4)^2},$$

$$D_{P,B} = \sqrt{(x-5)^2 + (y-0)^2 + (z-(-7))^2} = \sqrt{(x-5)^2 + y^2 + (z+7)^2}.$$

Since P is equidistant from A and from B, we have $D_{P,A} = D_{P,B}$. The radicals are squared away, then the binomials expanded by multiplication:

$$\sqrt{(x+2)^2 + (y-1)^2 + (z-4)^2} = \sqrt{(x-5)^2 + y^2 + (z+7)^2}.$$

$$(x+2)^2 + (y-1)^2 + (z-4)^2 = (x-5)^2 + y^2 + (z+7)^2.$$

$$x^2 + 4x + 4 + y^2 - 2y + 1 + z^2 - 8z + 16$$
$$= x^2 - 10x + 25 + y^2 + z^2 + 14z + 49.$$

Note that the squared terms cancel one another:

$$4x + 4 - 2y + 1 - 8z + 16 = -10x + 25 + 14z + 49.$$

The variable terms are collected to one side and the constant terms to the other:

$$14x - 2y - 22z = 53.$$

the equation $14x - 2y - 22z = 53$ is true upon substitution by all points that are equidistant from A and B. This forms a plane Q in R^3, which can be written as a set with z is isolated in terms of x and y:

$$Q = \left\{(x,y,z) \mid x \in R, y \in R, z = \tfrac{7}{11}x - \tfrac{1}{11}y - \tfrac{53}{22}\right\}.$$

To check, select arbitrary values for x and y. For example, let $x = 11$ and $y = 22$. This forces $z = \tfrac{57}{22}$, so a point on Q is $P = \left(11, 22, \tfrac{57}{22}\right)$. The distance from A to P, and from B to P, are

$$D_{A,P} = \sqrt{(-2-11)^2 + (1-22)^2 + \left(4 - \tfrac{57}{22}\right)^2} = \sqrt{13^2 + (-21)^2 + \left(\tfrac{31}{22}\right)^2} \approx 24.738,$$

$$D_{B,P} = \sqrt{(5-11)^2 + (0-22)^2 + \left(-7 - \tfrac{57}{22}\right)^2} = \sqrt{(-6)^2 + (-22)^2 + \left(-\tfrac{211}{22}\right)^2}$$
$$\approx 24.738.$$

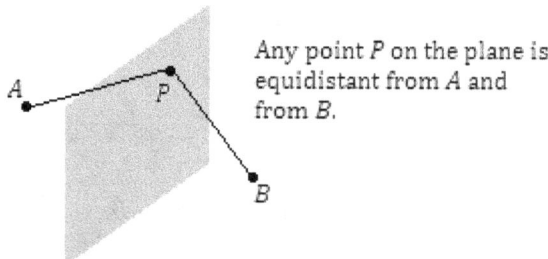

Any point P on the plane is equidistant from A and from B.

3. Triangles & Collinearity

Three points A, B and C form a **triangle** in that A, B and C are the vertices (corners) of the triangle, and that line segments $\overline{AB}$, $\overline{AC}$ and $\overline{BC}$ form the sides (edges).

Letting a, b and c represent the lengths of the sides of a triangle, and assuming c is the largest of the three values, the **triangle inequality** states that $c \leq a + b$, which simply states that the longest side of a triangle cannot be greater than the sum of the lengths of the two shorter sides:

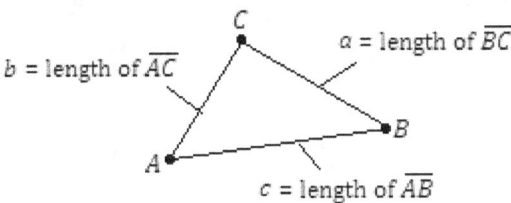

If $c = a + b$, then the length of the longest side is exactly the sum of the lengths of the two shorter sides, which can only happen when points A, B and C lie on a common line. In such a case, points A, B and C are **collinear**.

The three side-lengths of a triangle are related by the **law of cosines**:

$$c^2 = a^2 + b^2 - 2ab \cos \theta,$$

where c is assumed to be the length of the longest side and θ is the angle formed at point C, where side segments $\overline{AC}$ and $\overline{BC}$ meet. If $\theta = 90°$, then $\cos \theta = 0$, and we have the Pythagorean Formula, which relates the three side-lengths of a **right triangle**:

$$c^2 = a^2 + b^2.$$

♦ • ♦ • ♦ • ♦ • ♦

Example 3.1: Show that the points $A = (1,0,2)$, $B = (-2,3,1)$ and $C = (0,4,-2)$ are the vertices of a right triangle.

Solution: Find the lengths of the three sides of the triangle:

$$D_{A,B} = \sqrt{(1-(-2))^2 + (0-3)^2 + (2-1)^2} = \sqrt{3^2 + (-3)^2 + 1^2} = \sqrt{19},$$
$$D_{A,C} = \sqrt{(1-0)^2 + (0-4)^2 + (2-(-2))^2} = \sqrt{1^2 + (-4)^2 + 4^2} = \sqrt{33},$$
$$D_{B,C} = \sqrt{(-2-0)^2 + (3-4)^2 + (1-(-2))^2} = \sqrt{(-2)^2 + (-1)^2 + 3^2} = \sqrt{14}.$$

The length of the segment $\overline{AC}$ is the longest, and we use the Pythagorean Formula:

$$\left(\sqrt{33}\right)^2 = \left(\sqrt{19}\right)^2 + \left(\sqrt{14}\right)^2.$$

Since $33 = 19 + 14$ is a true statement, the triangle formed by A, B and C is a right triangle.

When sketching a triangle, the sides are named as is convenient. In the preceding example, segment $\overline{AC}$ would be given length b, if we followed the drawing on the previous page. This is fine, as long as in this case, we remember that b is the length of the longest side, and that c and a are the lengths of the two shorter sides.

◆ • ◆ • ◆ • ◆ • ◆

Example 3.2: Show that $A = (2,3,5)$, $B = (6,1,6)$ and $C = (14,-3,8)$ are collinear.

Solution: If A, B and C lie on the same line, then the largest distance between any of the three points will be equal to the sum of the two smaller distances.

The distances are:

$$\begin{aligned} D_{A,B} &= \sqrt{(6-2)^2 + (1-3)^2 + (6-5)^2} \\ &= \sqrt{4^2 + (-2)^2 + 1^2} \\ &= \sqrt{21}, \end{aligned}$$

$$\begin{aligned} D_{A,C} &= \sqrt{(14-2)^2 + (-3-3)^2 + (8-5)^2} \\ &= \sqrt{12^2 + (-6)^2 + 3^2} \\ &= \sqrt{189} \\ &= 3\sqrt{21}, \end{aligned}$$

$$\begin{aligned} D_{B,C} &= \sqrt{(14-6)^2 + (-3-1)^2 + (8-6)^2} \\ &= \sqrt{8^2 + (-4)^2 + 2^2} \\ &= \sqrt{84} = 2\sqrt{21}. \end{aligned}$$

Note that the distance between A and C is the longest, and that it is the sum of the distance between A and B, and the distance between B and C. That is, $3\sqrt{21} = 2\sqrt{21} + \sqrt{21}$, and so we conclude that A, B and C are collinear.

4. Spheres and Ellipsoids

A **sphere** is a set of ordered triples (x, y, z) that are of a fixed distance from a single fixed point (x_0, y_0, z_0), called the **center**, and the distance is called the **radius**, r. Using the distance formula, the simplified formula for a sphere can be written as

$$(x - x_0)^2 + (y - y_0)^2 + (z - z_0)^2 = r^2.$$

◆ ◆ ◆ ◆ ◆ ◆ ◆ ◆

Example 4.1: Find the equation of a sphere with center $(2, -1, 9)$ and radius 5.

Solution: The sphere is given by

$$(x - 2)^2 + (y - (-1))^2 + (z - 9)^2 = 5^2,$$

which simplifies to $(x - 2)^2 + (y + 1)^2 + (z - 9)^2 = 25$.

◆ ◆ ◆ ◆ ◆ ◆ ◆ ◆

Example 4.2: Find the equation of a sphere on which the two points $A = (4, 1, -1)$ and $B = (6, 7, 9)$ lie directly opposite one another (that is, the line through them forms a **diameter** of the sphere. Such points are called **antipodal points**).

Solution: The center is the midpoint of A and B:

$$M_{A,B} = \left(\frac{4+6}{2}, \frac{1+7}{2}, \frac{-1+9}{2}\right) = (5, 4, 4).$$

The distance from the midpoint to point A is:

$$D_{M,A} = \sqrt{(5-4)^2 + (4-1)^2 + \left(4-(-1)\right)^2} = \sqrt{1^2 + 3^2 + 5^2} = \sqrt{35}.$$

(This is also the distance from the midpoint to B.)

This is the radius, and since $r = \sqrt{35}$, then $r^2 = 35$. Thus, the sphere is

$$(x - 5)^2 + (y - 4)^2 + (z - 4)^2 = 35.$$

Example 4.3: Find the equation of the largest possible sphere with center (4,2,5) that is fully contained within the first octant (tangentially "touching" a coordinate plane is permissible).

Solution: The y-coordinate of 2 is the smallest of the three coordinates and is 2 units from the xz-coordinate plane. This will be the radius. Thus, the sphere is given by

$$(x-4)^2 + (y-2)^2 + (z-5)^2 = 4.$$

♦ • ♦ • ♦ • ♦ • ♦

Example 4.4: The sphere $(x+6)^2 + (y-1)^2 + (z-4)^2 = 100$ intersects the yz-coordinate plane, forming a circle. What is the radius of this circle?

Solution: The yz-coordinate plane is given by $x = 0$, so we substitute this into the equation of the sphere, and simplify:

$$\big((0)+6\big)^2 + (y-1)^2 + (z-4)^2 = 100$$
$$6^2 + (y-1)^2 + (z-4)^2 = 100$$
$$(y-1)^2 + (z-4)^2 = 64.$$

The intersection of the sphere with the yz-coordinate plane results in a circle of radius $\sqrt{64} = 8$.

♦ • ♦ • ♦ • ♦ • ♦

> A sphere may also be written as $x^2 + y^2 + z^2 + Dx + Ey + Fz = G$, in which case completing the square is needed to rewrite the sphere in simplified form.

Example 4.5: Find the center and radius of the sphere given by the equation $x^2 + 2x + y^2 - 6y + z^2 + 4z = 22$.

Solution: Complete the square three times:

$$\underbrace{x^2 + 2x + 1}_{(x+1)^2} + \underbrace{y^2 - 6y + 9}_{(y-3)^2} + \underbrace{z^2 + 4z + 4}_{(z+2)^2} = \underbrace{22 + 1 + 9 + 4}_{36}.$$

Simplified, we have

$$(x+1)^2 + (y-3)^2 + (z+2)^2 = 36.$$

Thus, the sphere has a center of $(-1,3,-2)$ and a radius of $r = \sqrt{36} = 6$.

Example 4.6: Explain why $x^2 + y^2 + z^2 + 4x + 6y + 10z + 50 = 0$ cannot represent a sphere.

Solution: Completing the square three times and simplifying, we have

$$x^2 + 4x + 4 + y^2 + 6y + 9 + z^2 + 10x + 25 = -50 + 4 + 9 + 25$$
$$(x + 2)^2 + (y + 3)^2 + (z + 5)^2 = -12.$$

The right side of the equation is negative, while the left side of the equation will always be a non-negative value, so this equation cannot have a solution in R^3. This equation is **inconsistent** (has no solutions).

An **axis intercept** in R^3 is found by setting two of the variables to 0. Thus, the x-axis intercept is given by the ordered triple $(x, 0, 0)$, the y-axis intercept is given by the ordered triple $(0, y, 0)$, and the z-axis intercept is given by the ordered triple $(0, 0, z)$. Not all axis intercepts will exist.

Example 4.7: Let $(x + 1)^2 + (y - 4)^2 + (z - 6)^2 = 41$. Find the axis intercepts if they exist.

Solution: When $x = 0$ and $y = 0$, we have

$$\left((0) + 1\right)^2 + \left((0) - 4\right)^2 + (z - 6)^2 = 41$$
$$1^2 + (-4)^2 + (z - 6)^2 = 41$$
$$1 + 16 + (z - 6)^2 = 41$$
$$(z - 6)^2 = 24$$
$$z - 6 = \pm\sqrt{24}$$
$$z = 6 \pm 2\sqrt{6}.$$

There are two z-axis intercepts, at $(0, 0, 6 + 2\sqrt{6})$ and $(0, 0, 6 - 2\sqrt{6})$.

When $x = 0$ and $z = 0$, we have

$$\left((0) + 1\right)^2 + (y - 4)^2 + \left((0) - 6\right)^2 = 41$$
$$1^2 + (y - 4)^2 + (-6)^2 = 41$$
$$1 + (y - 4)^2 + 36 = 41$$
$$(y - 4)^2 = 4$$
$$y - 4 = \pm 2$$
$$y = 4 \pm 2.$$

There are two y-axis intercepts, at $(0, 6, 0)$ and $(0, 2, 0)$.

When $y = 0$ and $z = 0$, we have

$$(x + 1)^2 + ((0) - 4)^2 + ((0) - 6)^2 = 41$$
$$(x + 1)^2 + (-4)^2 + (-6)^2 = 41$$
$$(x + 1)^2 + 16 + 36 = 41$$
$$(x + 1)^2 = -11.$$

Taking the square root of -11 results in a non-real value. Thus, there are no x-axis intercepts.

♦ • ♦ • ♦ • ♦ • ♦

Ellipsoids

An **ellipsoid** centered at the origin is written in the form

$$\frac{x^2}{a^2} + \frac{y^2}{b^2} + \frac{z^2}{c^2} = 1,$$

where $(\pm a, 0, 0)$ are the x-axis intercepts, $(0, \pm b, 0)$ are the y-axis intercepts, and $(0, 0, \pm c)$ are the z-axis intercepts. The **semi-principal axis radii** are a, b and c, respectively. The semi-principal diameters are $2a$, $2b$ and $2c$.

If the ellipsoid is centered at (x_0, y_0, z_0), the equation becomes

$$\frac{(x - x_0)^2}{a^2} + \frac{(y - y_0)^2}{b^2} + \frac{(z - z_0)^2}{c^2} = 1$$

♦ • ♦ • ♦ • ♦ • ♦

Example 4.8: Find the axis intercepts of the ellipsoid

$$\frac{x^2}{9} + y^2 + \frac{z^2}{12} = 1.$$

Solution: The x-axis intercepts are $(\pm 3, 0, 0)$, the y-axis intercepts are $(0, \pm 1, 0)$ and the z-axis intercepts are $(0, 0, \pm 2\sqrt{3})$. Note that this ellipsoid is centered at the origin.

The semi-principal radii are 3, 1 and $2\sqrt{3}$ units in the direction of the x-axis, y-axis and z-axis, respectively. The semi-principal diameters are twice these figures, or 6, 2 and $4\sqrt{3}$ units in the direction of the x-axis, y-axis and z-axis.

> Completing the square may be necessary to determine the ellipsoid's center and axis radii.

Example 4.9: Find the center, the semi-principal axis radii, and the axis intercepts of

$$x^2 + 2y^2 + 4z^2 + 2x - 8y + 24z = -5.$$

Solution: Group the terms by variable, and factor any constants from each grouping:

$$x^2 + 2x + 2y^2 - 8y + 4z^2 + 24z = -5$$
$$x^2 + 2x + 2(y^2 - 4y) + 4(z^2 + 6z) = -5.$$

Complete the square three times:

$$x^2 + 2x + 1 + 2(y^2 - 4y + 4) + 4(z^2 + 6z + 9) = -5 + 1 + 8 + 36$$

$$(x + 1)^2 + 2(y - 2)^2 + 4(z + 3)^2 = 40.$$

Note that the 8 on the right side is the "2 times 4" on the left side, and the 36 on the right is the "4 times 9" on the left. Divide now by 40:

$$\frac{(x + 1)^2}{40} + \frac{(y - 2)^2}{20} + \frac{(z + 3)^2}{10} = 1.$$

The ellipsoid's center is $(-1, 2, -3)$ and its semi-principal axis radii are $a = \sqrt{40} = 2\sqrt{10}$ in the direction parallel to the x-axis, $b = \sqrt{20} = 2\sqrt{5}$ in the direction parallel to the y-axis, and $c = \sqrt{10}$ in the direction parallel to the z-axis.

For the axis intercepts, we set two variables to 0, and solve for the third variable. For example, to find the z-axis intercepts, set $x = 0$ and $y = 0$. This can be done in the original equation:

$$x^2 + 2y^2 + 4z^2 + 2x - 8y + 24z = -5$$
$$(0)^2 + 2(0)^2 + 4z^2 + 2(0) - 8(0) + 24z = -5$$
$$4z^2 + 24z + 5 = 0.$$

Using the quadratic formula, we have

$$z = \frac{-24 \pm \sqrt{24^2 - 4(4)(5)}}{2(4)} = \frac{-24 \pm \sqrt{496}}{8} = \frac{-24 \pm 4\sqrt{31}}{8} = -3 \pm \frac{1}{2}\sqrt{31}.$$

Thus, the z-axis intercepts are $\left(0, 0, -3 \pm \frac{1}{2}\sqrt{31}\right)$. In a similar way, the y-axis intercepts are $(0, 4 \pm \sqrt{6}, 0)$. There are no x-axis intercepts (you verify).

5. Multivariable Functions

A function in R^3 has two independent variables, and a third variable dependent on the first two. If x and y represent the independent variables, and z the dependent variable, a **function in two variables** can be written $z = f(x, y)$. Depending on the situation, we can let y be the dependent variable, so that $y = f(x, z)$, or let x be the dependent variable, so that $x = f(y, z)$.

A function in three variables exists (is embedded) in R^4 and is written $w = f(x, y, z)$. Its points are called 4-tuples, written (x, y, z, w). In general, a function in n variables exists in R^{n+1}, has n independent variables and one dependent variable. A function with n independent variables is called a **multivariable function**, or an **n-variable function**. A point in R^n is called an **n-tuple**. Note that an n-variable function produces $(n + 1)$-tuples, since the final position will be the dependent variable. We often refer to 2-tuples as *pairs*, 3-tuples as *triples*, and so on.

The **domain** of an n-variable function (embedded in R^{n+1}) is the set of ordered n-tuples in R^n (the domain space) for which the function is defined. The **range** is the set of values in R^1 for which the dependent variable can assume. The visual representation of the set of points (ordered $(n + 1)$-tuples) for which a function is defined is called a **graph**. In R^3, the graph is often called a **surface**.

◆ • ◆ • ◆ • ◆ • ◆

Example 5.1: Given $z = f(x, y) = \frac{1}{x} + 2y$. Find $f\left(\frac{1}{3}, 4\right)$ and the domain of f.

Solution: We have

$$f\left(\frac{1}{3}, 4\right) = \frac{1}{1/3} + 2(4)$$
$$= 3 + 8$$
$$= 11.$$

This is an ordered triple $\left(\frac{1}{3}, 4, 11\right)$ on the graph of f. Since x is in the denominator, we must have $x \neq 0$. Thus, the domain is the set of x and y values for which $x \neq 0$. Using set-builder notation, we can write this as

$$\text{Dom } f = \{(x, y) \mid x \in R \text{ and } y \in R \text{ such that } x \neq 0\}.$$

The range can be inferred indirectly. For example, for any z-value, it is possible to find at least one ordered pair (x, y) that produces z. From this, we can state that the range of f is

$$\text{Ran } f = \{z \mid z \in R\}.$$

Example 5.2: Describe the graph of $y = 2x + 1$ as it appears in R^3.

Solution: In R^2, this is a line on an *xy*-coordinate axis system with *y*-intercept $(0,1)$ and a slope of 2. It is sketched below:

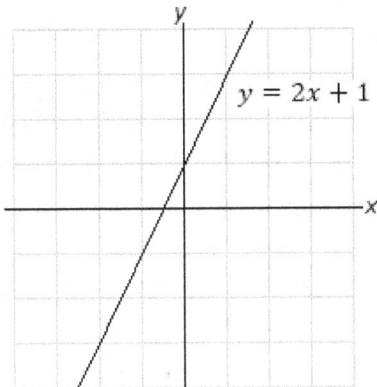

In R^3, we allow *z* to be any value. Thus, the graph of $y = 2x + 1$ in R^3 is the set of all ordered triples of the form $(x, 2x + 1, z)$. Note that we may choose *x* and *z* independently of one another. However, once *x* is chosen, *y* is then determined by the formula $y = 2x + 1$. Thus, in this example, we would let *x* and *z* be the independent variables, and *y* the dependent variable. The domain is $\{(x, z) \mid x \in R, z \in R\}$.

The graph of $y = 2x + 1$ in R^3 is a plane that extends into the positive and negative *z* directions:

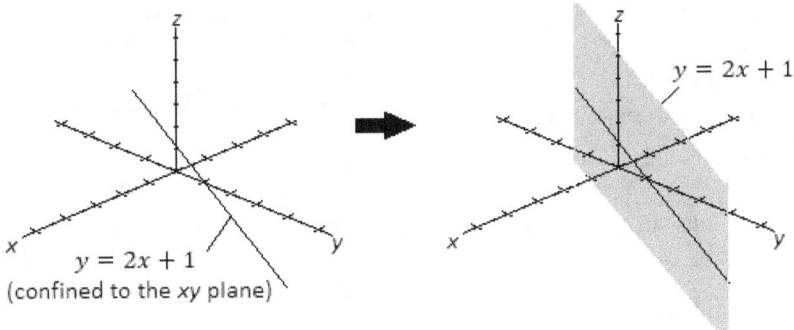

The graph of a function $y = f(x)$ in R^2 can be extended into R^3 by allowing the third variable *z* to be independent, with points of the form $(x, f(x), z)$. Such graphs are generically called **cylinders**. The above graph is technically a "linear cylinder". More commonly, these types of graphs are informally called **sheets**, with the term "cylinder" reserved for closed curves like circles and ellipses that are extended into the third dimension.

Example 5.3: Find the domain of $z = f(x,y) = \frac{1}{2x-y}$.

Solution: The expression $2x - y$ cannot be zero, $2x - y \neq 0$, or $y \neq 2x$. Using set-builder notation, this is

$$\text{Dom } f = \{(x,y) \mid \text{All } x \in R \text{ and } y \in R \text{ such that } y \neq 2x\}.$$

Thus, we may choose x and y independently of one another as long as $y \neq 2x$. For example, $f(3,1)$ is defined, but $f(2,4)$ is not defined. The range is inferred indirectly. If we set $z = 0$, then we have $\frac{1}{2x-y} = 0$. There are no ordered pairs (x,y) that solve this. However, if $z = k$, where $k \neq 0$, then $\frac{1}{2x-y} = k$ is always solvable. Thus, the range is

$$\text{Ran } f = \{z \mid z \in R \text{ except } z = 0\}.$$

◆ · ◆ · ◆ · ◆ · ◆

Determining domain is typically routine, in that we avoid zeros in the denominator, negative values inside an even-index root, and non-positive entries within a logarithm. The table below summarizes domains for common functions.

Type of function	Restrictions on the Domain
n-degree polynomials such as $x^2 + 3x - 1$ or $xy^3 + x^2y - 2x$.	No restrictions.
A radical expression such as $\sqrt[n]{x^3 - 2y}$, where n is an integer ≥ 2. (n is called the *index*)	No restrictions on the expression inside the radical if the index is odd. If the index is even, then the expression must be greater than or equal to 0.
Rational expression such as $\frac{x^2-1}{3x-y^2}$.	The denominator must not equal 0.
Exponential functions such as 2^x or x^y.	The base must be strictly greater than 0, and not equal to 1.
Logarithms such as $\ln(3y - 5x)$.	The expression inside the logarithm must be strictly greater than 0.
Sine and cosine functions.	No restrictions.
Tangent functions.	The expression must not equal $\pm \frac{n\pi}{2}$, where n is an odd integer.

Determining range is not as formulaic. We often use indirect means to infer the domain. For example, we might try setting the function equal to a particular z-value, and work backwards to see it it's possible to solve the equation. If not, then that z-value is outside the range. This method is highly inefficient. Often, the range is inferred by viewing the graph using software.

◆ • ◆ • ◆ • ◆ • ◆

Example 5.4: Find the domain and range of $z = g(x, y) = \sqrt{81 - x^2 - y^2}$.

Solution: The expression inside the radical must be non-negative. Thus, we have

$$81 - x^2 - y^2 \geq 0.$$

Rearranging the terms, the domain of g is $\{(x, y) \mid x^2 + y^2 \leq 81\}$.

The surface of g is a hemisphere of radius 9, and its domain is a filled-in circle of radius 9, centered at (0,0) on the xy-plane. The range of g is $\{z \mid 0 \leq z \leq 9\}$.

◆ • ◆ • ◆ • ◆ • ◆

Example 5.5: Let $z = h(x, y) = \sqrt{4 - (x - 3)^2 + (y + 1)^2} + 5$. Find the domain and range of h.

Solution: This surface is a hemisphere centered at $(3, -1, 5)$ with radius 2. It creates a "shadow" onto the xy-plane that is a circle centered at $(3, -1)$ with radius 2. These are the permissible ordered pairs (x, y) that will result in a real-value output z. Thus, the domain of this sphere is

$$\text{Dom } h = \{(x, y) \mid (x - 3)^2 + (y + 1)^2 \leq 4\}.$$

The range is

$$\text{Ran } h = \{z \mid 5 \leq z \leq 7\}.$$

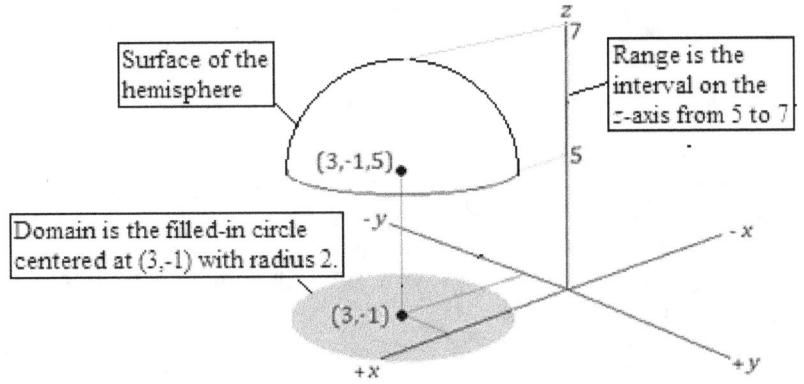

Given a multivariable function $z = f(x, y)$, we can set $x = 0$ and sketch its **trace** on the yz-plane, and then set $y = 0$ and sketch its trace on the xz-plane. From the two traces, it may be possible to infer the actual surface that results.

Example 5.6: Sketch $z = x^2 + y^2$.

Solution: When $x = 0$, we have $z = y^2$, which is a parabola opening in the positive z direction on the yz-plane. Similarly, when $y = 0$, we have another parabola $z = x^2$ opening in the positive z direction on the xz-plane. Together, the two parabola traces suggest that the surface of the function $z = x^2 + y^2$ is a parabolic bowl, or **paraboloid**.

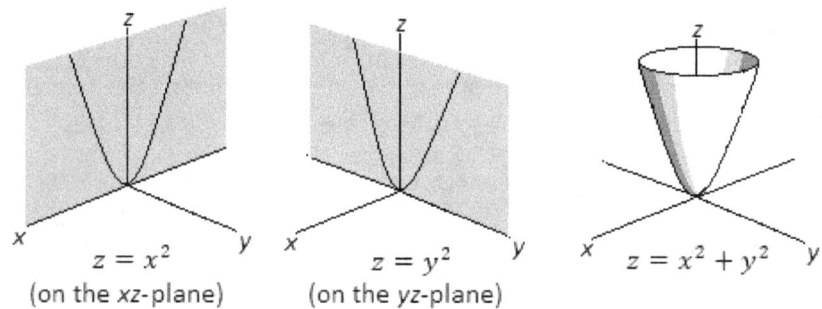

$z = x^2$ (on the xz-plane) $z = y^2$ (on the yz-plane) $z = x^2 + y^2$

Note that this paraboloid has a vertex at (0,0,0). If positive z is considered "up", then we say this paraboloid opens upward. The domain is $\{(x, y) | x \in R, y \in R\}$, and the range is $\{z | z \geq 0\}$.

♦ • ♦ • ♦ • ♦ • ♦

Example 5.7: Describe the surface of $z = -x^2 + 4x - y^2 - 2y$.

Solution: Completing the square twice gives

$$\begin{aligned} z &= -x^2 + 4x - y^2 - 2y \\ &= -(x^2 - 4x) - (y^2 + 2y) \\ &= -(x^2 - 4x + 4) - (y^2 + 2y + 1) + 4 + 1 \\ &= -(x - 2)^2 - (y + 1)^2 + 5. \end{aligned}$$

This is a paraboloid that has been shifted 2 units in the x-direction, -1 unit in the y direction, and 5 units in the z direction. The leading negatives in front of the quadratic terms suggest the paraboloid opens in the negative z direction. Thus, it has the identical shape as the paraboloid in the previous example, but it has a vertex at $(2, -1, 5)$ and opens "downward". The domain is $\{(x, y) | x \in R, y \in R\}$, and the range is $\{z | z \leq 5\}$.

Example 5.8: Describe the surface $z = \sqrt{x^2 + y^2}$.

Solution: Note that $x^2 + y^2 \geq 0$ for all x and y, so that the domain is $\{(x,y) | x \in R, y \in R\}$. Note also that the radical results in non-negative values for z, so that the range is $\{z | z \geq 0\}$.

We sketch traces. For example, let $y = 0$, so that means $z = \sqrt{x^2 + 0} = \pm x$. Similarly, when $x = 0$, we have $z = \sqrt{0 + y^2} = \pm y$. These are lines that form a "V" shape in their respective planes. The cross sections parallel to the xy-plane are circles, and together, these facts suggest that $z = \sqrt{x^2 + y^2}$ is a **cone**.

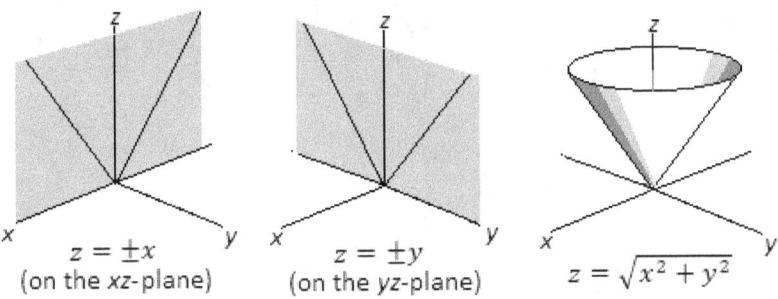

$z = \pm x$ (on the xz-plane) $z = \pm y$ (on the yz-plane) $z = \sqrt{x^2 + y^2}$

The cone opens in the positive z direction, indicating that the origin is a minimum. The surface given by $z = -\sqrt{x^2 + y^2}$ would be a cone opening in the negative z direction, where the origin would be a maximum, assuming that positive z is "up".

◆ • ◆ • ◆ • ◆ • ◆

Example 5.9: A cone with circular cross sections and the vertex at the origin opens in the positive z direction, passing through the point $(1,3,7)$. Find the equation of the cone.

Solution: The general equation of the cone is $z = a\sqrt{x^2 + y^2}$, where a can be determined by evaluating at a known point on the cone's surface. We have

$$7 = a\sqrt{1^2 + 3^2}$$
$$7 = a\sqrt{10}$$
$$a = \frac{7}{\sqrt{10}}.$$

Thus, the cone's equation is $z = \frac{7}{\sqrt{10}}\sqrt{x^2 + y^2} = 7\sqrt{\frac{x^2}{10} + \frac{y^2}{10}}$.

Example 5.10: A cone with circular cross sections and the vertex at the origin opens in the positive z direction, such that the angle at the vertex is $\frac{2\pi}{3}$ radians. Find the equation of the cone.

Solution: Viewing a trace of the cone, we can see the vertex angle. Note that the side of the cone is at an angle of $\frac{\pi}{3}$ radians (half of the vertex angle) from the positive z-axis. From this, we can determine a point on the cone's surface. In the images that follow, we set $y = 0$ and choose $z = 1$. Using a 30-60-90 triangle with shortest leg of length 1, the longer leg is of length $\sqrt{3}$. This is our x value, and the point is $(\sqrt{3}, 0, 1)$. Thus, we have

$$z = a\sqrt{x^2 + y^2}$$

$$1 = a\sqrt{\left(\sqrt{3}\right)^2 + 0^2}$$

$$1 = a\sqrt{3}$$

$$a = \frac{1}{\sqrt{3}} \text{ or } \frac{\sqrt{3}}{3}.$$

The cone's equation is $z = \frac{\sqrt{3}}{3}\sqrt{x^2 + y^2}$. The images and the final cone are below.

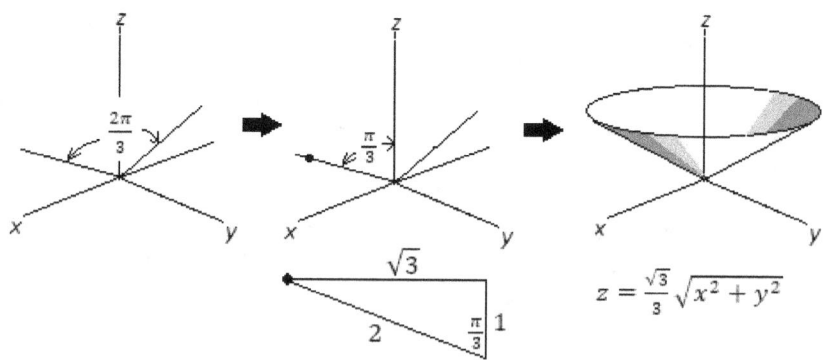

Using right triangles to help find a point on the cone's surface.

Example 5.11: Describe the surface $z = x^2 - y^2$.

Solution: When $y = 0$, the surface's trace on the xz-plane is $y = x^2$, a parabola that opens in the positive z direction. When $x = 0$, the trace on the yz-plane is $z = -y^2$, a parabola that opens in the negative z direction.

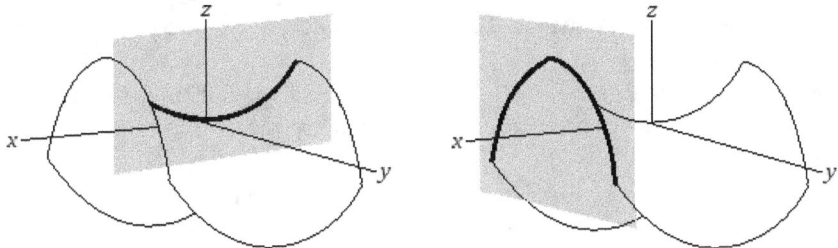

When a plane parallel to the xz-plane intersect the surface $z = x^2 - y^2$, it forms a parabola that opens up. When a plane parallel to the yz-plane intersects the surface, it forms a parabola that opens down.

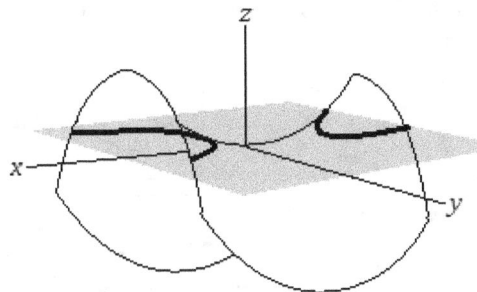

When a plane parallel to the xy-plane intersects
the surface, it forms a hyperbola.

The surface is called a **hyperbolic paraboloid**. It is shaped like a saddle and is informally called a saddle. The origin in this case is the saddle point.

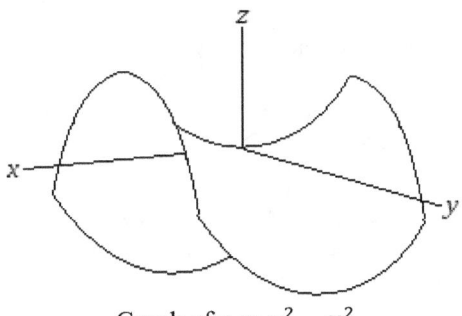

Graph of $z = x^2 - y^2$.

Example 5.12: Describe the surface $x^2 + y^2 - z^2 = 1$.

Solution: The surface's appearance can be inferred by setting each variable to 0, one at a time.

- When $x = 0$, this gives $y^2 - z^2 = 1$, which is a hyperbola in the yz-plane where the two halves open in the positive and negative y-directions.

- When $y = 0$, this gives $x^2 - z^2 = 1$, which is a hyperbola in the xz-plane where the two halves open in the positive and negative x-directions.

- When $z = 0$, this gives $x^2 + y^2 = 1$, which is a circle of radius 1 in the xy-plane, centered at the origin.

The resulting shape is called a *hyperboloid of one sheet*. It does not intersect the z-axis (the z-axis is the axis of symmetry of this surface). Any plane parallel to the xy-plane (that is, any plane with the equation $z = k$) will intersect this surface forming a circle. The surface is "narrowest" when $z = 0$.

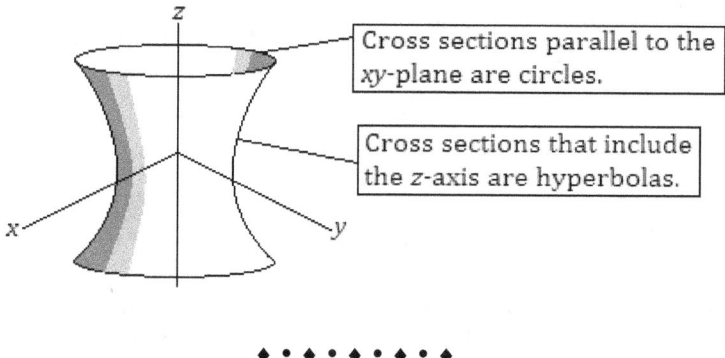

Hyperboloid of One Sheet

Cross sections parallel to the xy-plane are circles.

Cross sections that include the z-axis are hyperbolas.

♦ • ♦ • ♦ • ♦ • ♦

Example 5.13: Describe the surface $x^2 - y^2 - z^2 = 1$.

Solution: As in the previous example, the surface's appearance can be inferred by setting each variable to 0, one at a time.

- When $x = 0$, this gives $-y^2 - z^2 = 1$, which has no solution since the left side will always be 0 or negative, while the right side is 1. The surface will not intersect the yz-plane.

- When $y = 0$, this gives $x^2 - z^2 = 1$, which is a hyperbola in the xz-plane where the two halves open in the positive and negative x-directions.

- When $z = 0$, this gives $x^2 - y^2 = 1$, which is a hyperbola in the xy-plane where the two halves open in the positive and negative x-directions.

The resulting shape is called a *hyperboloid of two sheets*. Because it does not intersect the yz-plane, the surface is split into two symmetric halves. In fact, it is not difficult to show that the surface is not defined when $-1 < x < 1$.

Hyperboloid of Two Sheets

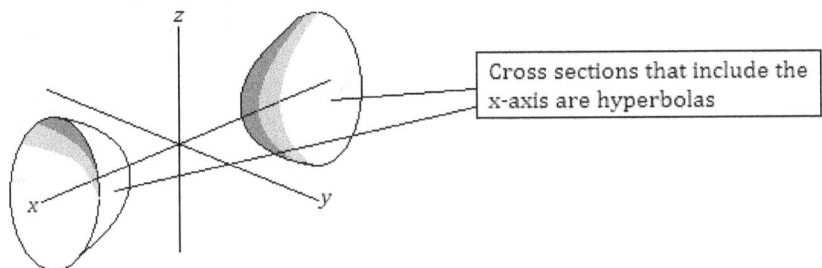

Cross sections that include the x-axis are hyperbolas

In this example, the x-axis is the axis of symmetry. Any plane that includes the x-axis will intersect the surface forming hyperbolas.

♦ • ♦ • ♦ • ♦ • ♦

It may be tempting to assume that a hyperboloid of one or two sheets can be "detected" by the number of quadratic terms with a negative coefficient. The next example illustrates how this initial assumption may not be correct.

Example 5.14: Describe the surface $x^2 - 2y^2 - 4z^2 + 2x - 24z + 5 = 0$.

Solution: Completing the square twice (on variables x and z), we have:

$$x^2 + 2x + 1 - 2y^2 - 4(z^2 - 6z + 9) = -5 + 1 - 36$$
$$(x + 1)^2 - 2y^2 - 4(z - 3)^2 = -40.$$

Dividing by -40, we have

$$-\frac{(x+1)^2}{40} + \frac{y^2}{20} + \frac{(z-3)^2}{10} = 1.$$

This is a hyperboloid of one sheet, centered at $(-1, 0, 3)$. Planes parallel to the yz-plane will intersect the surface and form ellipses. The axis of symmetry is a line parallel to the x-axis, passing through $(-1, 0, 3)$.

Common Graphs in R^3

Equation	Surface Description
$ax + by + cz = d$	Plane.
$(x - x_0)^2 + (y - y_0)^2 + (z - z_0)^2 = r^2$	Sphere of radius r and center (x_0, y_0, z_0).
$z = \pm\sqrt{r^2 - x^2 - y^2}$	Hemisphere with a circular base on the xy-plane and extending into the positive z direction if the root is positive, negative z direction if the root is negative.
$z = x^2 + y^2$ $z = -x^2 - y^2$	Paraboloid with vertex $(0,0,0)$ opening in the positive or negative z directions, respectively. The vertex is a minimum or maximum, respectively.
$z = a\sqrt{x^2 + y^2}$	Cone with vertex at $(0,0,0)$ and opening in the positive z direction when $a > 0$, or negative z direction when $a < 0$.
$z = (x - x_0)^2 + (y - y_0)^2 + z_0$ $z = -(x - x_0)^2 - (y - y_0)^2 + z_0$	Paraboloid with vertex at (x_0, y_0, z_0) opening in the positive z direction or negative z direction, respectively.
$z = x^2 - y^2$ or $z = -x^2 + y^2$	A hyperbolic paraboloid (saddle).
$\dfrac{(x - x_0)^2}{a^2} + \dfrac{(y - y_0)^2}{b^2} - \dfrac{(z - z_0)^2}{c^2} = 1$	Hyperboloid of one sheet
$\dfrac{(x - x_0)^2}{a^2} - \dfrac{(y - y_0)^2}{b^2} - \dfrac{(z - z_0)^2}{c^2} = 1$	Hyperboloid of two sheets.
$y = f(x)$	A "sheet" or "cylinder" in which the curve given by $y = f(x)$ extends into the positive and negative z directions and contains ordered pairs of the form $(x, f(x), z)$.

Analogs exist by isolating other variables. For example, $x = y^2 + z^2$ is a paraboloid that opens in the positive x direction.

◆ • ◆ • • ◆ • • ◆

Surfaces of the general form $Ax^2 + By^2 + Cz^2 + Dx + Ey + Fz + G = 0$ are called **quadric surfaces** in R^3, assuming that A, B and C are not all simultaneously 0 (If they are, then the equation represents a plane). The signs and values of A, B and C determine the type of surface; E, F and G govern shifts in the x, y and z directions simultaneously.

Assuming that the equation is **consistent** (has at least one solution), then some of the common quadric surfaces are spheres, ellipsoids, paraboloids, cones, hyperbolic paraboloids ("saddles"), hyperboloids of one sheet, and hyperboloids of two sheets.

Limits of Functions in R^3

Let $z = f(x, y)$ be a two-variable function in R^3. If x approaches a and y approaches b, then the **general limit** is written

$$\lim_{(x,y) \to (a,b)} f(x, y) = L.$$

For this limit to exist, it must be finite and true for *all* possible paths toward (a, b). If any pair of different paths result in a different limit value, or any one path results in an infinite or undefined limit, then the general limit does not exist.

♦ • ♦ • ♦ • ♦ • ♦

Example 5.15: Find the following limit:

$$\lim_{(x,y) \to (1,-2)} (2x^2 y).$$

Solution: For two-variable polynomial terms, the limit will exist and is found by direct evaluation:

$$\lim_{(x,y) \to (1,-2)} (2x^2 y) = 2(1)^2(-2) = -4.$$

♦ • ♦ • ♦ • ♦ • ♦

Example 5.16: Find the following limit:

$$\lim_{(x,y) \to (3,5)} \left(\frac{x}{y - 5} \right).$$

Solution: By direct evaluation, we have

$$\lim_{(x,y) \to (3,5)} \left(\frac{x}{y - 5} \right) = \frac{(3)}{(5) - 5} = \frac{3}{0}.$$

This is an undefined term. Thus, the limit fails to exist.

Undefined and Indeterminate: Recall that division by 0 is never allowed. However, depending on the numerator, letting a denominator approach 0 as a limit results in two different situations. If the numerator k is not zero (as a limit), then the expression $\frac{k}{0}$ is *undefined*, such as in Example 5.16. If the numerator is 0 as a limit, then the expression $\frac{0}{0}$ is *indeterminate*, which means that further investigation is needed to determine the limit if it exists. This is explored in Examples 5.17 and 5.18.

Example 5.17: Find the following limit:

$$\lim_{(x,y)\to(1,1)} \left(\frac{x^2 - xy}{x - y}\right).$$

Solution: Evaluation results in the indeterminate form $\frac{0}{0}$:

$$\lim_{(x,y)\to(1,1)} \left(\frac{x^2 - xy}{x - y}\right) = \frac{(1)^2 - (1)(1)}{(1) - (1)} = \frac{0}{0}.$$

However, we can factor the numerator, then simplify:

$$\frac{x^2 - xy}{x - y} = \frac{x(x - y)}{x - y} = x.$$

Re-evaluating the limit, we have

$$\lim_{(x,y)\to(1,1)} \left(\frac{x^2 - xy}{x - y}\right) = \lim_{(x,y)\to(1,1)} x = 1.$$

Note that the function $z = f(x, y) = \frac{x^2-xy}{x-y}$ is not defined when $y = x$.

♦ • ♦ • ♦ • ♦ • ♦

Example 5.18: Find the following limit:

$$\lim_{(x,y)\to(0,0)} \left(\frac{x^2 - y^2}{x^2 + y^2}\right).$$

Solution: Evaluation results in the indeterminate form $\frac{0}{0}$:

$$\lim_{(x,y)\to(0,0)} \left(\frac{x^2 - y^2}{x^2 + y^2}\right) = \frac{(0)^2 - (0)^2}{(0)^2 + (0)^2} = \frac{0}{0}.$$

The expression is not reducible by factoring. Instead, we try different paths in the xy-plane that approach the origin, (0,0). If we can show that two different paths result in two different limits, then the general limit fails to exist.

For the path along the positive x-axis towards (0,0), we have $y = 0$, so the expression $\frac{x^2-y^2}{x^2+y^2}$ simplifies to

$$\frac{x^2 - (0)^2}{x^2 + (0)^2} = \frac{x^2}{x^2} = 1 \text{ (assuming } x \neq 0\text{)}.$$

Thus, for this particular path, the limit is

$$\lim_{x \to 0} \left(\frac{x^2 - y^2}{x^2 + y^2}\right) = \lim_{x \to 0} 1 = 1.$$

For the path along the positive y-axis towards (0,0), we have $x = 0$, so the expression $\frac{x^2-y^2}{x^2+y^2}$ simplifies to

$$\frac{(0)^2 - y^2}{(0)^2 + y^2} = -\frac{y^2}{y^2} = -1 \text{ (assuming } y \neq 0\text{)}.$$

Thus, the limit for this particular path is

$$\lim_{y \to 0} \left(\frac{x^2 - y^2}{x^2 + y^2}\right) = \lim_{y \to 0}(-1) = -1.$$

Since two different paths lead to two different limit values, the general limit

$$\lim_{(x,y) \to (0,0)} \left(\frac{x^2 - y^2}{x^2 + y^2}\right)$$

does not exist. The function $z = f(x, y) = \frac{x^2-y^2}{x^2+y^2}$ is not defined at (0,0), nor does its limit exist as x and y approach (0,0).

Example 5.19: Find the following limit:

$$\lim_{(x,y) \to (0,0)} \left(\frac{xy}{x^2 + y^2}\right).$$

Solution: Direct evaluation results in the indeterminate form $\frac{0}{0}$:

$$\lim_{(x,y)\to(0,0)} \left(\frac{xy}{x^2+y^2}\right) = \frac{(0)(0)}{(0)^2+(0)^2} = \frac{0}{0}.$$

We try different paths: For a path along the x-axis ($y = 0$), we have

$$\frac{x(0)}{x^2+(0)^2} = 0, \quad \text{so that} \quad \lim_{x\to 0}\left(\frac{x(0)}{x^2+(0)^2}\right) = \lim_{x\to 0}(0) = 0.$$

For a path along the y-axis ($x = 0$), we have

$$\frac{(0)y}{(0)^2+y^2} = 0, \quad \text{so that} \quad \lim_{y\to 0}\left(\frac{(0)y}{(0)^2+y^2}\right) = \lim_{x\to 0}(0) = 0.$$

It might be tempting to infer that since the limit equals 0 along two paths, the general limit would exist and be 0 as well. This is false. Let's try a different path, along the line $y = x$:

$$\lim_{x\to 0}\left(\frac{x(x)}{x^2+(x)^2}\right) = \frac{x^2}{2x^2} = \frac{1}{2}.$$

We have shown two different paths result in different limit values. Thus, the general limit does not exist.

♦ • ♦ • ♦ • ♦ • ♦

6. Vectors

For purposes of applications in calculus and physics, a **vector** has both a direction and a magnitude (length) and is usually represented by an arrow. The start of the arrow is the vector's *foot*, and the end is its *head*. A vector is usually labelled in boldface, such as **v**. In an xy-axis system (R^2), a vector is written $\mathbf{v} = \langle v_1, v_2 \rangle$, which means that from the foot of **v**, move v_1 units in the x direction, and v_2 units in the y direction, to arrive at the vector's head. The values v_1 and v_2 are the vector's **components**. In R^3, a vector has three components and is written $\mathbf{v} = \langle v_1, v_2, v_3 \rangle$.

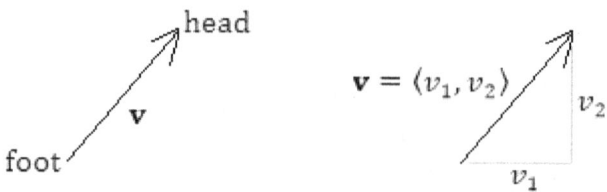

Given two points, $P_0 = (x_0, y_0)$ and $P_1 = (x_1, y_1)$ in R^2, a vector $\mathbf{P_0 P_1}$ can be drawn with its foot at P_0 and head at P_1, where $\mathbf{P_0 P_1} = \langle x_1 - x_0, y_1 - y_0 \rangle$. In R^3, the vector is expressed $\mathbf{P_0 P_1} = \langle x_1 - x_0, y_1 - y_0, z_1 - z_0 \rangle$.

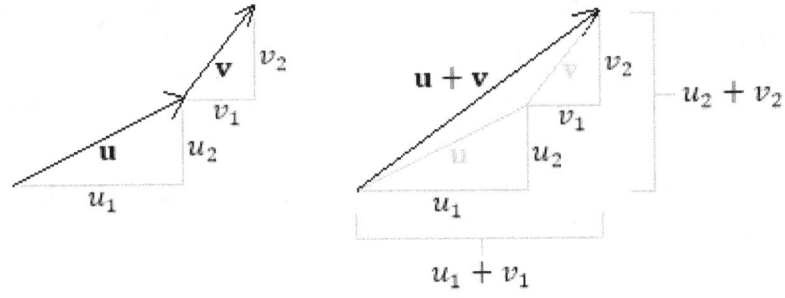

By comparison, a **scalar** is a number only, with no implied direction. Scalars are chosen from the set of real numbers R.

The **magnitude** of a vector **v** is found by the Pythagorean Formula:

$$|\mathbf{v}| = \sqrt{v_1^2 + v_2^2} \text{ (in } R^2\text{)} \quad \text{or} \quad \sqrt{v_1^2 + v_2^2 + v_3^2} \text{ (in } R^3\text{).}$$

The notation $|\mathbf{v}|$ represents the magnitude of **v** and is always a non-negative value. The expression $|\mathbf{v}|$ is a scalar.

To add two vectors $\mathbf{u} = \langle u_1, u_2 \rangle$ and $\mathbf{v} = \langle v_1, v_2 \rangle$, add the respective components:

$$\mathbf{u} + \mathbf{v} = \langle u_1 + v_1, u_2 + v_2 \rangle.$$

Place the foot of **v** at the head of **u**, then sketch a vector that begins at the foot of **u** and ends at the head of **v**. The vector $\mathbf{u} + \mathbf{v}$ is called the **resultant**.

Pay attention to notation. Parentheses ( ) are used to represent points, and angled brackets ⟨ ⟩ are used to represent vectors.

A vector may be multiplied by any real number c, called a **scalar multiple**. For example, if **u** is a vector, then $2\mathbf{u} = \mathbf{u} + \mathbf{u} = \langle 2u_1, 2u_2 \rangle$, which results in a vector $2\mathbf{u}$ that is twice the magnitude of $\mathbf{u}$. Scalars act as coefficients when multiplied to a vector. In general, for a vector $\mathbf{v}$ and a scalar c, the magnitude of $c\mathbf{v}$ is $|c\mathbf{v}| = |c||\mathbf{v}|$, where $|c|$ is the absolute value of c. Two non-zero vectors $\mathbf{u}$ and $\mathbf{v}$ are **parallel** if one can be written as a scalar multiple of the other, $\mathbf{u} = c\mathbf{v}$ for some non-zero scalar c.

There are two closure properties of vectors:

C1. If $\mathbf{u}$ and $\mathbf{v}$ are two vectors in R^2 (or R^3), then their vector sum $\mathbf{u} + \mathbf{v}$ is also in R^2 (or R^3).

C2. If $\mathbf{u}$ is a vector in R^2 (or R^3), then for any scalar c, its scalar multiple $c\mathbf{u}$ is also in R^2 (or R^3).

The structural properties of vectors are:

P1. *Commutativity of addition*: $\mathbf{u} + \mathbf{v} = \mathbf{v} + \mathbf{u}$. Vectors can be added in any order.

P2. *Associativity of addition*: $(\mathbf{u} + \mathbf{v}) + \mathbf{w} = \mathbf{u} + (\mathbf{v} + \mathbf{w})$.

P3. *Additive identity*: $\mathbf{0} = \langle 0,0 \rangle$ or $\langle 0,0,0 \rangle$, with the property that for any vector $\mathbf{u}$, the sum $\mathbf{u} + \mathbf{0} = \mathbf{0} + \mathbf{u} = \mathbf{u}$. Thus, $\mathbf{0}$ is called the **zero vector**, and is a single point with magnitude 0: $|\mathbf{0}| = 0$, and if $|\mathbf{v}| = 0$, then $\mathbf{v} = \mathbf{0}$.

P4. *Additive inverse*: For any non-zero vector $\mathbf{u}$, the vector $-\mathbf{u}$ exists with the property that $\mathbf{u} + (-\mathbf{u}) = \mathbf{0}$. Visually, $-\mathbf{u}$ has the same magnitude as $\mathbf{u}$, but points in the opposite direction. Subtraction of two vectors is now defined: $\mathbf{u} - \mathbf{v} = \mathbf{u} + (-\mathbf{v}) = \mathbf{u} + (-1\mathbf{v})$.

P5. *Distributivity of a scalar across vectors*: If c is a scalar, then $c(\mathbf{u} + \mathbf{v}) = c\mathbf{u} + c\mathbf{v}$.

P6. *Distributivity of a vector across scalars*: If c and d are scalars, then $(c + d)\mathbf{u} = c\mathbf{u} + d\mathbf{u}$.

P7. *Associativity and commutativity of scalar multiplication*: $cd\mathbf{u} = c(d\mathbf{u}) = (cd)\mathbf{u} = d(c\mathbf{u}) = dc\mathbf{u}$.

P8. *Multiplicative scalar identity*: $1\mathbf{v} = \mathbf{v}$.

Any set V for which the two closure properties and the eight structural properties are true for all elements in V and for all real-number scalars is called a **vector space**. Elements of a vector space are called vectors. Common vector spaces are R^n, where n is any non-negative integer. Thus, the *xy*-axis system R^2 is a vector space, where any ordered pair (a,b) can be thought of as a vector from $(0,0)$ to (a,b). In this manner, the elements of R^2 are vectors of the form $\langle a,b \rangle$, and all of the closure and structural properties listed above are met. Similarly, for R^3, the real line R^1 ($= R$), and even the trivial space $R^0 = \mathbf{0}$ are vector spaces.

Given any non-zero vector **v**, the **unit vector** of **v** is found by multiplying **v** by $\frac{1}{|\mathbf{v}|}$. The unit vector has magnitude 1. That is, $\left|\frac{\mathbf{v}}{|\mathbf{v}|}\right| = \frac{1}{|\mathbf{v}|}|\mathbf{v}| = 1$. The unit vector of **v** is any vector of length 1 parallel and in the same direction to **v**. Common notation for the unit vector of **v** is $\hat{\mathbf{v}}$ ("v-hat") or $\mathbf{v}_{unit}$.

In R^2, the vectors $\mathbf{i} = \langle 1,0 \rangle$ and $\mathbf{j} = \langle 0,1 \rangle$ are called the **standard orthonormal basis vectors**, which allows us to write a vector $\mathbf{v} = \langle v_1, v_2 \rangle = v_1\mathbf{i} + v_2\mathbf{j}$. In R^3, the standard orthonormal basis vectors are $\mathbf{i} = \langle 1,0,0 \rangle$, $\mathbf{j} = \langle 0,1,0 \rangle$ and $\mathbf{k} = \langle 0,0,1 \rangle$. The notation $\langle v_1, v_2 \rangle$ and $v_1\mathbf{i} + v_2\mathbf{j}$ to represent a vector in R^2 can be used interchangeably.

♦ • ♦ • ♦ • ♦ • ♦

Example 6.1: Sketch $\mathbf{u} = \langle 2,3 \rangle = 2\mathbf{i} + 3\mathbf{j}$.

Solution: From any starting point, move 2 units in the x (horizontal) direction, and 3 units in the y (vertical) direction. Below are five copies of the vector **u**.

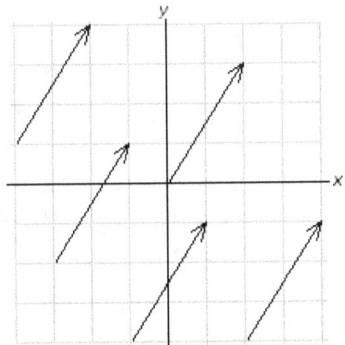

The foot can be placed anywhere. Multiple copies of the same vector can be drawn using different starting points. The position of a vector relative to a coordinate axis system is not relevant. If the direction and magnitude are not changed, it is considered to be the same vector.

♦ • ♦ • ♦ • ♦ • ♦

Example 6.2: Given $\mathbf{v} = \langle 4, -5 \rangle$, find $|\mathbf{v}|$, and the unit vector of **v**.

Solution: The magnitude of **v** is $|\mathbf{v}| = \sqrt{(4)^2 + (-5)^2} = \sqrt{16 + 25} = \sqrt{41}$.
The unit vector of **v** is $\hat{\mathbf{v}} = \frac{1}{\sqrt{41}}\langle 4, -5 \rangle = \left\langle \frac{4}{\sqrt{41}}, -\frac{5}{\sqrt{41}} \right\rangle$.

Example 6.3: Find a vector whose foot is $A = (-1, 4)$ and head is $B = (5, 2)$.

Solution: The vector is $\mathbf{AB} = \langle 5 - (-1), 2 - 4 \rangle = \langle 6, -2 \rangle$. Note that $\mathbf{BA} = -\mathbf{AB}$, has its foot at B, head at A, and is $\mathbf{BA} = -\langle 6, -2 \rangle = \langle -6, 2 \rangle$. Vectors $\mathbf{AB}$ and $\mathbf{BA}$ are parallel, but not in the same direction.

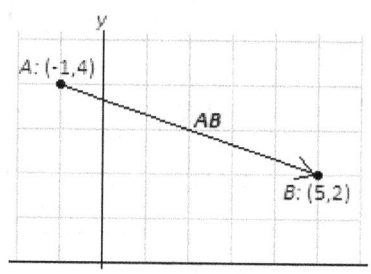

Example 6.4. Given $\mathbf{u} = \langle -2, 1 \rangle$ and $\mathbf{v} = \langle 1, 5 \rangle$, Find and sketch $\mathbf{u} + \mathbf{v}$, and $\mathbf{u} - \mathbf{v}$.

Solution:

$\mathbf{u} + \mathbf{v} = \langle -2 + 1, 1 + 5 \rangle = \langle -1, 6 \rangle$ and $\mathbf{u} - \mathbf{v} = \langle -2 - 1, 1 - 5 \rangle = \langle -3, -4 \rangle$.

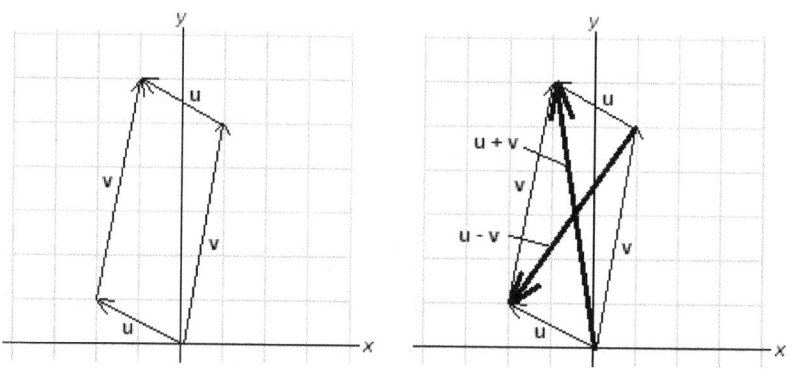

Vectors $\mathbf{u}$ and $\mathbf{v}$ form a parallelogram, with $\mathbf{u} + \mathbf{v}$ as one diagonal inside the parallelogram, and $\mathbf{u} - \mathbf{v}$ the other diagonal. If $\mathbf{u}$ and $\mathbf{v}$ have a common foot, then $\mathbf{u} - \mathbf{v}$ has its foot at the head of $\mathbf{v}$, and its head at the head of $\mathbf{u}$.

◆ • ◆ • ◆ • • ◆ • ◆

Example 6.5: Given $\mathbf{v} = \langle 3, 1 \rangle$, Find and sketch $2\mathbf{v}$ and $-3\mathbf{v}$.

Solution: We have $2\mathbf{v} = \langle 2(3), 2(1) \rangle = \langle 6, 2 \rangle$ and $-3\mathbf{v} = \langle -3(3), -3(1) \rangle = \langle -9, -3 \rangle$. These are shown on an xy-axis system on the next page.

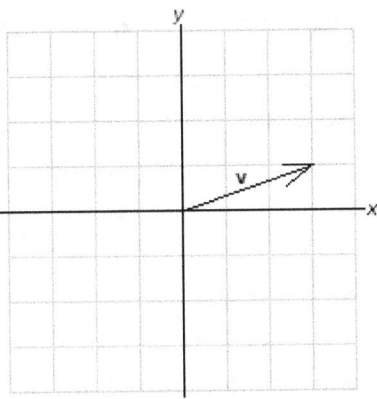

A representation of **v** = $\langle 3,1 \rangle$, with its foot at the origin.

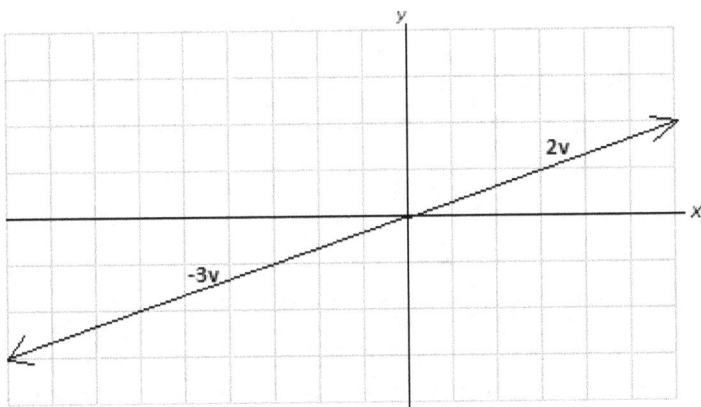

Representations of 2**v** and –3**v**.

Note that 2**v** has twice the magnitude of **v**, and –3**v** has three times the magnitude of **v**, pointing in the opposite direction. Vectors 2**v**, –3**v** and **v** are all parallel.

♦ • ♦ • ♦ • ♦ • ♦

Example 6.6: Given **u** = $\langle 1, -4, 2 \rangle$ and **v** = $\langle -5, 8, 3 \rangle$, find 4**u** – 3**v**.

Solution: We have

$$\begin{aligned} 4\mathbf{u} - 3\mathbf{v} &= 4\langle 1, -4, 2\rangle - 3\langle -5, 8, 3\rangle \\ &= \langle 4, -16, 8\rangle + \langle 15, -24, -9\rangle \\ &= \langle 19, -40, -1\rangle. \end{aligned}$$

Example 6.7: Given the vectors $\mathbf{u} = \langle 1, -4, 2 \rangle$ and $\mathbf{v} = \langle -5, 8, 3 \rangle$, find $\mathbf{w}$ where $3\mathbf{w} + 2\mathbf{u} = -\mathbf{v}$.

Solution: Using algebraic manipulation, solve for $\mathbf{w}$:

$$3\mathbf{w} + 2\mathbf{u} = -\mathbf{v}$$
$$3\mathbf{w} = -\mathbf{v} - 2\mathbf{u}$$
$$\mathbf{w} = -\frac{1}{3}\mathbf{v} - \frac{2}{3}\mathbf{u}.$$

Now, substitute $\mathbf{u}$ and $\mathbf{v}$, and simplify:

$$\mathbf{w} = -\frac{1}{3}\mathbf{v} - \frac{2}{3}\mathbf{u}$$
$$= -\frac{1}{3}\langle -5, 8, 3 \rangle - \frac{2}{3}\langle 1, -4, 2 \rangle$$
$$= \left\langle \frac{5}{3}, -\frac{8}{3}, -1 \right\rangle + \left\langle -\frac{2}{3}, \frac{8}{3}, -\frac{4}{3} \right\rangle$$
$$= \left\langle 1, 0, -\frac{7}{3} \right\rangle.$$

◆ • ◆ • ◆ • ◆ • ◆

Example 6.8: Let $\mathbf{v} = \langle 1, 2, 4 \rangle$, find a unit vector in the opposite direction of $\mathbf{v}$.

Solution: The magnitude is $|\mathbf{v}| = \sqrt{1^2 + 2^2 + 4^2} = \sqrt{21}$. A leading negative sign will cause the vector to point in the opposite direction. Thus,

$$-\hat{\mathbf{v}} = -\frac{1}{\sqrt{21}}\langle 1, 2, 4 \rangle = \left\langle -\frac{1}{\sqrt{21}}, -\frac{2}{\sqrt{21}}, -\frac{4}{\sqrt{21}} \right\rangle.$$

◆ • ◆ • ◆ • ◆ • ◆

Example 6.9: Find a vector in the opposite direction of $\mathbf{u} = \langle -2, 1 \rangle$ that is 4 times the length of $\mathbf{u}$.

Solution: Multiply by -4, where the 4 increases the magnitude by a factor of 4, and the negative reverses the direction: $-4\mathbf{u} = -4\langle -2, 1 \rangle = \langle 8, -4 \rangle$.

Example 6.10: Given $\mathbf{v} = \langle -1, 5, 2 \rangle$, find a vector $\mathbf{w}$ in the same direction as $\mathbf{v}$, with magnitude 3.

Solution: The unit vector is $\hat{\mathbf{v}} = \frac{1}{\sqrt{30}} \langle -1, 5, 2 \rangle$. Then multiply by 3:

$$\mathbf{w} = 3\hat{\mathbf{v}} = \frac{3}{\sqrt{30}} \langle -1, 5, 2 \rangle.$$

> In many situations, it is easier to leave any scalar multipliers at the front of the vector, rather than distributing it among the components.

◆ • ◆ • ◆ • ◆ • ◆

Example 6.11: A boat travels north at 30 miles per hour. Meanwhile, the current is moving toward the east at 5 miles per hour. If the boat's captain does not account for the current, the boat will drift to the east of its intended destination. After two hours, find (a) the boat's position as a vector, (b) the distance the boat travelled, and (c) the boat's position as a bearing.

Solution. Superimpose an *xy*-axis system, so that the positive *y*-axis North, and the positive *x*-axis is East. Thus, the boat's vector can be represented by $\mathbf{b} = \langle 0, 30 \rangle$, and the current's vector by $\mathbf{c} = \langle 5, 0 \rangle$.

a) The boat's position after two hours will be $2(\mathbf{b} + \mathbf{c}) = 2\langle 5, 30 \rangle = \langle 10, 60 \rangle$. From the boat's starting point, the boat moved 10 miles east and 60 miles north.

b) The boat travelled a distance of $|2(\mathbf{b} + \mathbf{c})| = 2\sqrt{5^2 + 30^2} = 2\sqrt{925}$, or about 60.83 miles.

c) Viewing a drawing below, we see that we can find the angle *t* using inverse trigonometry. Thus, $t = \tan^{-1}(10/60) \approx 9.46$ degrees East of North.

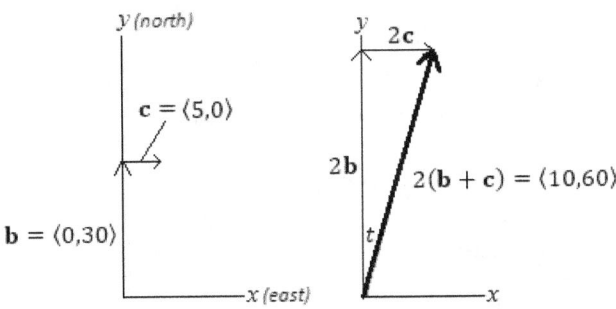

Example 6.12: A 10 kg mass hangs by two symmetric cables from a ceiling such that the cables meet at a 40-degree angle at the mass itself. Find the tension (in Newtons) on each cable.

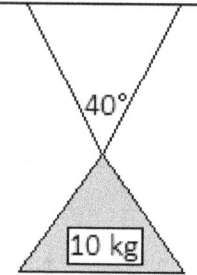

Solution: The force of the mass with respect to gravity is $F = ma = (10 \text{ kg})\left(9.8 \frac{m}{s^2}\right) = 98$ N. Let $|T|$ be the tension on one cable. We decompose $|T|$ into its vertical and horizontal components:

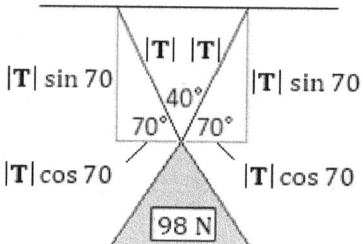

Note:

$|T|$ is not the length of the cable. It is the tension, as a scalar, within the cable.

The two horizontal components sum to 0, since the forces cancel one another, while the two vertical components support the 98 N downward force. Thus, we have

$$2|T| \sin 70 = 98.$$

Solving for $|T|$ we, we obtain

$$|T| = \frac{98}{2 \sin 70} = 52.14 \text{ N}.$$

Each cable has a tension of about 52.14 N. If the angle at which the cables meet was larger, the tensions would be greater. For example, if the cables were to meet at the mass at an angle of 150 degrees, then each cable would have a tension of

$$|T| = \frac{98}{2 \sin 15} = 189.32 \text{ N}$$

Example 6.13: A mass with a downward force of 100 N is being supported by two cables at full tension as shown in the diagram below. Let |T| represent the tension in one cable, and |U| the tension in the other cable. Find |T| and |U|.

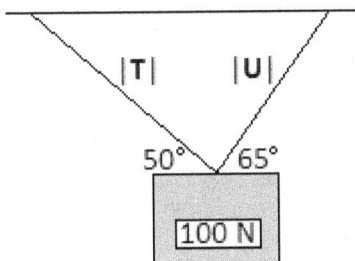

Solution: Decompose |T| and |U|:

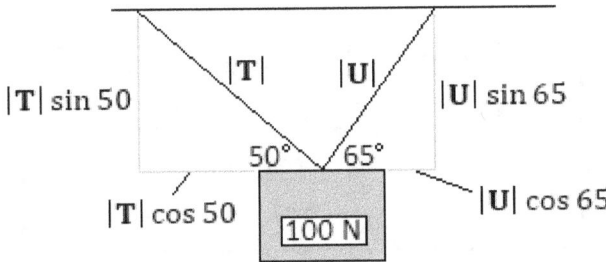

In doing so, we obtain two equations. The horizontal forces sum to 0, and since one force acts in the opposite direction of the other, we have |T| cos 50 − |U| cos 65 = 0. At the same time, the two vertical forces support the 100 N force, so we have |T| sin 50 + |U| sin 65 = 100. This is a system of two equations and two unknowns, |T| and |U|:

$$|T| \cos 50 - |U| \cos 65 = 0$$
$$|T| \sin 50 + |U| \sin 65 = 100.$$

Multiplying the top by sin 65 and the bottom by cos 65, we have

$$|T| \sin 65 \cos 50 - |U| \sin 65 \cos 65 = 0$$
$$|T| \sin 50 \cos 65 + |U| \sin 65 \cos 65 = 100 \cos 65.$$

Summing the two equations, we have

$$|T|(\sin 65 \cos 50 + \sin 50 \cos 65) = 100 \cos 65.$$

Now we can isolate |T|:

$$|T| = \frac{100 \cos 65}{\sin 65 \cos 50 + \sin 50 \cos 65} \approx 46.63 \text{ N}.$$

To find $|U|$, we use the first equation of the system, $|T| \cos 50 - |U| \cos 65 = 0$, solve for $|U|$, and substitute $|T| \approx 46.63$ N. We have

$$|U| = \frac{|T| \cos 50}{\cos 65} = \frac{46.63 \cos 50}{\cos 65} \approx 70.92 \text{ N}.$$

◆ • ◆ • • ◆ • • ◆

Example 6.14: An object with a weight of 75 lbs hangs at the center of a cable that is supported at opposite ends by poles 30 feet apart such that the cable meets the poles at the same height. Assume the weight sits in the center and causes a sag of 2 feet. Find the tension in each half of the cable.

Solution: A diagram helps us see the situation.

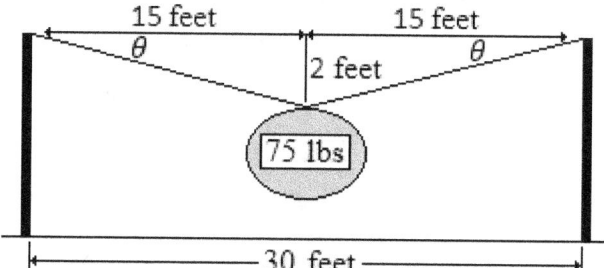

Since this is a symmetric system, the horizontal components of tension sum to 0, while the two vertical components support the 75 lb weight:

$$2|T| \sin \theta = 75; \quad \text{therefore} \quad |T| = \frac{75}{2 \sin \theta}.$$

The angle θ is not given, but we do know that $\sin \theta = \frac{2}{\sqrt{2^2+15^2}} = \frac{2}{\sqrt{229}}$, where $\sqrt{229}$ is the length of the hypotenuse of the right triangle, the hypotenuse being the cable under tension. Thus, we have

$$|T| = \frac{75}{2(2/\sqrt{229})} \approx 284.74 \text{ lbs}.$$

Each cable is supporting 284.78 lbs, which is plausible when considering the large angle at which the two halves of the cable meet at the object.

◆ • ◆ • • ◆ • • ◆

See an error? Have a suggestion?
Please see www.surgent.net/vcbook

7. The Dot Product

Let $\mathbf{u} = \langle u_1, u_2, u_3 \rangle$ and $\mathbf{v} = \langle v_1, v_2, v_3 \rangle$ be two vectors. The dot product of $\mathbf{u}$ and $\mathbf{v}$, written $\mathbf{u} \cdot \mathbf{v}$, is defined in two ways:

$\mathbf{u} \cdot \mathbf{v} = |\mathbf{u}||\mathbf{v}| \cos \theta$, where θ is the angle formed when the feet of $\mathbf{u}$ and $\mathbf{v}$ are placed together;

$\mathbf{u} \cdot \mathbf{v} = u_1 v_1 + u_2 v_2 + u_3 v_3$.

The two forms are equivalent and related to one another by the Law of Cosines. We often don't know the angle between the two vectors, so we tend to use the second formula. However, we may use the first formula to find the angle between vectors $\mathbf{u}$ and $\mathbf{v}$. Note that the dot product $\mathbf{u} \cdot \mathbf{v}$ results in a scalar value.

Common properties of the dot product are:

Commutativity: $\mathbf{u} \cdot \mathbf{v} = \mathbf{v} \cdot \mathbf{u}$.
Distributive property: $\mathbf{u} \cdot (\mathbf{v} + \mathbf{w}) = \mathbf{u} \cdot \mathbf{v} + \mathbf{u} \cdot \mathbf{w}$.
Scalar multiples can be combined as a product: $c\mathbf{u} \cdot d\mathbf{v} = cd\,(\mathbf{u} \cdot \mathbf{v})$.
Relation to magnitude: $\mathbf{u} \cdot \mathbf{u} = |\mathbf{u}|^2$.
Differentiation form: If $\mathbf{u}$ and $\mathbf{v}$ are vector-valued functions of a variable t then the derivative is $\frac{d}{dt}(\mathbf{u}(t) \cdot \mathbf{v}(t)) = \mathbf{u}(t) \cdot \frac{d}{dt}\mathbf{v}(t) + \mathbf{v}(t) \cdot \frac{d}{dt}\mathbf{u}(t)$. (vector-valued functions are discussed in Section 15).

The most useful feature of the dot product is its sign:

- If $\mathbf{u} \cdot \mathbf{v} > 0$, then the angle θ between vectors $\mathbf{u}$ and $\mathbf{v}$ is acute $\left(0 < \theta < \frac{\pi}{2}\right)$.

- If $\mathbf{u} \cdot \mathbf{v} < 0$, then the angle θ between vectors $\mathbf{u}$ and $\mathbf{v}$ is obtuse $\left(\frac{\pi}{2} < \theta < \pi\right)$.

- If $\mathbf{u} \cdot \mathbf{v} = 0$ (assuming $\mathbf{u}$ and $\mathbf{v}$ are non-zero vectors), then the angle θ between vectors $\mathbf{u}$ and $\mathbf{v}$ is right $\left(\theta = \frac{\pi}{2}\right)$, and the two vectors are **orthogonal** to one another.

Example 7.1: Let $\mathbf{u} = \langle 1, -2, -5 \rangle$ and $\mathbf{v} = \langle 3, 4, -2 \rangle$. Find (a) $\mathbf{u} \cdot \mathbf{v}$, and (b) the angle θ between $\mathbf{u}$ and $\mathbf{v}$.

Solution:

a) Use the second definition of the dot product:

$$\mathbf{u} \cdot \mathbf{v} = (1)(3) + (-2)(4) + (-5)(-2) = 3 - 8 + 10 = 5.$$

Since the dot product is positive, we know that the angle between $\mathbf{u}$ and $\mathbf{v}$ is acute.

b) Use the first definition of the dot product and solving for θ:

$$\theta = \cos^{-1}\left(\frac{\mathbf{u} \cdot \mathbf{v}}{|\mathbf{u}||\mathbf{v}|}\right) = \cos^{-1}\left(\frac{5}{\sqrt{30}\sqrt{29}}\right) \approx 1.4 \text{ radians, or } 80.34 \text{ degrees}.$$

◆ • ◆ • ◆ • ◆ • ◆

Example 7.2: Suppose 12 vectors of equal magnitude are arranged on an analog clock, such that all vectors have a common foot at the center. The vectors point to 12 o'clock, 1 o'clock, and so on.

a) Is the dot product of the vectors pointing to 2 o'clock and 4 o'clock be positive, negative, or zero? Why?
b) Is the dot product of the vectors pointing to 10 o'clock and 1 o'clock be positive, negative, or zero? Why?
c) Is the dot product of the vectors pointing to 5 o'clock and 9 o'clock be positive, negative, or zero? Why?

Solution:

a) The two vectors form an acute angle, so their dot product is positive.
b) The two vectors form a right angle, so their dot product is zero.
c) The two vectors form an obtuse angle, so their dot product is negative.

Be aware of notation. The dot product is defined between two vectors and is always written with the dot (·) symbol. Traditional scalar multiplication is written without the dot symbol. Thus, statements like $\mathbf{u} \cdot \mathbf{v}$ and $c\mathbf{u}$ are well defined, while statements like $\mathbf{uv}$ and $c \cdot \mathbf{u}$ are not defined.

Example 7.3: Let $\mathbf{u} = \langle 8,2,3 \rangle$ and $\mathbf{v} = \langle -3, k, 6 \rangle$. Find k so that $\mathbf{u}$ and $\mathbf{v}$ are orthogonal.

Solution: Since $\mathbf{u}$ and $\mathbf{v}$ are orthogonal, their dot product is 0:

$$\mathbf{u} \cdot \mathbf{v} = 0$$
$$(8)(-3) + (2)(k) + (3)(6) = 0$$
$$-24 + 2k + 18 = 0$$
$$2k - 6 = 0$$
$$k = 3.$$

Thus, the vectors $\mathbf{u} = \langle 8,2,3 \rangle$ and $\mathbf{v} = \langle -3,3,6 \rangle$ are orthogonal.

◆ • ◆ • ◆ • ◆ • ◆

Example 7.4: Suppose vector $\mathbf{u}$ has magnitude 6. What is $\mathbf{u} \cdot \mathbf{u}$?

Solution: Use the relationship $\mathbf{u} \cdot \mathbf{u} = |\mathbf{u}|^2$. Since $|\mathbf{u}| = 6$, then $\mathbf{u} \cdot \mathbf{u} = 6^2 = 36$.

◆ • ◆ • ◆ • ◆ • ◆

Example 7.5: Let $\mathbf{u} = \langle -1,4,6 \rangle$ and $\mathbf{v} = \langle 3,2,-4 \rangle$. Find all possible vectors $\mathbf{w}$ that are orthogonal to both $\mathbf{u}$ and $\mathbf{v}$.

Solution: Let $\mathbf{w} = \langle x, y, z \rangle$. Since $\mathbf{w}$ is orthogonal to $\mathbf{u}$, and orthogonal to $\mathbf{v}$, we know that $\mathbf{u} \cdot \mathbf{w} = 0$ and $\mathbf{v} \cdot \mathbf{w} = 0$.

$$\begin{array}{c} \mathbf{u} \cdot \mathbf{w} = 0 \\ \mathbf{v} \cdot \mathbf{w} = 0 \end{array} \text{ which gives } \begin{array}{c} -x + 4y + 6z = 0 \\ 3x + 2y - 4z = 0 \end{array}$$

This is a system of two equations in three variables. Let $x = t$ (a constant). Now we have a system of two equations in two variables. Here, we eliminate the y variable first:

$$\begin{array}{c} 4y + 6z = t \\ 2y - 4z = -3t \end{array} \xrightarrow{\binom{\text{Multiply bottom}}{\text{row by -2}}} \begin{array}{c} 4y + 6z = t \\ -4y + 8z = 6t \end{array} \xrightarrow{\binom{\text{Sum the two}}{\text{equations}}} 14z = 7t.$$

Thus, $z = \frac{1}{2}t$. Using the equation $4y + 6z = t$ where $z = \frac{1}{2}t$, we obtain $4y + 6\left(\frac{1}{2}t\right) = t$, which, after algebra, gives $y = -\frac{1}{2}t$. We now have vector $\mathbf{w}$ identified:

$$\mathbf{w} = \langle x, y, z \rangle = \left\langle t, -\frac{1}{2}t, \frac{1}{2}t \right\rangle = t\left\langle 1, -\frac{1}{2}, \frac{1}{2} \right\rangle.$$

There are infinitely many vectors **w** that are simultaneously orthogonal to **u** = ⟨−1,4,6⟩ and **v** = ⟨3,10,−2⟩, and they are all scalar multiples of the vector $\left\langle 1, -\frac{1}{2}, \frac{1}{2}\right\rangle$, or ⟨2,−1,1⟩ if we prefer a vector with integer components.

To verify, we show that the dot product of ⟨2,−1,1⟩ (or any non-zero scalar multiple thereof) with **u** and with **v** is 0:

$$\langle 2, -1, 1\rangle \cdot \langle -1, 4, 6\rangle = (2)(-1) + (-1)(4) + (1)(6) = -2 - 4 + 6 = 0$$

and that

$$\langle 2, -1, 1\rangle \cdot \langle 3, 2, -4\rangle = (2)(3) + (-1)(2) + (1)(-4) = 6 - 2 - 4 = 0$$

♦ • ♦ • ♦ • ♦ • ♦

Example 7.6: Let **u** = ⟨1,−4,2⟩. Find all possible vectors **v** orthogonal to **u**.

Solution: Let **v** = ⟨x, y, z⟩. Since **u** and **v** are orthogonal, we have **u** · **v** = 0:

$$\langle 1, -4, 2\rangle \cdot \langle x, y, z\rangle = 0, \quad \text{which gives} \quad x - 4y + 2z = 0.$$

Let $y = s$ and $z = t$. This is a system where x, y and z are written in terms of s and t:

$$x = 4s - 2t$$
$$y = s$$
$$z = t.$$

Now, "zero-fill" so that columns are evident:

$$\begin{aligned} x &= 4s - 2t \\ y &= 1s + 0t \\ z &= 0s + 1t \end{aligned} \quad \text{which can be written as} \quad \begin{bmatrix} x \\ y \\ z \end{bmatrix} = s\begin{bmatrix} 4 \\ 1 \\ 0 \end{bmatrix} + t\begin{bmatrix} -2 \\ 0 \\ 1 \end{bmatrix}.$$

Thus, any vector **v** that is orthogonal to **u** = ⟨1,−4,2⟩ is of the form

$$\mathbf{v} = s\langle 4, 1, 0\rangle + t\langle -2, 0, 1\rangle.$$

Here, s and t are parameter variables, chosen independently of one another. For example, if $s = -4$ and $t = 9$, then

$$\mathbf{v} = -4\langle 4, 1, 0\rangle + 9\langle -2, 0, 1\rangle = \langle -16, -4, 0\rangle + \langle -18, 0, 9\rangle = \langle -34, -4, 9\rangle.$$

It is easy to verify that **v** = ⟨−34,−4,9⟩ is orthogonal to **u** = ⟨1,−4,2⟩:

$$\langle 1, -4, 2\rangle \cdot \langle -34, -4, 9\rangle = -34 + 16 + 18 = 0.$$

Thus, we have a "formula" that can generate all possible vectors **v** that are orthogonal to **u** = ⟨1, −4, 2⟩. Try it with any other choices of s and t.

This is generalized and discussed in Section 9, The Cross Product.

◆ • ◆ • ◆ • ◆ • ◆

8. Projections

Given two vectors **u** and **v**, the **orthogonal projection** (or **projection**) of **u** onto **v** is given by

$$\text{proj}_\mathbf{v}\, \mathbf{u} = \frac{\mathbf{u} \cdot \mathbf{v}}{\mathbf{v} \cdot \mathbf{v}} \mathbf{v}. \quad \left(\text{The expression } \frac{\mathbf{u} \cdot \mathbf{v}}{\mathbf{v} \cdot \mathbf{v}} \text{ is a scalar multiplier of } \mathbf{v}. \right)$$

Think of a right triangle: **u** is the hypotenuse, and $\text{proj}_\mathbf{v}\, \mathbf{u}$ is the adjacent leg of the triangle that points in the direction of **v**. The opposite leg, called $\text{norm}_\mathbf{v}\, \mathbf{u}$, is found by vector summation:

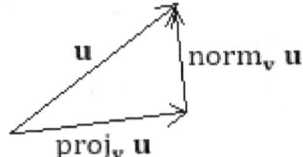

Since $\text{proj}_\mathbf{v}\, \mathbf{u} + \text{norm}_\mathbf{v}\, \mathbf{u} = \mathbf{u}$, then $\text{norm}_\mathbf{v}\, \mathbf{u} = \mathbf{u} - \text{proj}_\mathbf{v}\, \mathbf{u}$.

Often, we need to write a vector **u** in terms of a vector parallel to **v**, and another vector orthogonal (or normal) to **v**. This is called **decomposition** of a vector, and is done using projections.

> Take care to perform the projection operations in the correct order. When we say "project **u** onto **v**", it is vector **v** that is being altered.

Example 8.1: Find the projection of **u** = ⟨2, 5⟩ onto **v** = ⟨4, 1⟩.

Solution: The projection of **u** onto **v** is

$$\text{proj}_\mathbf{v}\, \mathbf{u} = \frac{\mathbf{u} \cdot \mathbf{v}}{\mathbf{v} \cdot \mathbf{v}} \mathbf{v} = \frac{\langle 2,5 \rangle \cdot \langle 4,1 \rangle}{\langle 4,1 \rangle \cdot \langle 4,1 \rangle} \langle 4,1 \rangle$$

$$= \frac{(2)(4) + (5)(1)}{(4)(4) + (1)(1)} \langle 4,1 \rangle$$

$$= \frac{13}{17} \langle 4,1 \rangle \approx \langle 3.06, 0.76 \rangle.$$

The graphical representation is on the next page.

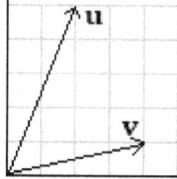 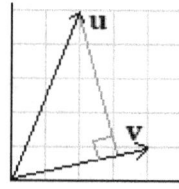 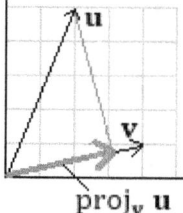

In the left image, vectors **u** and **v** are plotted. In the center image, a perpendicular is dropped from **u** onto **v**. In the right image, $\text{proj}_v\,\mathbf{u}$ is the vector "on" **v** that forms the adjacent leg of a right triangle, with **u** as the hypotenuse.

◆ • ◆ • • ◆ • ◆ • ◆

Example 8.2: Decompose $\mathbf{u} = \langle 2,5 \rangle$ into two vectors, one parallel to $\mathbf{v} = \langle 4,1 \rangle$, and another normal to $\mathbf{v} = \langle 4,1 \rangle$.

Solution: The vector parallel to **v** is found by projecting $\mathbf{u} = \langle 2,5 \rangle$ onto $\mathbf{v} = \langle 4,1 \rangle$. From the previous example, we have $\text{proj}_v\,\mathbf{u} = \frac{13}{17}\langle 4,1 \rangle$. The vector normal to **v** is found by vector subtraction

$$\text{norm}_v\,\mathbf{u} = \mathbf{u} - \text{proj}_v\,\mathbf{u}$$
$$= \langle 2,5 \rangle - \frac{13}{17}\langle 4,1 \rangle$$
$$= \left\langle -\frac{18}{17}, \frac{72}{17} \right\rangle$$
$$= \frac{18}{17}\langle -1,4 \rangle.$$

Viewing this as a right triangle, $\text{proj}_v\,\mathbf{u} = \frac{13}{17}\langle 4,1 \rangle$ and $\text{norm}_v\,\mathbf{u} = \frac{18}{17}\langle -1,4 \rangle$ are the two legs of a right triangle, with $\mathbf{u} = \langle 2,5 \rangle$ being the hypotenuse.

◆ • ◆ • • ◆ • ◆ • ◆

Example 8.3: Jimmy is standing at the point $A = (1,2)$ and wants to visit his friend who lives at the point $B = (7,3)$, all units in miles. Jimmy starts walking in the direction given by $\mathbf{v} = \langle 4,1 \rangle$. If he continues to walk in this direction, he will miss point B. Suppose Jimmy is allowed one right-angle turn. At what point should Jimmy make this right angle turn so that he arrives at point B?

Solution: We sketch a diagram to get a sense of Jimmy's location and direction of travel, as well as to see where his right-angle turn should be made.

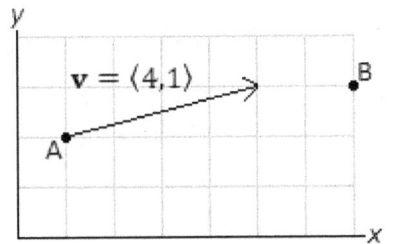 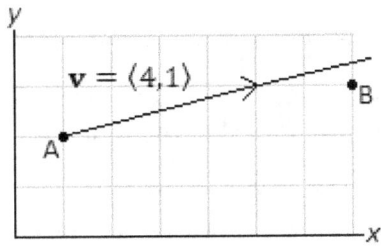

Left: The sketch of Jimmy's position vector, with his starting point at A, his intended finish point B, and his direction vector **v**. **Right:** If Jimmy walks in the direction of **v**, he will bypass B.

The diagram suggests a right triangle, with the vector $\mathbf{AB} = \langle 6,1 \rangle$ forming the hypotenuse, and $\mathbf{v} = \langle 4,1 \rangle$ defining the direction of the adjacent leg (relative to Jimmy's initial position). Thus, we find the projection of **AB** onto **v**:

$$\text{proj}_\mathbf{v} \mathbf{AB} = \frac{\mathbf{AB} \cdot \mathbf{v}}{\mathbf{v} \cdot \mathbf{v}} \mathbf{v}$$

$$= \frac{\langle 6,1 \rangle \cdot \langle 4,1 \rangle}{\langle 4,1 \rangle \cdot \langle 4,1 \rangle} \langle 4,1 \rangle$$

$$= \frac{(6)(4) + (1)(1)}{(4)(4) + (1)(1)} \langle 4,1 \rangle$$

$$= \frac{25}{17} \langle 4,1 \rangle.$$

This vector, $\frac{25}{17} \langle 4,1 \rangle = \langle \frac{100}{17}, \frac{25}{17} \rangle$, can be placed so its foot is at $A = (1,2)$. Thus, its head will be at $\left(1 + \frac{100}{17}, 2 + \frac{25}{17}\right) = \left(\frac{117}{17}, \frac{59}{17}\right)$, or about $(6.88, 3.47)$, which agrees well with the diagram. This is where Jimmy should make his right-angle turn.

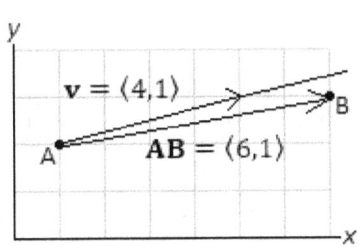 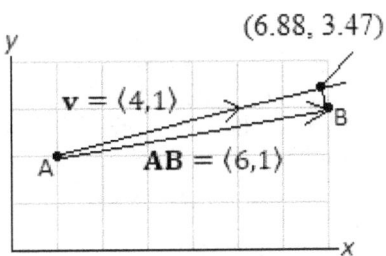

Jimmy will walk a distance of $|\text{proj}_\mathbf{v} \mathbf{AB}| = \left|\langle \frac{100}{17}, \frac{25}{17} \rangle\right| = \sqrt{\left(\frac{100}{17}\right)^2 + \left(\frac{25}{17}\right)^2}$, or about 6.06 miles in the direction of **v**, and then walk $|\text{norm}_\mathbf{v} \mathbf{AB}| = \sqrt{\left(7 - \frac{117}{17}\right)^2 + \left(3 - \frac{59}{17}\right)^2}$, about 0.49 miles orthogonal to **v**.

9. The Cross Product

Let $\mathbf{u} = \langle u_1, u_2, u_3 \rangle$ and $\mathbf{v} = \langle v_1, v_2, v_3 \rangle$ be two vectors in R^3. The **cross product** of $\mathbf{u}$ and $\mathbf{v}$ is denoted $\mathbf{u} \times \mathbf{v}$ and is defined by the formula

$$\mathbf{u} \times \mathbf{v} = \langle u_2 v_3 - u_3 v_2, u_3 v_1 - u_1 v_3, u_1 v_2 - u_2 v_1 \rangle.$$

The cross product is a vector that is simultaneously orthogonal to $\mathbf{u}$ and to $\mathbf{v}$.

$$(\mathbf{u} \times \mathbf{v}) \cdot \mathbf{u} = 0 \text{ and } (\mathbf{u} \times \mathbf{v}) \cdot \mathbf{v} = 0.$$

The calculation of the cross product is summarized as the determinant of a 3 by 3 matrix:

$$\mathbf{u} \times \mathbf{v} = \begin{vmatrix} \mathbf{i} & \mathbf{j} & \mathbf{k} \\ u_1 & u_2 & u_3 \\ v_1 & v_2 & v_3 \end{vmatrix} = \begin{vmatrix} u_2 & u_3 \\ v_2 & v_3 \end{vmatrix} \mathbf{i} - \begin{vmatrix} u_1 & u_3 \\ v_1 & v_3 \end{vmatrix} \mathbf{j} + \begin{vmatrix} u_1 & u_2 \\ v_1 & v_2 \end{vmatrix} \mathbf{k}.$$

Don't forget the negative in front of the $\mathbf{j}$ term.

The cross product vector, $\mathbf{u} \times \mathbf{v}$, points in a direction as defined by the "right hand rule". Curling one's fingers on their right hand from $\mathbf{u}$ to $\mathbf{v}$, the cross product vector will point in the direction of the thumb pointing away from the palm.

Common properties of the cross product are:

- Switching the order of the vectors results in a factor of -1: $\mathbf{u} \times \mathbf{v} = -(\mathbf{v} \times \mathbf{u})$. The cross product is *not* commutative.

- Scalars can be grouped to the front: $c\mathbf{u} \times d\mathbf{v} = cd(\mathbf{u} \times \mathbf{v})$.

- Distributive property: $\mathbf{u} \times (\mathbf{v} + \mathbf{w}) = \mathbf{u} \times \mathbf{v} + \mathbf{u} \times \mathbf{w}$.

- The magnitude of the cross product is $|\mathbf{u} \times \mathbf{v}| = |\mathbf{u}||\mathbf{v}| \sin \theta$. It is the area of the parallelogram formed by $\mathbf{u}$ and $\mathbf{v}$. (A proof of this is given at the end of the book)

A corollary to the last property is that if $\mathbf{u}$ and $\mathbf{v}$ are parallel, then $\mathbf{u} \times \mathbf{v} = \mathbf{0}$. If vectors $\mathbf{u}$ and $\mathbf{v}$ are in R^2, then the cross product is defined if the vectors are rewritten as $\mathbf{u} = \langle u_1, u_2, 0 \rangle$ and $\mathbf{v} = \langle v_1, v_2, 0 \rangle$. In this case, the result is $\mathbf{u} \times \mathbf{v} = \langle 0, 0, u_1 v_2 - u_2 v_1 \rangle$.

Example 9.1: In the image below, vectors **u**, **v** and **w** are drawn. State whether the cross products (a) **u** × **v**, (b) **u** × **w** and (c) **w** × **v** are positive (coming out of the paper), negative (going into the paper) or **0**.

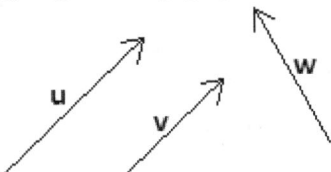

Solution: (a) Since **u** and **v** are parallel, **u** × **v** = **0**; (b) Placing the feet of **u** and **w** together, curl your fingers from **u** to **w**. Your thumb will point upwards, out of the paper, so **u** × **w** is positive. (c) placing the feet of **w** and **v** together, you have to invert your hand in order to curl your fingers from **w** towards **v**. Thus, your thumb would point down, or into the paper, and **w** × **v** is negative.

◆ • ◆ • ◆ • ◆ • ◆

Example 9.2: Let $\mathbf{u} = \langle -1, 4, 6 \rangle$ and $\mathbf{v} = \langle 3, 2, -4 \rangle$. Find $\mathbf{u} \times \mathbf{v}$.

Solution: We have

$$\mathbf{u} \times \mathbf{v} = \begin{vmatrix} \mathbf{i} & \mathbf{j} & \mathbf{k} \\ -1 & 4 & 6 \\ 3 & 2 & -4 \end{vmatrix} = \begin{vmatrix} 4 & 6 \\ 2 & -4 \end{vmatrix} \mathbf{i} - \begin{vmatrix} -1 & 6 \\ 3 & -4 \end{vmatrix} \mathbf{j} + \begin{vmatrix} -1 & 4 \\ 3 & 2 \end{vmatrix} \mathbf{k}$$

$$= ((4)(-4) - (6)(2))\mathbf{i} - ((-1)(-4) - (6)(3))\mathbf{j} + ((-1)(2) - (4)(3))\mathbf{k}$$

$$= -28\mathbf{i} + 14\mathbf{j} - 14\mathbf{k}, \quad \text{or} \quad \langle -28, 14, -14 \rangle.$$

Check that this is correct by showing $(\mathbf{u} \times \mathbf{v}) \cdot \mathbf{u} = 0$ and $(\mathbf{u} \times \mathbf{v}) \cdot \mathbf{v} = 0$:

$$(\mathbf{u} \times \mathbf{v}) \cdot \mathbf{u} = \langle -28, 14, -14 \rangle \cdot \langle -1, 4, 6 \rangle$$

$$= (-28)(-1) + (14)(4) + (-14)(6)$$

$$= 28 + 56 - 84 = 0;$$

$$(\mathbf{u} \times \mathbf{v}) \cdot \mathbf{v} = \langle -28, 14, -14 \rangle \cdot \langle 3, 2, -4 \rangle$$

$$= (-28)(3) + (14)(2) + (-14)(-4)$$

$$= -84 + 28 + 56 = 0.$$

Thus, $\mathbf{u} \times \mathbf{v} = \langle -28, 14, -14 \rangle$ is orthogonal to **u** and orthogonal to **v**. In Example 7.5, we showed that there are infinitely many vectors that are orthogonal to $\mathbf{u} = \langle -1, 4, 6 \rangle$ and $\mathbf{v} = \langle 3, 2, -4 \rangle$, all having the form $t \langle 1, -\frac{1}{2}, \frac{1}{2} \rangle$. Note that $\langle -28, 14, -14 \rangle$ is one such vector, when $t = -28$.

Example 9.3: Let $\mathbf{u} = \langle -1,4,6 \rangle$ and $\mathbf{v} = \langle 3,2,-4 \rangle$. Find the area of the parallelogram formed by $\mathbf{u}$ and $\mathbf{v}$, then find the area of the triangle formed by $\mathbf{u}$ and $\mathbf{v}$.

Solution: From the previous example, we have $\mathbf{u} \times \mathbf{v} = \langle -28, 14, -14 \rangle$. Thus, the area of the parallelogram formed by $\mathbf{u}$ and $\mathbf{v}$ is

$$|\mathbf{u} \times \mathbf{v}| = |\langle -28, 14, -14 \rangle|$$
$$= \sqrt{(-28)^2 + 14^2 + (-14)^2}$$
$$= \sqrt{1176}, \text{ or about } 34.29 \text{ units}^2.$$

The area of the triangle formed by $\mathbf{u}$ and $\mathbf{v}$ is half this quantity, $\frac{1}{2}\sqrt{1176}$, or about 17.145 units².

◆ • ◆ • • ◆ • ◆ • ◆

Example 9.4: Find the area of the triangle formed by the points $A = (1,3,-2)$, $B = (4,0,3)$ and $C = (6,-3,5)$.

Solution: We form two vectors from among the three points (any pair of vectors will suffice). The vector from A to B is $\mathbf{AB} = \langle 3, -3, 5 \rangle$, and the vector from A to C is $\mathbf{AC} = \langle 5, -6, 7 \rangle$. The cross product of $\mathbf{AB}$ and $\mathbf{AC}$ is

$$\mathbf{AB} \times \mathbf{AC} = \begin{vmatrix} \mathbf{i} & \mathbf{j} & \mathbf{k} \\ 3 & -3 & 5 \\ 5 & -6 & 7 \end{vmatrix} = \langle 9, 4, -3 \rangle.$$

(Check that this is correct by verifying that $(\mathbf{AB} \times \mathbf{AC}) \cdot \mathbf{AB} = 0$ and $(\mathbf{AB} \times \mathbf{AC}) \cdot \mathbf{AC} = 0$.)

The area of the triangle is half the magnitude of $\mathbf{AB} \times \mathbf{AC}$:

$$\text{Area} = \frac{1}{2}\sqrt{9^2 + 4^2 + (-3)^2} = \frac{1}{2}\sqrt{106} \approx 5.15 \text{ units}^2.$$

Other vectors formed from points A, B and C will work. For example, try this with vectors $\mathbf{BA}$ and $\mathbf{CB}$.

10. The Scalar Triple Product

Let $\mathbf{u} = \langle u_1, u_2, u_3 \rangle$, $\mathbf{v} = \langle v_1, v_2, v_3 \rangle$ and $\mathbf{w} = \langle w_1, w_2, w_3 \rangle$ be three vectors in R^3. The **scalar triple product** of $\mathbf{u}$, $\mathbf{v}$ and $\mathbf{w}$ is given by

$$\mathbf{u} \cdot (\mathbf{v} \times \mathbf{w}) = \begin{vmatrix} u_1 & u_2 & u_3 \\ v_1 & v_2 & v_3 \\ w_1 & w_2 & w_3 \end{vmatrix}.$$

The scalar triple product is a scalar quantity. Its absolute value is the volume of the **parallelepiped** (a "tilted" box) formed by $\mathbf{u}$, $\mathbf{v}$ and $\mathbf{w}$. The ordering of the vectors is not important. For example,

$$\mathbf{u} \cdot (\mathbf{v} \times \mathbf{w}), \qquad \mathbf{v} \cdot (\mathbf{u} \times \mathbf{w}), \qquad \text{and} \qquad \mathbf{w} \cdot (\mathbf{u} \times \mathbf{v}),$$

result in values that differ by at most the factor -1.

◆ • ◆ • • ◆ • • ◆

Example 10.1: Find the volume of the parallelepiped with sides represented by the vectors $\mathbf{u} = \langle -1, 3, 2 \rangle$, $\mathbf{v} = \langle 4, 2, -5 \rangle$ and $\mathbf{w} = \langle 0, -3, 1 \rangle$.

Solution: We have

$$\mathbf{u} \cdot (\mathbf{v} \times \mathbf{w}) = \begin{vmatrix} -1 & 3 & 2 \\ 4 & 2 & -5 \\ 0 & -3 & 1 \end{vmatrix}$$

$$= -1 \begin{vmatrix} 2 & -5 \\ -3 & 1 \end{vmatrix} - 3 \begin{vmatrix} 4 & -5 \\ 0 & 1 \end{vmatrix} + 2 \begin{vmatrix} 4 & 2 \\ 0 & -3 \end{vmatrix}$$

$$= -1(2 - 15) - 3(4 - 0) + 2(-12 - 0)$$

$$= -23.$$

Taking the absolute value, the volume is 23 cubic units.

Example 10.2: Four ordered triples, $A = (3,2,1)$, $B = (-2,5,4)$, $C = (6,10,1)$ and $D = (0,7,-4)$ form a tetrahedron, a four-sided solid in which each face is a triangle. Find the volume of this tetrahedron.

Solution: The volume of a tetrahedron is one-sixth the volume of the parallelepiped that contains it. We find three vectors, and then find their scalar triple product.

$$AB = \langle -5,3,3 \rangle, \quad AC = \langle 3,8,0 \rangle, \quad AD = \langle -3,5,-5 \rangle.$$

Next, we have:

$$AB \cdot (AC \times AD) = \begin{vmatrix} -5 & 3 & 3 \\ 3 & 8 & 0 \\ -3 & 5 & -5 \end{vmatrix} = 362.$$

Thus, the volume of the tetrahedron is $\frac{1}{6}(362) = \frac{181}{3} = 60\frac{1}{3}$ cubic units.

◆ • ◆ • • ◆ • • ◆

Example 10.3: Find the volume of the parallelepiped with sides represented by the vectors $\mathbf{u} = \langle 2,1,3 \rangle$, $\mathbf{v} = \langle -5,2,1 \rangle$ and $\mathbf{w} = \langle -3,3,4 \rangle$. Interpret the result.

Solution: We have

$$\mathbf{u} \cdot (\mathbf{v} \times \mathbf{w}) = \begin{vmatrix} 2 & 1 & 3 \\ -5 & 2 & 1 \\ -3 & 3 & 4 \end{vmatrix}$$

$$= 2\begin{vmatrix} 2 & 1 \\ 3 & 4 \end{vmatrix} - 1\begin{vmatrix} -5 & 1 \\ -3 & 4 \end{vmatrix} + 3\begin{vmatrix} -5 & 2 \\ -3 & 3 \end{vmatrix}$$

$$= 2(8-3) - (-20-(-3)) + 3(-15-(-6))$$

$$= 2(5) - (-17) + 3(-9)$$

$$= 10 + 17 - 27$$

$$= 0.$$

The volume of the parallelepiped formed by $\mathbf{u}$, $\mathbf{v}$ and $\mathbf{w}$ is 0. The object has no thickness, which would imply that the three vectors are **coplanar** (lying on a common plane).

11. Work & Torque

Work is defined as force F (in Newtons) applied to move an object a distance of d (in meters). It is the product of F and d:

$$W = Fd$$

The standard metric unit for work is Joules, which is equivalent to Newton-meters.

If the force is not applied in the same direction that the object will move, then we need to find the component of **F** (written now as a vector) that is parallel to the direction **d**, also now written as a vector. Placing the feet of **F** and **d** together, the component of **F** in the direction of **d** is given by $|\mathbf{F}|\cos\theta$. If $|\mathbf{d}|$ is the length of vector **d** (not necessarily the distance moved by the object in the direction of **d**), then the work is given by

$$W = |\mathbf{F}||\mathbf{d}|\cos\theta.$$

This is the dot product of **F** and **d**. Thus, work can be defined as a dot product,

$$W = \mathbf{F} \cdot \mathbf{d}.$$

Work is a scalar value. It may be a negative value, which can be interpreted that the object is moving against the force being applied to it. For example, walking into a headwind at an angle.

◆ • ◆ • • ◆ • • ◆

Example 11.1: A force of 10 Newtons is applied in the direction of $\langle 1,1 \rangle$ to an object that moves in the direction of the positive x-axis for 5 meters. Find the work performed on this object.

Solution: Using geometry, the component of the force in the direction of the positive x-axis is $|\mathbf{F}|\cos\left(\frac{\pi}{4}\right) = 10\left(\frac{\sqrt{2}}{2}\right) = 5\sqrt{2}$. The object moves 5 meters, so the work performed is

$$W = Fd = (5\sqrt{2})(5) = 25\sqrt{2} \text{ Joules.}$$

Using vectors, the force vector is $\mathbf{F} = \left\langle \frac{10}{\sqrt{2}}, \frac{10}{\sqrt{2}} \right\rangle$ and the direction vector is $\mathbf{d} = \langle 5,0 \rangle$. Thus, the work performed is

$$W = \mathbf{F} \cdot \mathbf{d} = \left\langle \frac{10}{\sqrt{2}}, \frac{10}{\sqrt{2}} \right\rangle \cdot \langle 5,0 \rangle = \frac{50}{\sqrt{2}} = 50\left(\frac{\sqrt{2}}{2}\right) = 25\sqrt{2} \text{ Joules.}$$

Example 11.2: A force of 50 N is applied in the direction of the positive x-axis, moving an object 10 meters up an inclined plane of 30 degrees. Find the work.

Solution: Using geometry, the component of the force that is parallel to the inclined ramp is $50 \cos 30 = 50 \left(\frac{\sqrt{3}}{2}\right) = 25\sqrt{3}$ N. This component of the force moves the object 10 m, so the total work performed is

$$W = Fd = (25\sqrt{3})(10) = 250\sqrt{3} \text{ Joules.}$$

Using vectors, the force is given by $\mathbf{F} = \langle 50, 0 \rangle$ while the 10-mtere ramp inclined at 30 degrees has is described by the vector $\mathbf{d} = \langle 10 \cos 30, 10 \sin 30 \rangle$. Thus the work performed is

$$\begin{aligned} W &= \mathbf{F} \cdot \mathbf{d} \\ &= \langle 50, 0 \rangle \cdot \langle 10 \cos 30, 10 \sin 30 \rangle \\ &= 500 \cos 30 \\ &= 250\sqrt{3} \text{ Joules.} \end{aligned}$$

◆ • ◆ • ◆ • ◆ • ◆

Torque describes the force resulting from a pivoting motion. For example, when a wrench is turned around a pivot point (*e.g.* a bolt), it creates a force in the direction of the bolt. If the wrench is described as a vector $\mathbf{r}$ (where the length of the wrench is $|\mathbf{r}|$) and the force applied to the wrench as another vector $\mathbf{F}$ (with magnitude $|\mathbf{F}|$), then the torque τ (tau) is orthogonal to both $\mathbf{r}$ and $\mathbf{F}$, defined by the cross product:

$$\tau = \mathbf{r} \times \mathbf{F}.$$

This means that torque τ is a vector. However, if both $\mathbf{r}$ and $\mathbf{F}$ lie in the xy-plane, then τ is a vector of the form $\langle 0, 0, k \rangle$, and its magnitude is $|\tau| = k$. This scalar value is usually given as the torque in place of its vector form.

It makes intuitive sense that the force $\mathbf{F}$ applied to a wrench $\mathbf{r}$ (or anything that turns on a pivot) should be orthogonal to $\mathbf{r}$. If $\mathbf{F}$ is not orthogonal to $\mathbf{r}$, then we look for the component of $\mathbf{F}$ that is orthogonal to $\mathbf{r}$. This component of $\mathbf{F}$ is what "turns the wrench", so to speak.

In the following images, vectors $\mathbf{r}$ and $\mathbf{F}$ are both drawn with their feet at a common point, the pivot point. The angle between $\mathbf{r}$ and $\mathbf{F}$ is θ. Naturally, we would not apply the force $\mathbf{F}$ at the pivot itself. Recall that a vector can be moved at will, so we locate $\mathbf{F}$ so that its foot is at $\mathbf{r}$'s head. Now we look for the component of the magnitude of $\mathbf{F}$ that is orthogonal to $\mathbf{r}$. We see that it is $|\mathbf{F}| \sin \theta$. If $|\mathbf{r}|$ is the length of the wrench, then the area given by $|\mathbf{F}||\mathbf{r}| \sin \theta$ is interpreted as the torque, as a scalar. This is $|\tau| = |\mathbf{r} \times \mathbf{F}|$.

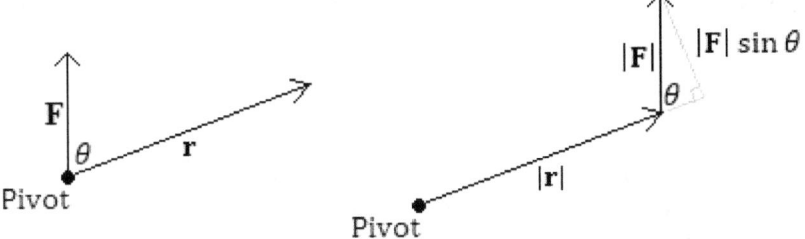

Left: Initial configuration. **Right:** F has been moved to r's head, and the component orthogonal to r is identified.

Also note that in this example, **r** × **F** points out of the page. This force **F** will turn **r** in a counter-clockwise direction, and in the usual sense of a wrench turning a bolt, this will loosen the bolt – *i.e.* it will emerge from this page.

Since **F** is in Newtons, then **r** must be in meters to maintain consistency in units. The units for torque are Newton-meters, abbreviated Nm.

♦ • ♦ • ♦ • ♦ • ♦

Example 11.3: Find the torque around the pivot shown in the following diagram.

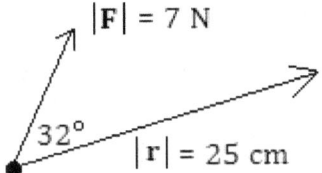

Solution: We convert the magnitude of **r** into meters. Thus, the torque is

$$|\tau| = |\mathbf{r} \times \mathbf{F}|$$
$$= |\mathbf{F}||\mathbf{r}| \sin \theta$$
$$= (7)(0.25) \sin 32° \quad (25 \text{ cm} = 0.25 \text{ m})$$
$$= 0.927 \text{ Nm}.$$

Example 11.4: Find the torque around the pivot shown in the following diagram.

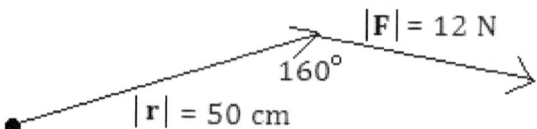

Solution: Note the position of the force vector **F** relative to **r**. Although it is reasonable to assume that to turn **r** around its pivot, we would apply the force **F** at its head, the angle given in the diagram must be handled carefully. The desired angle θ is defined by placing the feet of **F** and **r** at a common point. In other words, the equation $|\tau| = |\mathbf{F}||\mathbf{r}| \sin \theta = (12)(0.5) \sin 160°$ is *incorrect*.

Instead, redraw the diagram slightly, placing the foot of **F** at the pivot to better see the necessary angle:

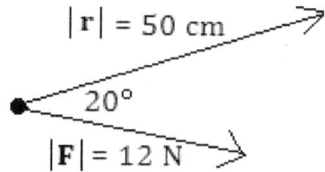

The desired angle is 20°. Thus, the torque is $|\tau| = |\mathbf{F}||\mathbf{r}| \sin \theta = (12)(0.5) \sin 20° \approx 2.05$ Nm.

12. Lines in R^3

Given a point $P_0 = (x_0, y_0, z_0)$ and a direction vector $\mathbf{v}_1 = \langle a, b, c \rangle$ in R^3, a line L that passes through P_0 and is parallel to $\mathbf{v}$ is written parametrically as a function of t:

$$x(t) = x_0 + at, \qquad y(t) = y_0 + bt, \qquad z(t) = z_0 + ct$$

Using vector notation, the same line is written

$$\begin{aligned}\langle x, y, z \rangle &= \mathbf{v}_0 + t\mathbf{v}_1 \\ &= \langle x_0, y_0, z_0 \rangle + t\langle a, b, c \rangle \\ &= \langle x_0 + at, y_0 + bt, z_0 + ct \rangle,\end{aligned}$$

where $\mathbf{v}_0$ is the vector whose head is located at $P_0 = (x_0, y_0, z_0)$.

◆ ◆ ◆ ◆ ◆ ◆ ◆ ◆ ◆

Example 12.1: Find the parametric equation of a line passing through $P_0 = (2, -1, 3)$ and parallel to the vector $\mathbf{v}_1 = \langle 5, 8, -4 \rangle$.

Solution: The line is represented parametrically by

$$x(t) = 2 + 5t, \qquad y(t) = -1 + 8t, \qquad z(t) = 3 - 4t,$$

or in vector notation as

$$\langle x, y, z \rangle = \langle 2, -1, 3 \rangle + t\langle 5, 8, -4 \rangle = \langle 2 + 5t, -1 + 8t, 3 - 4t \rangle.$$

Note that when $t = 0$, we obtain the vector $\langle 2, -1, 3 \rangle$. If the foot of this vector is placed at the origin, then its head is the ordered triple $P_0 = (2, -1, 3)$.

◆ ◆ ◆ ◆ ◆ ◆ ◆ ◆ ◆

A **line segment** from a point $P_0 = (x_0, y_0, z_0)$ to a point $P_1 = (x_1, y_1, z_1)$ over $a \leq t \leq b$ has the form

$$\langle x, y, z \rangle = \langle x_0, y_0, z_0 \rangle + \frac{t - a}{b - a} \langle x_1 - x_0, y_1 - y_0, z_1 - z_0 \rangle,$$

Note that $\langle x_1 - x_0, y_1 - y_0, z_1 - z_0 \rangle$ is the direction vector of the line.

Example 12.2: Find the parametric equation of the line segment from $P_0 = (4, 2, -1)$ to $P_1 = (7, -3, -2)$.

Solution: The direction vector $\mathbf{v}_1$ is found by subtracting P_0 from P_1:

$$\begin{aligned}\mathbf{v}_1 &= P_1 - P_0 \\ &= \langle 7-4, -3-2, -2-(-1)\rangle \\ &= \langle 3, -5, -1\rangle.\end{aligned}$$

Thus, the line can be written

$$\begin{aligned}\langle x, y, z\rangle &= \langle 4, 2, -1\rangle + t\langle 3, -5, -1\rangle \\ &= \langle 4+3t, 2-5t, -1-t\rangle, \quad \text{for } 0 \le t \le 1.\end{aligned}$$

The bounds are such that $t = 0$ gives $P_0 = (4, 2, -1)$ and $t = 1$ gives $P_1 = (7, -3, -2)$. Since there is a direction implied by increasing t, this is called a *directed line segment*.

> When a line segment between two points is constructed in this manner, the bounds on t are always $0 \le t \le 1$.

◆ ◆ ◆ ◆ ◆ ◆ ◆ ◆ ◆ ◆

Example 12.3: Find the parametric equation of the line segment connecting $P_0 = (4, 2, -1)$ and $P_1 = (7, -3, -2)$ such that $t = 0$ gives P_0 and that $t = 5$ gives P_1.

Solution: The difference in t-values is $b - a = 5 - 0 = 5$. Thus, the line segment is

$$\begin{aligned}\langle x, y, z\rangle &= \langle 4, 2, -1\rangle + \frac{t}{5}\langle 3, -5, -1\rangle \\ &= \langle 4, 2, -1\rangle + t\left\langle \frac{3}{5}, -1, -\frac{1}{5}\right\rangle \\ &= \left\langle 4 + \frac{3}{5}t, 2-t, -1-\frac{1}{5}t\right\rangle, \quad \text{for } 0 \le t \le 5.\end{aligned}$$

Note that $t = 0$ gives P_0 and that $t = 5$ gives P_1.

Example 12.4: Find the parametric equation of a line segment connecting $P_0 = (4, 2, -1)$ and $P_1 = (7, -3, -2)$ such that $t = 4$ gives P_0 and $t = 10$ gives P_1.

Solution: The starting point occurs when $t = 4$, indicating in a horizontal shift. The difference in t-values is $b - a = 10 - 4 = 6$. Thus, the line segment is

$$\langle x, y, z \rangle = \langle 4,2,-1 \rangle + \frac{t-4}{10-4}\langle 3,-5,-1 \rangle$$

$$= \langle 4,2,-1 \rangle + \frac{1}{6}\langle 3(t-4), -5(t-4), -1(t-4) \rangle$$

$$= \langle 4,2,-1 \rangle + \frac{1}{6}\langle 3t-12, -5t+20, -t+4 \rangle \quad \left\{ \begin{array}{l} \text{Clearing} \\ \text{parentheses} \end{array} \right.$$

$$= \left\langle 4 - \frac{12}{6}, 2 + \frac{20}{6}, -1 + \frac{4}{6} \right\rangle + \frac{1}{6}\langle 3t, -5t, -t \rangle \quad \left\{ \begin{array}{l} \text{Combining} \\ \text{constants} \end{array} \right.$$

$$= \left\langle 2, \frac{16}{3}, -\frac{1}{3} \right\rangle + \left\langle \frac{1}{2}t, -\frac{5}{6}t, -\frac{1}{6}t \right\rangle, \quad \text{for } 4 \leq t \leq 10.$$

Note that $t = 4$ gives P_0 and that $t = 10$ gives P_1.

◆ ◆ ◆ ◆ ◆ ◆ ◆ ◆ ◆

> When more than one line is being considered,
> use different parameter variables.

Example 12.5: Let L_1: $\langle x, y, z \rangle = \langle 1, 2, 5 \rangle + t\langle 2, 4, -3 \rangle$ and L_2: $\langle x, y, z \rangle = \langle 6, 1, -2 \rangle + s\langle 4, 8, -6 \rangle$ be two lines defined parametrically. Are lines L_1 and L_2 parallel? If so, are lines L_1 and L_2 the same line?

Solution: The direction vector for line L_1 is $\mathbf{v}_1 = \langle 2, 4, -3 \rangle$ and the direction vector for line L_2 is $\mathbf{v}_2 = \langle 4, 8, -6 \rangle$. Since $\mathbf{v}_2 = 2\mathbf{v}_1$ (or equivalently, $\mathbf{v}_1 = \frac{1}{2}\mathbf{v}_2$), the two vectors are parallel. Thus, so are the lines.

Choose a point from one line and show that the other line passes through it. In this example, choose point $P_0 = (1, 2, 5)$ from line L_1. Does L_2 pass through $P_0 = (1, 2, 5)$? We substitute the coordinates in P_0 for x, y and z, and attempt to solve for a unique value of s that would indicate line L_2 passes through a point in line L_1:

$$1 = 6 + 4s, \qquad 2 = 1 + 8s, \qquad 5 = -2 - 6s.$$

From the first equation, we get $s = -5/4$, but from the second equation, we get $s = 1/8$. Since s is not unique, we conclude it is impossible that L_2 passes through $P_0 = (1,2,5)$. Thus, lines L_1 and L_2 represent two different parallel lines.

◆ ◆ ◆ ◆ ◆ ◆ ◆ ◆ ◆ ◆

Example 12.6: Show that the lines $L_1: \langle 2,3,0 \rangle + t\langle 1,-2,5 \rangle$ and $L_2: \langle 6,-5,20 \rangle + s\langle -3,6,-15 \rangle$ are the same line.

Solution: Their direction vectors are $\mathbf{v}_1 = \langle 1,-2,5 \rangle$ and $\mathbf{v}_2 = \langle -3,6,-15 \rangle$. Since $\mathbf{v}_2 = -3\mathbf{v}_1$, the two lines are parallel. Had the direction vectors not been parallel, the parametric equations cannot possibly represent the same line.

Now, choose a point from one line, and see if it is possible to find a unique value for the parameter variable in the other line. From line L_1, choose the point $P_0 = (2,3,0)$ and then attempt to find a unique value for s in L_2:

$$2 = 6 - 3s, \qquad 3 = -5 + 6s, \qquad 0 = 20 - 15s.$$

From the first equation, $s = 4/3$. From the second equation, $s = 4/3$ as well, and from the third equation, $s = 4/3$. Since we were able to find a unique value s, we conclude that line L_2 passes through a point that is in line L_1. Since the lines are already parallel, this would force the two lines to be **coincident**, that is, the same line.

◆ ◆ ◆ ◆ ◆ ◆ ◆ ◆ ◆ ◆

Example 12.7: Find the point of intersection of lines $L_1: \langle x,y,z \rangle = \langle 1,2,-1 \rangle + t\langle 2,-3,4 \rangle$ and $L_2: \langle x,y,z \rangle = \langle 1,8,9 \rangle + s\langle 4,-12,-2 \rangle$.

Solution: The direction vectors are $\mathbf{v}_1 = \langle 2,-3,4 \rangle$ and $\mathbf{v}_2 = \langle 4,-12,-2 \rangle$. Since they are not scalar multiples of one another, the two lines are not parallel. To see if they intersect, set the equations for x equal to one another, and for y, and for z:

$$\begin{aligned} x: & \quad 1 + 2t = 1 + 4s \\ y: & \quad 2 - 3t = 8 - 12s \\ z: & \quad -1 + 4t = 9 - 2s. \end{aligned}$$

Simplifying, this is a system of two variables in three equations:

$$2t - 4s = 0$$
$$-3t + 12s = 6$$
$$4t + 2s = 10.$$

One of two things happens: either a solution in s and t is found, in which case there is an intersection point, or a solution in s and t cannot be found, in which case there is no intersection point. From the first two equations, we solve a system:

$$\begin{array}{l} 2t - 4s = 0 \\ -3t + 12s = 6 \end{array} \xrightarrow{\binom{\text{Multiply}}{\text{top row by 3}}} \begin{array}{l} 3(2t - 4s) = 0 \\ -3t + 12s = 6 \end{array} \xrightarrow{\binom{\text{Distribute}}{\text{then add}}} \begin{array}{l} 6t - 12s = 0 \\ -3t + 12s = 6 \end{array} \rightarrow 3t = 6.$$

Thus, $t = 2$, and back-substituting, $s = 1$. Does this solve the third equation? Substitute and simplify:

$$4(2) + 2(1) = 8 + 2 = 10.$$

This is a true statement. We were able to show that when $t = 2$, we generate the point $(5, -4, 7)$ on line L_1, and when $s = 1$, we generate the same point $(5, -4, 7)$ on line L_2. Thus, the two lines intersect at this point.

◆ ◆ ◆ ◆ ◆ ◆ ◆ ◆ ◆

> Two non-parallel and non-intersecting lines in R^3 are called **skew lines**.

Example 12.8: Show that the lines $L_1: \langle x, y, z \rangle = \langle 1, 2, -1 \rangle + t \langle 2, -3, 4 \rangle$ and $L_2: \langle x, y, z \rangle = \langle 1, 8, 5 \rangle + s \langle 4, -12, -2 \rangle$ are skew lines.

Solution: These are the same lines as from the previous example, except that a small change has been made to the equation for z in line L_2. From the previous example, we established that since the direction vectors are not scalar multiples of one another, then the lines are not parallel. Next, set the equations for x equal to one another, and for y, and for z:

$$\begin{array}{ll} x: & 1 + 2t = 1 + 4s \\ y: & 2 - 3t = 8 - 12s \\ z: & -1 + 4t = 5 - 2s. \end{array}$$

This simplifies to

$$\begin{aligned} 2t - 4s &= 0 \\ -3t + 12s &= 6 \\ 4t + 2s &= 6. \end{aligned}$$

From the previous example, solving the system formed by the first two equations gave $t = 2$ and $s = 1$. However, when substituting these values in the third equation, we get $4(2) + 2(1) = 6$, which is false. There is no solution of this system in s and t. Thus, the two lines do not intersect, and are skew.

♦ • ♦ • ♦ • ♦ • ♦

13. Planes in R^3

Given a point $P_0 = (x_0, y_0, z_0)$ and a vector $\mathbf{n} = \langle a, b, c \rangle$, a **plane** that passes through P_0 and is normal (orthogonal) to $\mathbf{n}$ has the equation

$$a(x - x_0) + b(y - y_0) + c(z - z_0) = 0,$$

which simplifies to $ax + by + cz = d$, where $d = ax_0 + by_0 + cz_0$.

♦ • ♦ • ♦ • ♦ • ♦

Example 13.1: Find the equation of the plane passing through $P_0 = (-3,9,1)$ and normal to $\mathbf{n} = \langle 7,3,-5 \rangle$.

Solution: The plane has the equation

$$7(x - (-3)) + 3(y - 9) - 5(z - 1) = 0.$$

Simplifying, we have

$$\begin{aligned} 7(x + 3) + 3(y - 9) - 5(z - 1) &= 0 \\ 7x + 21 + 3y - 27 - 5z + 5 &= 0 \\ 7x + 3y - 5z &= 1. \end{aligned}$$

Example 13.2: State a vector that is normal to the plane $2x - 4y + 3z = 12$.

Solution: There are infinitely-many possible vectors. One is $\langle 2, -4, 3 \rangle$, which is found by reading the coefficients of x, y and z. Any non-zero multiple of $\langle 2, -4, 3 \rangle$ is also a vector normal to the plane.

♦ ● ♦ ● ● ♦ ● ● ♦

Example 13.3: Let $-2x + 7y + 4z = -7$ be a plane in R^3. Find the equation of the line normal to the plane that passes through the point $(1, -3, 5)$ on the plane.

Solution: The line is parallel to (in the direction of) $\mathbf{n} = \langle -2, 7, 4 \rangle$, so the line is given by

$$\langle x, y, z \rangle = \langle 1, -3, 5 \rangle + t \langle -2, 7, 4 \rangle = \langle 1 - 2t, -3 + 7t, 5 + 4t \rangle.$$

Example 13.4: Find the point of intersection of the line $\langle x, y, z \rangle = \langle 4 - t, 2 + 3t, 3 - 5t \rangle$ and the plane $6x + 2y - 3z = 79$.

Solution: Substitute the equations for x, y and z into the plane, and solve for t:

$$6(4 - t) + 2(2 + 3t) - 3(3 - 5t) = 79$$
$$24 - 6t + 4 + 6t - 9 + 15t = 79$$
$$15t = 60$$
$$t = 4.$$

Now substitute $t = 4$ into the equations for the line:

$$\langle x, y, z \rangle = \langle 4 - (4), 2 + 3(4), 3 - 5(4) \rangle = \langle 0, 14, -17 \rangle.$$

When referenced from the origin, its head lies at the point $(0, 14, -17)$.

♦ ● ♦ ● ● ♦ ● ● ♦

Example 13.5: Find the equation of the plane passing through the points $A = (1, 3, 4)$, $B = (-3, 2, 6)$ and $C = (1, 0, -6)$.

Solution: From the three points, form two vectors. For example, vectors **AB** and **AC**:

$$\mathbf{AB} = \langle -4, -1, 2 \rangle, \quad \mathbf{AC} = \langle 0, -3, -10 \rangle$$

Next, find a vector **n** normal to **AB** and **AC** by finding the cross product **AB** × **AC**:

$$\mathbf{n} = \mathbf{AB} \times \mathbf{AC} = \langle 16, -40, 12 \rangle$$

Any non-zero multiple of **n** will suffice, so divide through by 4, getting **n** = $\langle 4, -10, 3 \rangle$. Using any one of the three given points, we now find the equation of the plane. Using $A = (1,3,4)$ and **n** = $\langle 4, -10, 3 \rangle$ gives

$$4(x-1) - 10(y-3) + 3(z-4) = 0$$
$$4x - 4 - 10y + 30 + 3z - 12 = 0$$
$$4x - 10y + 3z = -14.$$

Verify this is correct by substituting the coordinates for points B and C into the equation:

$B = (-3,2,6)$: $\quad 4(-3) - 10(2) + 3(6) = -12 - 20 + 18 = -14$, true.
$C = (1,0,-6)$: $\quad 4(1) - 10(0) + 3(-6) = 4 - 18 = -14$, true.

◆ • ◆ • ◆ • ◆ • ◆

Example 13.6: Show that the equation of a plane passing through the axis-intercepts $(a, 0, 0)$, $(0, b, 0)$ and $(0, 0, c)$ is $\frac{x}{a} + \frac{y}{b} + \frac{z}{c} = 1$.

Solution: Form two vectors:

The vector from $(a, 0, 0)$ to $(0, b, 0)$ is $\mathbf{u} = \langle 0-a, b-0, 0-0 \rangle = \langle -a, b, 0 \rangle$.

The vector from $(0, b, 0)$ to $(0, 0, c)$ is $\mathbf{v} = \langle 0-0, 0-b, c-0 \rangle = \langle 0, -b, c \rangle$.

Their cross product is $\mathbf{u} \times \mathbf{v} = \langle bc, ac, ab \rangle$, and using $(a, 0, 0)$ as a point in the plane, the plane's equation is

$$bc(x-a) + ac(y-0) + ab(z-0) = 0$$
$$bcx - abc + acy + abz = 0$$
$$bcx + acy + abz = abc.$$

Now, divide through by abc:

$$\frac{bcx}{abc} + \frac{acy}{abc} + \frac{abz}{abc} = \frac{abc}{abc}, \text{ which simplifies as } \frac{x}{a} + \frac{y}{b} + \frac{z}{c} = 1.$$

This is a handy formula for finding the equation of a plane given its axis-intercepts.

◆ ◆ ◆ ◆ ◆ ◆ ◆ ◆ ◆

Example 13.7: Find the acute angle formed by the intersection of the planes $x + 3y - 2z = 5$ and $4x - y + 5z = -2$.

Solution: The respective normal vectors of each plane are $\mathbf{n}_1 = \langle 1,3,-2 \rangle$ and $\mathbf{n}_2 = \langle 4,-1,5 \rangle$. The angle between these two planes is the same as the angle between the two normal vectors:

$$\theta = \cos^{-1}\left(\frac{\mathbf{n}_1 \cdot \mathbf{n}_2}{|\mathbf{n}_1||\mathbf{n}_2|}\right) = \cos^{-1}\left(\frac{-9}{\sqrt{14 \cdot 42}}\right) \approx 111.79 \text{ degrees}$$

However, planes always intersect at an acute angle (except in the case where they are orthogonal). The preferred answer is the supplement: $180° - 111.79° = 68.21°$.

◆ ◆ ◆ ◆ ◆ ◆ ◆ ◆ ◆

Example 13.8: Find the acute angle formed by vector $\mathbf{v} = \langle 1,4,3 \rangle$ and plane $2x + y + 5z = 4$.

Solution: The plane's normal vector is $\mathbf{n} = \langle 2,1,5 \rangle$, so the angle between $\mathbf{v}$ and $\mathbf{n}$ is

$$\theta = \cos^{-1}\left(\frac{\mathbf{v} \cdot \mathbf{n}}{|\mathbf{v}||\mathbf{n}|}\right) = \cos^{-1}\left(\frac{21}{\sqrt{26 \cdot 30}}\right) \approx 41.24°.$$

Visually, $\mathbf{v}$ and $\mathbf{n}$ are on the same side of the plane, as their angle is acute. Note that $41.24°$ is the angle between $\mathbf{v}$ and $\mathbf{n}$, so the angle between $\mathbf{v}$ and the plane is the complement, $90° - 41.24° = 48.76°$.

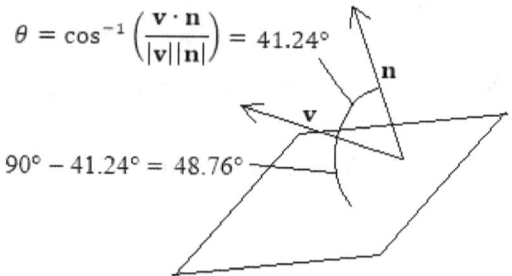

Find the angle between **v** and **n** first, then find the complement.
(Example 13.8)

♦ • ♦ • ♦ • ♦ • ♦

Example 13.9: Find the acute angle formed by vector $\mathbf{v} = \langle 2,1,-8 \rangle$ and plane $x + 2y + z = 4$.

Solution: The plane's normal vector is $\mathbf{n} = \langle 1,2,1 \rangle$, so the angle between **v** and **n** is

$$\theta = \cos^{-1}\left(\frac{\mathbf{v} \cdot \mathbf{n}}{|\mathbf{v}||\mathbf{n}|}\right) = \cos^{-1}\left(\frac{-4}{\sqrt{69 \cdot 6}}\right) \approx 101.34°.$$

In this example, the vector **n** is on one side of the plane, and **v** on the opposite side. Since the angle from **n** to the plane is 90°, the remainder, 101.34° − 90° = 11.34°, is the desired angle.

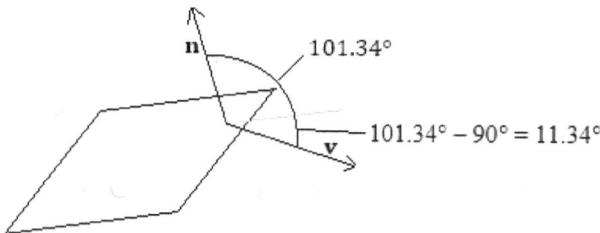

Vectors **v** and **n** lie on opposite sides of the plane, so the acute angle **v** makes with the plane is found by subtracting 90 degrees.

Another way to view this is to use $\mathbf{n} = \langle -1,-2,-1 \rangle$, which would have given us

$$\theta = \cos^{-1}\left(\frac{\mathbf{v} \cdot \mathbf{n}}{|\mathbf{v}||\mathbf{n}|}\right) = \cos^{-1}\left(\frac{4}{\sqrt{69 \cdot 6}}\right) \approx 78.66°.$$

Thus, **v** and **n** are on the same side of the plane, and the complement gives us $90° - 78.66° = 11.34°$, the desired angle.

♦ • ♦ • ♦ • ♦ • ♦

Example 13.10: Find the parametric equation of the line formed by the intersection of the planes $x - 3y + 2z = -16$ and $2x + y - 7z = 18$.

Solution: The planes are not parallel since their normal vectors, $\mathbf{n}_1 = \langle 1, -3, 2 \rangle$ and $\mathbf{n}_2 = \langle 2, 1, -7 \rangle$, are not scalar multiples of one another. Had the planes been parallel, they would not intersect (unless they were the same coincident plane). To find the parametric equation of the line, we need a point on this line of intersection, and a direction vector parallel to the line of intersection.

For a point on the line of intersection, set one variable to 0, and solve for the other two variables. Let $x = 0$ in both equations. This results in a system

$$-3y + 2z = -16$$
$$y - 7z = 18.$$

The solution is $y = 4$ and $z = -2$. Thus, $(0, 4, -2)$ is a point on the line of intersection of the two planes.

For the direction vector, we note that it will be orthogonal to both normal vectors of each plane. Thus, we can find a vector parallel to the line of intersection by finding the cross product of the two normal vectors. We get $\mathbf{n}_1 \times \mathbf{n}_2 = \langle 19, 11, 7 \rangle$. Therefore, the line of intersection of the two planes is given parametrically by

$$\langle x, y, z \rangle = \langle 19t, 4 + 11t, -2 + 7t \rangle.$$

♦ • ♦ • ♦ • ♦ • ♦

Example 13.11: Find the equation of the plane passing through $P_0 = (7, 3, -11)$ and parallel to the plane $2x - y + 6z = 1$.

Solution: The two planes are parallel, so they share the same normal vector $\mathbf{n} = \langle 2, -1, 6 \rangle$. Thus, the plane's equation is

$$2(x - 7) - 1(y - 3) + 6(z - (-11)) = 0$$
$$2x - 14 - y + 3 + 6z + 66 = 0$$
$$2x - y + 6z = -55.$$

Parametric Representation of a Plane

A plane can also be represented parametrically in vector form. Assume that a plane passes through $P_0 = (x_0, y_0, z_0)$ and that $\mathbf{v}_1$ and $\mathbf{v}_2$ are two (non-parallel) vectors that lie *within* the plane. The plane can then be written as

$$\langle x, y, z \rangle = \mathbf{v}_0 + s\mathbf{v}_1 + t\mathbf{v}_2,$$

where s and t are independent parameter variables that may assume any real-number value, and that $\mathbf{v}_0$ is the vector whose head points to P_0 when $s = 0$ and $t = 0$.

Example 13.12: Write the plane $3x + 2y + z = 5$ in parametric form.

Solution: Any of the three variables can be isolated, leaving one dependent variable and two independent variables. In this case, isolate z:

$$z = 5 - 3x - 2y.$$

Now, let $x = s$ and $y = t$. This gives

$$\begin{aligned} x &= s \\ y &= t \\ z &= 5 - 3s - 2t. \end{aligned}$$

Observe what happens when we "zero-fill" the missing terms:

$$\begin{aligned} x &= 0 + 1s + 0t \\ y &= 0 + 0s + 1t \\ z &= 5 - 3s - 2t. \end{aligned}$$

The vectors $\mathbf{v}_0$, $\mathbf{v}_1$ and $\mathbf{v}_2$ can be inferred from the columns in the previous step:

$$\begin{bmatrix} x \\ y \\ z \end{bmatrix} = \begin{bmatrix} 0 \\ 0 \\ 5 \end{bmatrix} + s\begin{bmatrix} 1 \\ 0 \\ -3 \end{bmatrix} + t\begin{bmatrix} 0 \\ 1 \\ -2 \end{bmatrix}.$$

Thus, the plane $3x + 2y + z = 5$ written in parametric form is

$$\langle x, y, z \rangle = \mathbf{v}_0 + s\mathbf{v}_1 + t\mathbf{v}_2 = \langle 0, 0, 5 \rangle + s\langle 1, 0, -3 \rangle + t\langle 0, 1, -2 \rangle.$$

Example 13.13: Rewrite the plane given by $\langle x,y,z \rangle = \langle 0,0,2 \rangle + s\langle 1,0,4 \rangle + t\langle 0,1,-7 \rangle$ into the form $ax + by + cz = d$.

Solution: Reading across component by component, we have

$$x = 0 + 1s + 0t$$
$$y = 0 + 0s + 1t$$
$$z = 2 + 4s - 7t.$$

In this case, we have $x = s$ and $y = t$, so the last equation can be written as

$$z = 2 + 4x - 7y.$$

Thus, the plane $\langle x,y,z \rangle = \langle 0,0,2 \rangle + s\langle 1,0,4 \rangle + t\langle 0,1,-7 \rangle$ can be written as $-4x + 7y + z = 2$.

◆ ◆ ◆ ◆ ◆ ◆ ◆ ◆ ◆ ◆

Example 13.14: Rewrite the plane given by $\langle x,y,z \rangle = \langle 1,3,4 \rangle + s\langle 2,-1,5 \rangle + t\langle -3,8,1 \rangle$ into the form $ax + by + cz = d$.

Solution: Reading across component by component, we have

$$x = 1 + 2s - 3t \quad \textbf{(1)}$$
$$y = 3 - s + 8t \quad \textbf{(2)}$$
$$z = 4 + 5s + t. \quad \textbf{(3)}$$

We need to rewrite s and t in terms of x and y. From the first two equations **(1)** and **(2)**, we have

$$2s - 3t = x - 1$$
$$-s + 8t = y - 3.$$

Eliminate s by multiplying the bottom row by 2, then adding:

$$2s - 3t = x - 1$$
$$-2s + 16t = 2y - 6,$$

which gives $13t = x + 2y - 7$. Thus, $t = \frac{1}{13}(x + 2y - 7)$. In a similar way, t can be eliminated, in which case we have $s = \frac{1}{13}(8x + 3y - 17)$.

These are substituted into equation **(3)**:

$$z = 4 + 5s + t = 4 + 5\left(\frac{1}{13}(8x + 3y - 17)\right) + \left(\frac{1}{13}(x + 2y - 7)\right).$$

This simplifies to

$$z = \frac{41}{13}x + \frac{17}{13}y - \frac{40}{13}.$$

Multiplying by 13 to clear fractions, we have $13z = 41x + 17y - 40$. Thus, the plane can be written as $41x + 17y - 13z = 40$.

◆ • ◆ • ◆ • ◆ • ◆ • ◆

Example 13.15: Find the equation of the plane passing through $P_0 = (-3,9,1)$ and normal to $\mathbf{n} = \langle 7,3,-5 \rangle$. (This is a repeat of Example 13.1)

Solution: We need two vectors, $\mathbf{v}_1$ and $\mathbf{v}_2$, in the plane. These two vectors will be orthogonal to $\mathbf{n}$. For example, if we choose the x-component of $\mathbf{v}_1$ to be 0, then $\mathbf{v}_1 = \langle 0,5,3 \rangle$ is orthogonal to $\mathbf{n}$ and hence, in the plane. In a similar way, we could let the y-component of $\mathbf{v}_2$ be 0, and so $\mathbf{v}_2 = \langle 5,0,7 \rangle$ is orthogonal to $\mathbf{n}$ and also in the plane.

Noting that $\mathbf{v}_0$ corresponds to point $P_0 = (-3,9,1)$, we have

$$\langle x, y, z \rangle = \mathbf{v}_0 + s\mathbf{v}_1 + t\mathbf{v}_2 = \langle -3,9,1 \rangle + s\langle 0,5,3 \rangle + t\langle 5,0,7 \rangle.$$

Is this the same as $7x + 3y - 5z = 1$, the result from Example 13.1? Let's find out. Reading component-wise, we have

$$x = -3 + 0s + 5t$$
$$y = 9 + 5s + 0t$$
$$z = 1 + 3s + 7t.$$

Solving for t in the first equation gives $t = \frac{1}{5}(x + 3)$, and doing similar in the second equation gives $s = \frac{1}{5}(y - 9)$. These are substituted into the third equation:

$$z = 1 + 3s + 7t = 1 + 3\left(\frac{1}{5}(y - 9)\right) + 7\left(\frac{1}{5}(x + 3)\right).$$

Clearing parentheses, we have

$$z = \frac{7}{5}x + \frac{3}{5}y - \frac{1}{5}.$$

Multiplying by 5, we have $5z = 7x + 3y - 1$. Rearranging terms, we have $7x + 3y - 5z = 1$.

♦ ♦ ♦ ♦ ♦ ♦ ♦ ♦ ♦

14. Distances in R^3

We use projections to find the shortest distances between a point and a line, between two non-intersecting lines, between a point and a plane, and between two parallel planes. The shortest distance is defined to be a line that meets the objects orthogonally.

♦ ♦ ♦ ♦ ♦ ♦ ♦ ♦ ♦

Example 14.1: Find the shortest distance between the line $\langle x, y, z \rangle = \langle 1, 2, -1 \rangle + t \langle 2, -3, 4 \rangle$ and the point $Q = (4, 8, 3)$.

Solution: Choose any point P on the line. For example, when $t = 0$, we have $P = (1, 2, -1)$. Then find the vector from P to Q, which is $\mathbf{u} = \langle 4 - 1, 8 - 2, 3 - (-1) \rangle = \langle 3, 6, 4 \rangle$. Meanwhile, the directional vector of the line is $\mathbf{v} = \langle 2, -3, 4 \rangle$. The projection of $\mathbf{u}$ onto $\mathbf{v}$ is:

$$\text{proj}_\mathbf{v} \mathbf{u} = \frac{\mathbf{u} \cdot \mathbf{v}}{\mathbf{v} \cdot \mathbf{v}} \mathbf{v} = \frac{4}{29} \mathbf{v} = \left\langle \frac{8}{29}, -\frac{12}{29}, \frac{16}{29} \right\rangle.$$

The normal vector is found by subtracting $\text{proj}_\mathbf{v} \mathbf{u}$ from $\mathbf{u}$:

$$\text{norm}_\mathbf{v} \mathbf{u} = \mathbf{u} - \text{proj}_\mathbf{v} \mathbf{u} = \langle 3, 6, 4 \rangle - \left\langle \frac{8}{29}, -\frac{12}{29}, \frac{16}{29} \right\rangle = \left\langle \frac{79}{29}, \frac{186}{29}, \frac{100}{29} \right\rangle.$$

The magnitude of this normal vector is the distance from the point Q to the line:

$$|\text{norm}_\mathbf{v} \mathbf{u}| = \sqrt{\left(\frac{79}{29}\right)^2 + \left(\frac{186}{29}\right)^2 + \left(\frac{100}{29}\right)^2} \approx 7.78 \text{ units.}$$

Example 14.2: Find the shortest distance between point $Q = (1,4,3)$ and plane $x - 3y + 2z = 6$.

Solution: Pick any point P in the plane by choosing values for two of the variables and solving for the third. If $x = 0$ and $y = 0$, we get $z = 3$, so a point in the plane is $P = (0,0,3)$. The vector $\mathbf{u}$ from P to Q is $\mathbf{u} = \langle 1,4,0 \rangle$. The normal vector to the plane is $\mathbf{n} = \langle 1,-3,2 \rangle$. Project $\mathbf{u}$ onto $\mathbf{n}$:

$$\text{proj}_\mathbf{n}\mathbf{u} = \frac{\mathbf{u} \cdot \mathbf{n}}{\mathbf{n} \cdot \mathbf{n}} \mathbf{n} = -\frac{11}{14} \langle 1,-3,2 \rangle.$$

The magnitude of this vector is the distance from $Q = (1,4,3)$ to the plane $x - 3y + 2z = 6$:

$$|\text{proj}_\mathbf{n}\mathbf{u}| = \frac{11}{14}\sqrt{1^2 + (-3)^2 + 2^2} = \frac{11}{14}\sqrt{14} \approx 2.94 \text{ units}.$$

♦ ♦ ♦ ♦ ♦ ♦ ♦ ♦ ♦

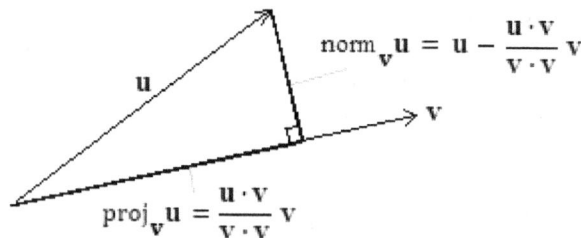

Decomposing a vector $\mathbf{u}$ into a right triangle with one leg parallel to vector $\mathbf{v}$, and the other orthogonal to $\mathbf{v}$.

A Pictorial Guide to Finding the Shortest Distance Between a Point and a Line

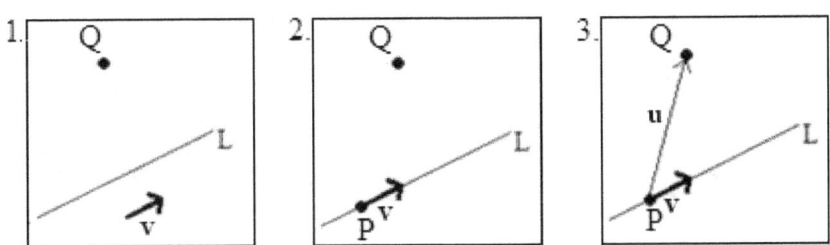

Steps: (1) Start with a line L, a point Q not on the line, and a vector **v** parallel to L. (2) Pick any point P on the line. If it helps, translate **v** onto the line with foot at P. (3) Create vector **u** from P to Q.

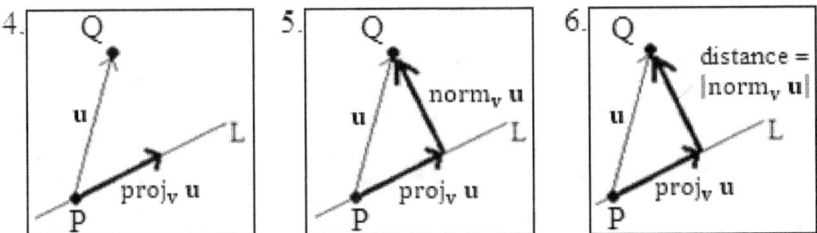

(4) Find the projection of **u** onto **v**. (5) Find the normal vector to complete the right triangle. (6) The desired distance from Q to L is the magnitude of this normal vector.

♦ • ♦ • • ♦ • ♦

When a vector **u** is projected onto a vector **v**, this produces a vector called proj_v **u**. This can be viewed as one leg of a right triangle, with **u** being the hypotenuse. The other leg is called norm_v **u**. In vector addition, we have:

$$\mathbf{u} = \text{proj}_v \mathbf{u} + \text{norm}_v \mathbf{u} \quad \text{(Remember, these are vectors.)}$$

Their magnitudes are related by the Pythagorean Theorem:

$$|\mathbf{u}|^2 = |\text{proj}_v \mathbf{u}|^2 + |\text{norm}_v \mathbf{u}|^2 \quad \text{(These are scalars.)}$$

A Pictorial Guide to Finding the Shortest Distance Between a Point and a Plane

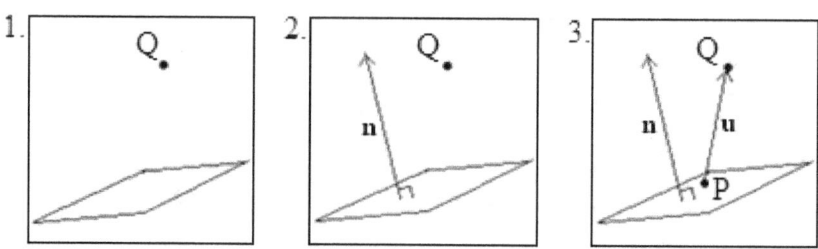

Steps: (1) Start with a plane and a point Q not on the plane. (2) From the plane's equation, determine the normal vector **n** to the plane. (3) Pick any point P on the plane and draw vector **u** from P to Q.

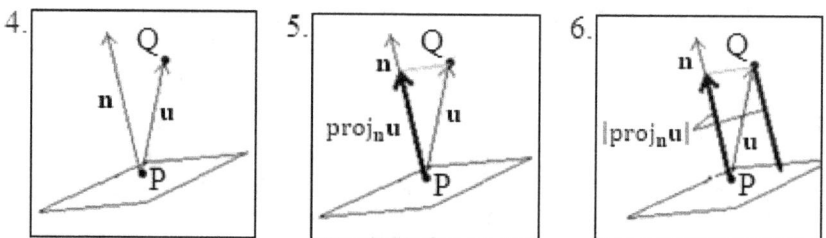

(4) If it helps, place **n**'s foot at P, this forms part of a right triangle with **u** as the hypotenuse. (5) Find the projection of **u** onto **n**. (6) The magnitude of the projection is the distance from Q to the plane.

◆ ◆ ◆ ◆ ◆ ◆ ◆ ◆ ◆ ◆

Example 14.3: Find the shortest distance between the parallel planes

$$x - 3y + 2z = 6 \text{ and } -2x + 6y - 4z = 1.$$

Solution: Note that the vector $\mathbf{n} = \langle 1, -3, 2 \rangle$ is orthogonal to both planes. Choose any point on each plane. From the plane $x - 3y + 2z = 6$, we use the point $A = (0,0,3)$, and from the plane $-2x + 6y - 4z = 1$, we use $B = \left(-\frac{1}{2}, 0, 0\right)$. The vector **v** from A to B is $\mathbf{v} = \langle \frac{1}{2}, 0, 3 \rangle$. We then project **v** onto **n**:

$$\text{proj}_\mathbf{n} \mathbf{v} = \frac{\mathbf{v} \cdot \mathbf{n}}{\mathbf{n} \cdot \mathbf{n}} \mathbf{n} = \frac{13}{28} \langle 1, -3, 2 \rangle.$$

The magnitude of this vector is the shortest distance between the two planes:

$$|\text{proj}_n \mathbf{v}| = \sqrt{\left(\frac{13}{28}\right)^2 + \left(-\frac{39}{28}\right)^2 + \left(\frac{26}{28}\right)^2} \approx 1.737 \text{ units.}$$

Example 14.4: Find the shortest distance between the lines L_1: $\langle x, y, z \rangle = \langle 1, -2, 6 \rangle + t\langle 1, -5, 4 \rangle$ and L_2: $\langle x, y, z \rangle = \langle 0, 8, 1 \rangle + s\langle 4, -7, 3 \rangle$

Solution: These lines are skew. Thus, they will not intersect. Two skew lines in R^3 can always be placed within two parallel planes, so that this example is nearly identical to the previous example. From the first line, we have a point $A = (1, -2, 6)$ and from the second line, we have a point $B = (0, 8, 1)$, and the vector $\mathbf{v}$ from A to B is $\mathbf{v} = \langle -1, 10, -5 \rangle$.

The normal vector $\mathbf{n}$ to the two planes that each contain one of the lines is found by finding the cross product of the two direction vectors of each line. From the first line, we have $\mathbf{v}_1 = \langle 1, -5, 4 \rangle$ and from the second line, we have $\mathbf{v}_2 = \langle 4, -7, 3 \rangle$. Their cross product is

$$\mathbf{n} = \mathbf{v}_1 \times \mathbf{v}_2 = \langle 13, 13, 13 \rangle.$$

However, we can use any non-zero multiple of $\mathbf{n}$ for this example, so $\mathbf{n} = \langle 1, 1, 1 \rangle$ will suffice, since the projection step will extend $\mathbf{n}$ to the desired length. Now project $\mathbf{v}$ onto $\mathbf{n}$:

$$\text{proj}_n \mathbf{v} = \frac{\mathbf{v} \cdot \mathbf{n}}{\mathbf{n} \cdot \mathbf{n}} \mathbf{n} = \frac{4}{3} \langle 1, 1, 1 \rangle.$$

The distance is the magnitude of this projection vector:

$$|\text{proj}_n \mathbf{v}| = \frac{4}{3} \sqrt{1^2 + 1^2 + 1^2} \approx 2.309 \text{ units.}$$

◆ ◆ ◆ ◆ ◆ ◆ ◆ ◆ ◆

See an error? Have a suggestion?
Please see www.surgent.net/vcbook

15. Vector Valued Functions

Up to this point, we have presented vectors with constant components, for example, ⟨1,2⟩ and ⟨2, −5,4⟩. We now allow the components of a vector to be functions of a common variable. For example, $\mathbf{r}(t) = \langle 2t + 1, t^2 + 3 \rangle$ presents a function whose input is a scalar t, and whose output is a vector in R^2. Such a function is called a **vector-valued function** and t is called a **parameter variable**. The common notation is to write $\mathbf{r}(t) = \langle x(t), y(t) \rangle$ for vector-valued functions in R^2, and $\mathbf{r}(t) = \langle x(t), y(t), z(t) \rangle$ for vector-valued functions in R^3. The number of parameter variables can be greater than one.

♦ • ♦ • ♦ • ♦ • ♦

Example 15.1: Sketch $\mathbf{r}(t) = \langle 2t + 1, t^2 + 3 \rangle$ for $-1 \leq t \leq 2$.

Solution: Let's build an input-output table:

t	$\mathbf{r}(t) = \langle 2t + 1, t^2 + 3 \rangle$
−1	$\mathbf{r}(-1) = \langle 2(-1) + 1, (-1)^2 + 3 \rangle = \langle -1, 4 \rangle$
−0.5	$\mathbf{r}(-0.5) = \langle 2(-0.5) + 1, (-0.5)^2 + 3 \rangle = \langle 0, 3.25 \rangle$
0	$\mathbf{r}(0) = \langle 2(0) + 1, (0)^2 + 3 \rangle = \langle 1, 3 \rangle$
0.5	$\mathbf{r}(0.5) = \langle 2(0.5) + 1, (0.5)^2 + 3 \rangle = \langle 2, 3.25 \rangle$
1	$\mathbf{r}(1) = \langle 2(1) + 1, (1)^2 + 3 \rangle = \langle 3, 4 \rangle$
1.5	$\mathbf{r}(1.5) = \langle 2(1.5) + 1, (1.5)^2 + 3 \rangle = \langle 4, 5.25 \rangle$
2	$\mathbf{r}(2) = \langle 2(2) + 1, (2)^2 + 3 \rangle = \langle 5, 7 \rangle$

We then sketch vectors for each t such that its foot is at the origin:

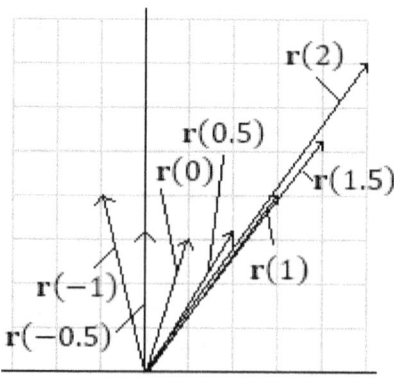

This looks like a mess, but it is a truthful and literal representation of $\mathbf{r}(t) = \langle 2t + 1, t^2 + 3 \rangle$ for certain values of t in the interval $-1 \leq t \leq 2$. However, when representing the graph of a vector valued function, it is common to only show the position at the head of the vector, and the curve that results.

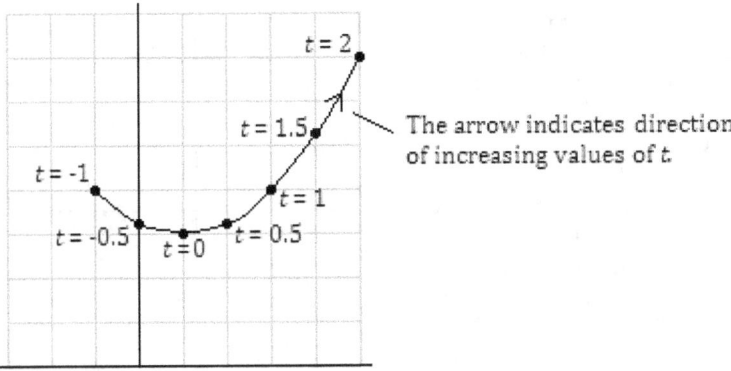

This image is much cleaner and we see that the path traced out by the heads of the vectors given by $\mathbf{r}(t) = \langle 2t + 1, t^2 + 3 \rangle$ for $-1 \leq t \leq 2$ forms a parabola. Note that some of the t values are stated at certain points. It is common to place an arrow on this path to show the direction of increasing value of the variable t.

Example 15.2: Sketch $\mathbf{r}(t) = \langle a \cos t, a \sin t \rangle$, for $0 \leq t \leq 2\pi$, and describe the curve that is traced out by the vectors.

Solution: We build an input-output table:

t	$\mathbf{r}(t) = \langle a \cos t, a \sin t \rangle$
0	$\mathbf{r}(0) = \langle a \cos 0, a \sin 0 \rangle = \langle a, 0 \rangle$
$\pi/4$	$\mathbf{r}(\pi/4) = \langle a \cos(\pi/4), a \sin(\pi/4) \rangle = \langle a\sqrt{2}/2, a\sqrt{2}/2 \rangle$
$\pi/2$	$\mathbf{r}(\pi/2) = \langle a \cos(\pi/2), a \sin(\pi/2) \rangle = \langle 0, a \rangle$
$3\pi/4$	$\mathbf{r}(3\pi/4) = \langle a \cos(3\pi/4), a \sin(3\pi/4) \rangle$
	$= \langle -a\sqrt{2}/2, a\sqrt{2}/2 \rangle$
π	$\mathbf{r}(\pi) = \langle a \cos(\pi), a \sin(\pi) \rangle = \langle -a, 0 \rangle$
$5\pi/4$	$\mathbf{r}(5\pi/4) = \langle a \cos(5\pi/4), a \sin(5\pi/4) \rangle$
	$= \langle -a\sqrt{2}/2, -a\sqrt{2}/2 \rangle$
$3\pi/2$	$\mathbf{r}(3\pi/2) = \langle a \cos(3\pi/2), a \sin(3\pi/2) \rangle = \langle 0, -a \rangle$
$7\pi/4$	$\mathbf{r}(7\pi/4) = \langle a \cos(7\pi/4), a \sin(7\pi/4) \rangle$
	$= \langle a\sqrt{2}/2, -a\sqrt{2}/2 \rangle$
2π	$\mathbf{r}(2\pi) = \langle a \cos(2\pi), a \sin(2\pi) \rangle = \langle a, 0 \rangle$

The curve is below. The vectors are not actually drawn. Instead, the curve formed by the placement of each vector's head is drawn.

The curve is a circle of radius a, centered at the origin. The bounds $0 \leq t \leq 2\pi$ ensure that exactly one revolution of the circle is sketched.

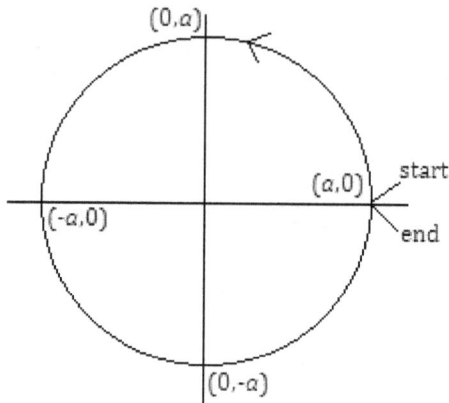

Note that certain points on the path are given by ordered pairs. Remember that these are the heads of the vectors, which are not drawn. Thus, the point $(0, a)$ represents the head of the vector $\langle 0, a \rangle$ when $t = \pi/2$. The arrow shows the direction of increasing t, and the circle "starts" at the point $(a, 0)$ and ends at this same point, one revolution later.

There is more than one way to define a circle of radius a. For example, $\mathbf{r}(t) = \langle a \sin t, a \cos t \rangle$, for $0 \leq t \leq 2\pi$ traces the same circle, but this time starting at $(0, a)$ and in the clockwise direction.

♦ • ♦ • ♦ • ♦ • ♦

Example 15.3: Rewrite the function $y = f(x) = x^3$ from (0,0) to (3,27) as a vector-valued function.

Solution: Any function of the form $y = f(x)$ can be rewritten as a vector-valued function by letting $x(t) = t$ and $y(t) = f(t)$. Thus, the function $y = f(x) = x^3$ from (0,0) to (3,27) can be re-written as

$$\mathbf{r}(t) = \langle t, t^3 \rangle \text{ for } 0 \leq t \leq 3.$$

Note that $\mathbf{r}(0) = \langle 0,0 \rangle$ and that $\mathbf{r}(3) = \langle 3,27 \rangle$. These are vectors whose heads lie at the points (0,0) and (3,27) respectively.

♦ • ♦ • ♦ • ♦ • ♦

Example 15.4: Find the domain of $\mathbf{r}(t) = \left\langle \ln t, 2t, \frac{1}{3-t} \right\rangle$.

Solution: The domain is the largest subset of the real numbers for which all three component functions are defined simultaneously. Note that $x(t) = \ln t$ is defined only if $x > 0$, that $y(t) = 2t$ is defined for all real numbers t, and that

$z(t) = \frac{1}{3-t}$ is not defined when $t = 3$. Thus, the domain of **r** is given by all non-negative values of t, excluding $t = 3$. Thus, Dom **r** = $\{t|(0,3) \cup (3, \infty)\}$.

◆ • ◆ • • ◆ • • ◆

Example 15.5: Find the domain of $\mathbf{r}(t) = \left\langle \frac{2}{t}, \sqrt{4 - 3t}, e^t \right\rangle$.

Solution: The first component $x(t) = \frac{2}{t}$ requires that $t \neq 0$, and the second component $y(t) = \sqrt{4 - 3t}$ requires that $4 - 3t \geq 0$, or $t \leq \frac{4}{3}$. There are no restrictions on t implied by $z(t) = e^t$. The domain of **r** is $\left\{ t \middle| (-\infty, 0) \cup \left(0, \frac{4}{3}\right] \right\}$.

◆ • ◆ • • ◆ • • ◆

Example 15.6: Find a vector valued function that describes the line segment in R^3 from $(1, -2, 5)$ to $(3, 1, -4)$.

Solution: Find the direction vector:

$$\mathbf{v} = \langle 3 - 1, 1 - (-2), -4 - 5 \rangle = \langle 2, 3, -9 \rangle.$$

Using $(1, -2, 5)$ as the initial point, we have $\langle 1, -2, 5 \rangle + t\langle 2, 3, -9 \rangle$ as the line segment using vector notation. As a vector-valued function, we have

$$\mathbf{r}(t) = \langle 1 + 2t, -2 + 3t, 5 - 9t \rangle \text{ for } 0 \leq t \leq 1.$$

Note that $\mathbf{r}(0) = \langle 1, -2, 5 \rangle$, a vector whose head lies at the point $(1, -2, 5)$, and that $\mathbf{r}(1) = \langle 3, 1, -4 \rangle$, a vector whose head lies at the point $(3, 1, -4)$.

◆ • ◆ • • ◆ • • ◆

Example 15.7: Describe $\mathbf{r}(t) = \langle 2\cos t, 2\sin t, t \rangle$ for $t \geq 0$.

Solution: This is a curve in R^3. Look at two of the components at a time:

The components $x(t) = 2\cos t$ and $y(t) = 2\sin t$ trace a circle of radius 2 repeatedly in the xy-plane since t increases without bound. The components $x(t) = 2\cos t$ and $z(t) = t$ trace a cosine wave "upward" in the xz-plane, e.g. assuming that x is the horizontal axis and z the vertical axis. The components $y(t) = 2\sin t$ and $z(t) = t$ trace a sine wave "upward" in the yz-plane.

The curve is a *helix*, which looks like a coiled spring. This helix has a radius of 2 centered around the positive z-axis, "wrapping" around the z-axis (but never touching it) as t increases in value.

Example 15.8: In R^3, the circular cylinder $x^2 + y^2 = 25$ is intersected by the plane $y + z = 4$. Find a vector-valued function $\mathbf{r}(t) = \langle x(t), y(t), z(t) \rangle$ that describes the curve formed by the intersection of these two surfaces.

Solution: There are many possible vector-valued functions that describe this curve. One possible way is to note that we can write $x(t) = 5\cos t$ and $y(t) = 5\sin t$ for $0 \le t \le 2\pi$. Then, since $y + z = 4$, we have $z = 4 - y$, so that $z(t) = 4 - 5\sin t$. The curve of intersection is given by

$$\mathbf{r}(t) = \langle 5\cos t, 5\sin t, 4 - 5\sin t \rangle, \quad \text{for } 0 \le t \le 2\pi.$$

♦ • ♦ • ♦ • ♦ • ♦

The number of parameter variables of a vector-valued function describe the "type" of graph that will result. For example, a vector-valued function of one parameter variable will result in a curve, as demonstrated in the previous examples. A vector-valued function of two variables results in a surface, as the next two examples show.

Example 15.9: A circular cylinder of radius 2 is centered at the origin such that the x-axis is the axis of symmetry of the cylinder. Describe this surface parametrically, using u and v as the parameter variables.

Solution: Since the x-axis is the axis of symmetry, we infer that the circular cross sections lie on planes parallel to the yz-plane. For example, a circle of radius 2 on the yz-plane ($x = 0$) is described by $y^2 + z^2 = 4$. Using parameter variable u, we can describe the circle by letting $y = 2\cos u$ and $z = 2\sin u$, where the 2 represents the circle's radius. Note that the circular cross-sections depend only on variable u. Thus, we can let $x = v$, representing the extension of the circle into the positive and negative x direction, with no restrictions on v. The cylinder is described parametrically as

$$\mathbf{r}(u, v) = \langle v, 2\cos u, 2\sin u \rangle, \quad 0 \le u \le 2\pi, \quad -\infty < v < \infty.$$

♦ • ♦ • ♦ • ♦ • ♦

Example 15.10: Describe the cone $z = \sqrt{x^2 + y^2}$ parametrically using variables u and v.

Solution: Observe that cross sections of this surface with a plane $z = k$ results in a circle of radius k. Thus, if we let $z = u$, we can then define $x = u\cos v$ and $y = u\sin v$, which result in circles of radius u. Thus, we have $\mathbf{r}(u, v) = \langle u\cos v, u\sin v, u \rangle$, where $0 \le v \le 2\pi$ and $u \ge 0$.

16. Vector Valued Functions: Limits & Continuity

The same notions of limits and continuity hold true for vector-valued functions. For example, the **limit** of $r(t) = \langle x(t), y(t), z(t)\rangle$ as $t \to a$ is given by

$$\lim_{t \to a} r(t) = \langle \lim_{t \to a} x(t), \lim_{t \to a} y(t), \lim_{t \to a} z(t)\rangle,$$

assuming that all three limits exist.

Similarly, a vector-valued function $r(t) = \langle x(t), y(t), z(t)\rangle$ is **continuous** at $t = a$ if

- The limit as $t \to a$ exists,
- The vector $r(a)$ exists (that is, a is in the domain of r), and
- $\lim_{t \to a} r(t) = r(a)$.

◆ • ◆ • ◆ • ◆ • ◆

Example 16.1: Let $r(t) = \langle t^2, e^t, \frac{1}{t+3}\rangle$, find $\lim_{t \to 2} r(t)$. Is r continuous at $t = 2$?

Solution: The limit is $\lim_{t \to 2} r(t) = \langle \lim_{t \to 2} t^2, \lim_{t \to 2} e^t, \lim_{t \to 2} (\frac{1}{t+3})\rangle = \langle 4, e^2, \frac{1}{5}\rangle.$

Note that $r(2) = \langle 4, e^2, \frac{1}{5}\rangle$. Since all three conditions of continuity are met, the curve traced out by $r(t) = \langle t^2, e^t, \frac{1}{t+3}\rangle$ is continuous at $t = 2$.

In this example, the limit of r as $t \to -3$ does not exist since the limit fails to exist for the expression $\frac{1}{t+3}$. This curve is not continuous when $t = -3$. It is continuous everywhere else.

◆ • ◆ • ◆ • ◆ • ◆

Example 16.2: Given $r(t) = \langle 2t + 1, \frac{t^2-9}{t-3}, t^2\rangle$, find $\lim_{t \to 3} r(t)$. Is r continuous at $t = 3$?

Solution: Note that the domain of r excludes the value $t = 3$. However, the limit does exist as $t \to 3$, since $\lim_{t \to 3} r(t) = \langle \lim_{t \to 3} (2t + 1), \lim_{t \to 3} (\frac{t^2-9}{t-3}), \lim_{t \to 3} t^2\rangle = \langle 7, 6, 9\rangle.$

The middle expression simplifies as $\frac{t^2-9}{t-3} = \frac{(t+3)(t-3)}{t-3} = t+3$, then the limit is taken. However, the value $t = 3$ is still excluded from the domain, so **r** is not continuous at $t = 3$. There is a deleted point in the curve when $t = 3$. But **r** is continuous everywhere else.

◆ • ◆ • ◆ • ◆ • ◆

17. Vector Valued Functions: Differentiation

Given a vector-valued function $\mathbf{r}(t) = \langle x(t), y(t), z(t) \rangle$, the derivative of **r** with respect to t is given by

$$\mathbf{r}'(t) = \frac{d}{dt}\mathbf{r}(t)$$

$$= \frac{d}{dt}\langle x(t), y(t), z(t) \rangle$$

$$= \left\langle \frac{d}{dt}x(t), \frac{d}{dt}y(t), \frac{d}{dt}z(t) \right\rangle$$

$$= \langle x'(t), y'(t), z'(t) \rangle,$$

assuming that the derivatives exist. Note that $\mathbf{r}'(t) = \langle x'(t), y'(t), z'(t) \rangle$ is itself a vector-valued function. Visually, the vectors given by $\mathbf{r}'(t)$ can be shifted in such a way so that they are tangent to the curve traced out by $\mathbf{r}(t) = \langle x(t), y(t), z(t) \rangle$.

In a physical setting, if $\mathbf{r}(t) = \langle x(t), y(t), z(t) \rangle$ represents the displacement of an object, then $\mathbf{v}(t) = \mathbf{r}'(t) = \langle x'(t), y'(t), z'(t) \rangle$ represents the object's velocity and the magnitude, $|\mathbf{r}'(t)|$, is the object's speed. Acceleration is $\mathbf{a}(t) = \mathbf{v}'(t) = \mathbf{r}''(t) = \langle x''(t), y''(t), z''(t) \rangle$.

◆ • ◆ • ◆ • ◆ • ◆

Example 17.1: An object moves through R^3 along a path defined by $\mathbf{r}(t) = \langle t^3, 2t^2 + t, 5t \rangle$ where all dimensions are in meters. Find the object's velocity and its speed when $t = 4$ seconds.

Solution: The derivative of $\mathbf{r}(t) = \langle t^3, 2t^2 + t, 5t \rangle$ is $\mathbf{r}'(t) = \langle 3t^2, 4t + 1, 5 \rangle$. Thus, when $t = 4$ seconds, the object has a velocity of $\mathbf{r}'(4) = \langle 3(4)^2, 4(4) + 1, 5 \rangle = \langle 48, 17, 5 \rangle$. The object's speed at $t = 4$ seconds is $|\mathbf{r}'(4)| = \sqrt{48^2 + 17^2 + 5^2} \approx 51.2$ meters per second.

Example 17.2: An object moves through R^2 along a path defined by $\mathbf{r}(t) = \langle t, -4.9t^2 + 24t \rangle$, where the first component is the horizontal displacement in meters, and the second component is vertical displacement in meters, and where t is in seconds. Find the maximum height that this object achieves.

Solution: Note that the object traces a downward-opening parabolic arc in R^2. The object will achieve its maximum height when the vertical component of velocity of the object is temporarily 0. Thus, we differentiate: $\mathbf{v}(t) = \mathbf{r}'(t) = \langle 1, -9.8t + 24 \rangle$.

We then set the vertical component of velocity to 0, and solve:

$$-9.8t + 24 = 0 \quad \text{gives} \quad t = \frac{24}{9.8} \approx 2.449 \text{ seconds}.$$

This is the *time* at which the object achieves its maximum height. When we substitute $t = 2.449$ into $\mathbf{r}$, we have

$$\mathbf{r}(2.449) = \langle 2.449, -4.9(2.449)^2 + 24(2.449) \rangle = \langle 2.449, 29.388 \rangle.$$

The object achieves a maximum height of about 29.388 meters above the ground after 2.449 seconds in flight. The object has moved 2.449 meters horizontally in this same period of time.

◆ • ◆ • ◆ • ◆ • ◆

Example 17.3: An object moves through R^2 along a path defined by $\mathbf{r}(t) = \langle t^3, t^2 + 2t \rangle$, where the components are in meters and t is in seconds. What is the minimum speed of the object?

Solution: The derivative is $\mathbf{r}'(t) = \langle 3t^2, 2t + 2 \rangle$, so that the speed can be now stated as a function in variable t: $s(t) = |\mathbf{r}'(t)| = \sqrt{(3t^2)^2 + (2t + 2)^2} = \sqrt{9t^4 + 4t^2 + 8t + 4}$. We now minimize $s(t)$:

$$\frac{d}{dt}s(t) = \frac{d}{dt}\sqrt{9t^4 + 4t^2 + 8t + 4} = \frac{36t^3 + 8t + 8}{2\sqrt{9t^4 + 4t^2 + 8t + 4}}.$$

This expression is 0 when the numerator is 0. Using a calculator, we find that $36t^3 + 8t + 8 = 0$ when $t = -0.485$ seconds. This can be verified to be a minimum by using either the first or second derivative test. Thus, the object's minimum speed occurs when $t = -0.485$ seconds and is

$$s(-0.485) = \sqrt{9(-0.485)^4 + 4(-0.485)^2 + 8(-0.485) + 4} \approx 1.249$$
meters per second.

Example 17.4: An object moves through R^2 along a path defined by $\mathbf{r}(t) = \langle 2t^2 + 1,\ t^4 \rangle$. Find the equation of the tangent line in vector form when $t = 5$.

Solution. The derivative is $\mathbf{r}'(t) = \langle 4t, 4t^3 \rangle$. Thus, when $t = 5$, the object is moving (instantaneously) in the direction of $\mathbf{r}'(5) = \langle 4(5), 4(5)^3 \rangle = \langle 20, 500 \rangle$. This is the object's direction vector. Furthermore, at $t = 5$, the object's location is $\mathbf{r}(5) = \langle 2(5)^2 + 1,\ (5)^4 \rangle = \langle 51, 625 \rangle$. Thus, the object's tangent line in vector form when $t = 5$ is $\langle 51, 625 \rangle + t\langle 20, 500 \rangle$, or equivalently, $\langle 51 + 20t, 625 + 500t \rangle$.

♦ • ♦ • ♦ • ♦ • ♦

Example 17.5: An object moves through R^3 along a path defined by $\mathbf{r}(t) = \langle t + 3,\ t^2 + t,\ 5t \rangle$. Find the equation of the tangent line to this path when the object is at $(7,20,20)$.

Solution. As in the previous example, we need both a direction vector and a position vector. The location $(7,20,20)$ corresponds to a position vector $\langle 7,20,20 \rangle$, and setting this equal to $\mathbf{r}(t) = \langle t + 3,\ t^2 + t,\ 5t \rangle$, we can deduce that $t = 4$. The derivative is $\mathbf{r}'(t) = \langle 1, 2t + 1, 5 \rangle$, so the direction vector is $\mathbf{r}'(4) = \langle 1, 2(4) + 1, 5 \rangle = \langle 1,9,5 \rangle$.

Thus, the object's tangent line in vector form at this instant is $\langle 7, 20, 20 \rangle + t\langle 1, 9, 5 \rangle$, or equivalently, $\langle 7 + t, 20 + 9t, 20 + 5t \rangle$.

Example 17.6: An object revolves around the origin in a circular orbit. The circle is of radius 5 meters and the object completes a revolution every 10 seconds. Assume the object moves counter-clockwise and that is started on the positive x-axis. Find this object's position (displacement), velocity, speed and acceleration at time t.

Solution: Let's assume that $0 \le t \le 10$ seconds represents one revolution of the object. Then, the object's displacement is given by

$$\mathbf{r}(t) = \left\langle 5\cos\left(\frac{2\pi t}{10}\right),\ 5\sin\left(\frac{2\pi t}{10}\right) \right\rangle = \left\langle 5\cos\left(\frac{\pi t}{5}\right),\ 5\sin\left(\frac{\pi t}{5}\right) \right\rangle.$$

The leading coefficient 5 represents the radius, and note that when $t = 10$, the arguments within the sine and cosine operators are both $\frac{\pi}{5}(10) = 2\pi$, the usual period of the sine and cosine functions.

The velocity is

$$\mathbf{v}(t) = \mathbf{r}'(t) = \left\langle -5\sin\left(\frac{\pi t}{5}\right)\left(\frac{\pi}{5}\right),\ 5\cos\left(\frac{\pi t}{5}\right)\left(\frac{\pi}{5}\right) \right\rangle = \left\langle -\pi\sin\left(\frac{\pi t}{5}\right),\ \pi\cos\left(\frac{\pi t}{5}\right) \right\rangle.$$

The chain rule was used followed by simplification. Note that $\mathbf{r}(t) \cdot \mathbf{v}(t) = 0$. This is always true for objects moving in a circular path: the (tangential) velocity vector is orthogonal to the displacement vector.

The object's speed is

$$|\mathbf{v}(t)| = |\mathbf{r}'(t)| = \sqrt{\left(-\pi \sin\left(\frac{\pi t}{5}\right)\right)^2 + \left(\pi \cos\left(\frac{\pi t}{5}\right)\right)^2} = \pi \text{ meters per second.}$$

This makes sense: the circumference of the object's path is $2\pi(5) = 10\pi$ meters. If it takes the object 10 seconds to complete one revolution at π meters per second, then it will have travelled a distance of 10π meters in that revolution.

The acceleration is

$$\mathbf{a}(t) = \mathbf{v}'(t) = \mathbf{r}''(t) = \left\langle -\frac{\pi^2}{5}\cos\left(\frac{\pi t}{5}\right), -\frac{\pi^2}{5}\sin\left(\frac{\pi t}{5}\right)\right\rangle.$$

Note that the acceleration vector is always opposite the displacement vector for an object in circular motion.

♦ • • ♦ • • ♦ • • ♦

18. Vector Valued Functions: Integration

Given a vector-valued function $\mathbf{r}(t) = \langle x(t), y(t), z(t) \rangle$, the indefinite integral of $\mathbf{r}$ with respect to t is given by

$$\int \mathbf{r}(t)\, dt = \left\langle \int x(t)\, dt, \int y(t)\, dt, \int z(t)\, dt \right\rangle + \langle a, b, c \rangle,$$

where $\langle a, b, c \rangle$ is a vector composed of the constants of integration of the components of $\mathbf{r}$.

♦ • • ♦ • • ♦ • • ♦

Example 18.1: Find $\int \mathbf{r}(t)\, dt$, where $\mathbf{r}(t) = \langle 3t^2, \frac{1}{t}, \sin(3t) \rangle$, where $t > 0$.

Solution: We have

$$\int \mathbf{r}(t)\, dt = \left\langle \int 3t^2\, dt, \int \left(\frac{1}{t}\right) dt, \int \sin(3t)\, dt \right\rangle$$
$$= \left\langle t^3, \ln t, -\frac{1}{3}\cos(3t) \right\rangle + \langle a, b, c \rangle.$$

Example 18.2: Find $\mathbf{r}(t) = \int \mathbf{r}'(t)\, dt$, where $\mathbf{r}'(t) = \langle e^{2t}, \sqrt{t}, \sin t \rangle$, and $\mathbf{r}(0) = \langle 0,0,0 \rangle$.

Solution: Note that $\mathbf{r}(t) = \int \mathbf{r}'(t)\, dt + \mathbf{k}$, where $\mathbf{k} = \langle a, b, c \rangle$ is a constant vector. We have

$$\mathbf{r}(t) = \int \mathbf{r}'(t)\, dt$$
$$= \left\langle \int e^{2t}\, dt, \int \sqrt{t}\, dt, \int \sin(t)\, dt \right\rangle + \mathbf{k}$$
$$= \left\langle \frac{1}{2} e^{2t}, \frac{2}{3} t^{3/2}, -\cos t \right\rangle + \langle a, b, c \rangle.$$

Since $\mathbf{r}(0) = \langle 0,0,0 \rangle$, we have

$$\langle 0,0,0 \rangle = \left\langle \frac{1}{2} e^{2(0)}, \frac{2}{3}(0)^{3/2}, -\cos(0) \right\rangle + \langle a, b, c \rangle$$
$$\langle 0,0,0 \rangle = \left\langle \frac{1}{2}, 0, -1 \right\rangle + \langle a, b, c \rangle.$$

This forces $a = -\frac{1}{2}$, $b = 0$ and $c = 1$. Thus, $\mathbf{r}(t) = \left\langle \frac{1}{2} e^{2t}, \frac{2}{3} t^{3/2}, -\cos t \right\rangle +$  $\langle -\frac{1}{2}, 0, 1 \rangle$, or simplified as $\mathbf{r}(t) = \left\langle \frac{1}{2}(e^{2t} - 1), \frac{2}{3} t^{3/2}, 1 - \cos t \right\rangle$. Don't confuse $\mathbf{r}(0) = \langle 0,0,0 \rangle$ as being the constant vector $\langle a, b, c \rangle$.

◆ • ◆ • ◆ • ◆ • ◆

Example 18.3: An object's acceleration is given by $\mathbf{a}(t) = \langle 0, t \rangle$, where t is in seconds and the components are meters per seconds-squared. Find $\mathbf{v}(t)$ and $\mathbf{r}(t)$ such that $\mathbf{v}(1) = \langle 2,5 \rangle$ and $\mathbf{r}(1) = \langle -1, 3 \rangle$.

Solution: Integrating acceleration, we obtain velocity:

$$\mathbf{v}(t) = \int \mathbf{a}(t)\, dt = \int \langle 0, t \rangle\, dt = \left\langle k_1, \frac{1}{2} t^2 + k_2 \right\rangle.$$

To find $\mathbf{k} = \langle k_1, k_2 \rangle$, note that $\mathbf{v}(1) = \langle 2,5 \rangle$:

$$\langle 2,5 \rangle = \left\langle k_1, \frac{1}{2}(1)^2 + k_2 \right\rangle.$$

This forces $k_1 = 2$ and $k_2 = \frac{9}{2}$, so that $\mathbf{v}(t) = \left\langle 2, \frac{1}{2} t^2 + \frac{9}{2} \right\rangle$.

Next, we have

$$r(t) = \int v(t)\, dt = \int \left\langle 2, \frac{1}{2}t^2 + \frac{9}{2} \right\rangle dt = \left\langle 2t + m_1, \frac{1}{6}t^3 + \frac{9}{2}t + m_2 \right\rangle.$$

To find $\mathbf{m} = \langle m_1, m_2 \rangle$, we note that $\mathbf{r}(1) = \langle -1, 3 \rangle$:

$$\langle -1, 3 \rangle = \left\langle 2(1) + m_1, \frac{1}{6}(1)^3 + \frac{9}{2}(1) + m_2 \right\rangle.$$

This forces $m_1 = -3$ and $m_2 = -\frac{5}{3}$. Therefore, $\mathbf{r}(t) = \left\langle 2t - 3, \frac{1}{6}t^3 + \frac{9}{2}t - \frac{5}{3} \right\rangle$.

For a definite integral, evaluate the components in the usual manner:

Example 18.4: Find $\int_0^2 \mathbf{r}(t)\, dt$, where $\mathbf{r}(t) = \left\langle t^2, e^{2t}, \frac{t}{t^2+1} \right\rangle$.

Solution: Integrate. Note that u-du substitution is used for the latter two components.

$$\int_0^2 \mathbf{r}(t)\, dt = \left\langle \int_0^2 t^2\, dt, \int_0^2 e^{2t}\, dt, \int_0^2 \left(\frac{t}{t^2+1} \right) dt \right\rangle$$

$$= \left\langle \left[\frac{1}{3}t^3 \right]_0^2, \left[\frac{1}{2}e^{2t} \right]_0^2, \left[\frac{1}{2}\ln(t^2+1) \right]_0^2 \right\rangle$$

$$= \left\langle \frac{8}{3}, \frac{e^4 - 1}{2}, \frac{1}{2}\ln 5 \right\rangle.$$

19. Arc Length

Let vector-valued function $r(t) = \langle x(t), y(t), z(t)\rangle$ be defined over the closed interval $a \le t \le b$ and differentiable over the open interval $a < t < b$. Visually, this means that r is a smooth curve, with no discontinuities or corners.

The **arc length** s of the curve r over the interval $a \le t \le b$ is given by the definite integral

$$s = \int_a^b \sqrt{(x'(t))^2 + (y'(t))^2 + (z'(t))^2}\, dt.$$

Note that the integrand $\sqrt{(x'(t))^2 + (y'(t))^2 + (z'(t))^2}$ is the same as $|r'(t)|$. Thus, we can write the integral as

$$s = \int_a^b |r'(t)|\, dt.$$

◆ • ◆ • ◆ • ◆ • ◆

Example 19.1: Find the length of the curve traced by $r(t) = \langle 2\cos t, 2\sin t\rangle$ for $0 \le t \le \pi$.

Solution: Find the derivative: $r'(t) = \langle -2\sin t, 2\cos t\rangle$. Then, using the arc length formula, we have

$$\begin{aligned}
s &= \int_0^\pi \sqrt{(-2\sin t)^2 + (2\cos t)^2}\, dt \\
&= \int_0^\pi \sqrt{4\sin^2 t + 4\cos^2 t}\, dt \\
&= \int_0^\pi \sqrt{4(\sin^2 t + \cos^2 t)}\, dt \\
&= 2\int_0^\pi dt = 2\pi.
\end{aligned}$$

The arc length is 2π units. This can be verified using geometry: r traces a semicircle of radius 2. The circumference of a circle of radius 2 is $2\pi(2) = 4\pi$, and half of this figure is 2π.

Example 19.2: Find the arc length of the curve traced by $r(t) = \langle 4t, 2t^2, 2\ln t \rangle$ between the points $(8, 8, 2\ln 2)$ and $(20, 50, 2\ln 5)$.

Solution: The derivative is $r'(t) = \langle 4, 4t, \frac{2}{t} \rangle$. Furthermore, the bounds of t can be inferred from the points. The point $(8, 8, 2\ln 2)$ suggests that $t = 2$ and the point $(20, 50, 2\ln 5)$ suggests that $t = 5$. We have

$$s = \int_2^5 \sqrt{4^2 + (4t)^2 + (2/t)^2} \, dt$$

$$= \int_2^5 \sqrt{16 + 16t^2 + \frac{4}{t^2}} \, dt$$

$$= \int_2^5 \sqrt{\frac{16t^2 + 16t^4 + 4}{t^2}} \, dt$$

$$= \int_2^5 \sqrt{\frac{(4t^2 + 2)^2}{t^2}} \, dt$$

$$= \int_2^5 \left(\frac{4t^2 + 2}{t} \right) dt$$

$$= \int_2^5 \left(4t + \frac{2}{t} \right) dt$$

$$= [2t^2 + 2\ln t]_2^5$$

$$= (50 + 2\ln 5) - (8 + 2\ln 2)$$

$$= 42 + 2\ln\left(\frac{5}{2}\right) \approx 43.832 \text{ units.}$$

Example 19.3: Find the arc length of the curve traced by $r(t) = \langle t^2, 3t, 4t^3 \rangle$ for $1 \leq t \leq 3$.

Solution. The derivative is $r'(t) = \langle 2t, 3, 12t^2 \rangle$. Thus, the arc length is given by

$$s = \int_1^3 \sqrt{(2t)^2 + 3^2 + (12t^2)^2} \, dt = \int_1^3 \sqrt{144t^4 + 4t^2 + 9} \, dt.$$

Using a calculator or any numerical method of integrating, we find that the arc length is

$$\int_1^3 \sqrt{144t^4 + 4t^2 + 9} \, dt \approx 104.58 \text{ units.}$$

Example 19.4: Find the length of the helix traced by $\mathbf{r}(t) = \langle 2\cos t, 2\sin t, t\rangle$ for $0 \le t \le 2\pi$.

Solution: The derivative is $\mathbf{r}'(t) = \langle -2\sin t, 2\cos t, 1\rangle$. We have

$$\begin{aligned}
s &= \int_0^{2\pi} \sqrt{(-2\sin t)^2 + (2\cos t)^2 + 1^2}\, dt \\
&= \int_0^{2\pi} \sqrt{4\sin^2 t + 4\cos^2 t + 1}\, dt \\
&= \int_0^{2\pi} \sqrt{4(\sin^2 t + \cos^2 t) + 1}\, dt \\
&= \int_0^{2\pi} \sqrt{5}\, dt \\
&= 2\pi\sqrt{5} \text{ units.}
\end{aligned}$$

♦ • ♦ • ♦ • ♦ • ♦

Arc Length as a Function

Consider the arc length formula, $s = \int_a^b |\mathbf{r}'(t)|\, dt$, and allow the upper bound to be a variable rather than a fixed value. If we allow the upper bound to be t, and use a dummy variable in the integrand, we have arc length s as a function of t:

$$s(t) = \int_a^t |\mathbf{r}'(u)|\, du.$$

Differentiating both sides with respect to t, we have

$$\frac{d}{dt} s(t) = \frac{d}{dt} \int_a^t |\mathbf{r}'(u)|\, du.$$

Using the Fundamental Theorem of Calculus, we have

$$\frac{d}{dt} \int_a^t |\mathbf{r}'(u)|\, du = |\mathbf{r}'(t)|.$$

Thus, we have

$$\frac{ds}{dt} = |\mathbf{r}'(t)|, \quad \text{or equivalently,} \quad ds = |\mathbf{r}'(t)|\, dt.$$

★ ★ This formula is *extremely* useful later on! Do *not* forget it! ★ ★

20. Unit Tangent and Unit Normal Vectors

Consider an object that moves along a differentiable (smooth, no discontinuities or corners) curve traced by $\mathbf{r}(t) = \langle x(t), y(t), z(t) \rangle$. At each point on the curve, the tangent vector is given by $\mathbf{r}'(t) = \langle x'(t), y'(t), z'(t) \rangle$. The magnitude of the tangent vector, $|\mathbf{r}'(t)|$, can be interpreted as the object's speed. For most curves the speed of an object can vary. In a rough sense, the speed of an object dictates the segmentation of the curve.

♦ • ♦ • • ♦ • • ♦

Example 20.1: Sketch the curve traced by $\mathbf{r}(t) = \langle t, t^2 \rangle$ for $0 \le t \le 4$.

Solution: The curve is shown below. It is a parabola $y = x^2$ from (0,0) to (4,16). The values for integer values of t are shown on the graph.

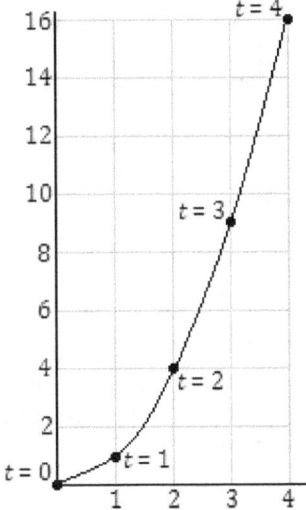

The segments of the curve between consecutive integer values of t vary in length. If t is a unit of time, then the object traverses each segment in the same amount of time. Thus, the object must move faster in order to traverse longer segments. The segmentation of the curve in terms of a unit time interval t is not consistent. The table on the following page shows the object's position, velocity and speed for integer values of t:

t	$r(t) = \langle t, t^2 \rangle$	$r'(t) = \langle 1, 2t \rangle$	$\lvert r'(t) \rvert = \sqrt{1 + 4t^2}$
0	$\langle 0,0 \rangle$	$\langle 1,0 \rangle$	1
1	$\langle 1,1 \rangle$	$\langle 1,2 \rangle$	$\sqrt{5}$
2	$\langle 2,4 \rangle$	$\langle 1,4 \rangle$	$\sqrt{17}$
3	$\langle 3,9 \rangle$	$\langle 1,6 \rangle$	$\sqrt{37}$
4	$\langle 4,16 \rangle$	$\langle 1,8 \rangle$	$\sqrt{65}$

To control the speed of the object, we can force all tangent vectors to have a length of 1 unit. This is called the **unit tangent** vector, and is given by

$$\mathbf{T}(t) = \frac{\mathbf{r}'(t)}{\lvert \mathbf{r}'(t) \rvert}.$$

This means that $\lvert \mathbf{T}(t) \rvert = 1$ for all t in the domain of $\mathbf{r}'$.

♦ • ♦ • ♦ • ♦ • ♦

Example 20.2: Find $\mathbf{T}(t)$, where $\mathbf{r}(t) = \langle t, t^2 \rangle$.

Solution: From the previous example, we have $\mathbf{r}'(t) = \langle 1, 2t \rangle$ and $\lvert \mathbf{r}'(t) \rvert = \sqrt{1 + 4t^2}$. Thus,

$$\mathbf{T}(t) = \frac{\mathbf{r}'(t)}{\lvert \mathbf{r}'(t) \rvert} = \frac{\langle 1, 2t \rangle}{\sqrt{1 + 4t^2}} = \left\langle \frac{1}{\sqrt{1 + 4t^2}}, \frac{2t}{\sqrt{1 + 4t^2}} \right\rangle.$$

You should verify that $\lvert \mathbf{T}(t) \rvert = 1$. If the object moves along this curve at a constant speed of 1 unit of distance per unit of time, then this will force the segmentation of the curve into equal-sized segments, so that it can traverse the same length each time, per unit of time. This is often called the ds segmentation.

♦ • ♦ • ♦ • ♦ • ♦

Example 20.3: Find $\mathbf{T}(t)$, where $\mathbf{r}(t) = \langle 3 \cos t, 3 \sin t, t \rangle$.

Solution: We have

$$\mathbf{T}(t) = \frac{\mathbf{r}'(t)}{\lvert \mathbf{r}'(t) \rvert} = \frac{\langle -3 \sin t, 3 \cos t, 1 \rangle}{\sqrt{10}} = \left\langle \frac{-3 \sin t}{\sqrt{10}}, \frac{3 \cos t}{\sqrt{10}}, \frac{1}{\sqrt{10}} \right\rangle.$$

Note that in this case, the speed of the object is always $\sqrt{10}$ units of distance per unit of time.

The **unit normal** vector is given by

$$\mathbf{N}(t) = \frac{\mathbf{T}'(t)}{|\mathbf{T}'(t)|}.$$

The vector **N** has a length of 1 unit. It is orthogonal to **T** (that is, $\mathbf{N} \cdot \mathbf{T} = 0$). For an object moving along a differentiable curve, **T** will point in the object's (tangential) direction of travel, and **N** will point orthogonal to **T**, representing one component of acceleration. It generally points "inward" to concave side of the curve.

◆ • ◆ • ◆ • ◆ • ◆

Example 20.4: Find $\mathbf{N}(t)$, where $\mathbf{r}(t) = \langle t, t^2 \rangle$.

Solution: From Example 20.2, we have

$$\mathbf{T}(t) = \left\langle \frac{1}{\sqrt{1+4t^2}}, \frac{2t}{\sqrt{1+4t^2}} \right\rangle.$$

We now find $\mathbf{T}'(t)$:

$$\mathbf{T}'(t) = \left\langle \frac{-4t}{(1+4t^2)^{3/2}}, \frac{2}{(1+4t^2)^{3/2}} \right\rangle.$$

Now, we need $|\mathbf{T}'(t)|$:

$$|\mathbf{T}'(t)| = \sqrt{\left(\frac{-4t}{(1+4t^2)^{3/2}}\right)^2 + \left(\frac{2}{(1+4t^2)^{3/2}}\right)^2}.$$

This simplifies after many steps to

$$|\mathbf{T}'(t)| = \frac{2}{1+4t^2}.$$

Thus, the unit normal **N** is given by

$$\mathbf{N}(t) = \frac{\mathbf{T}'(t)}{|\mathbf{T}'(t)|} = \frac{1}{\left(\frac{2}{1+4t^2}\right)} \left\langle \frac{-4t}{(1+4t^2)^{3/2}}, \frac{2}{(1+4t^2)^{3/2}} \right\rangle$$

$$= \left\langle \frac{-2t}{\sqrt{1+4t^2}}, \frac{1}{\sqrt{1+4t^2}} \right\rangle.$$

Note the similarities in **T** and **N** and note also that $\mathbf{N} \cdot \mathbf{T} = 0$.

Example 20.5: Find $\mathbf{N}(t)$, where $\mathbf{r}(t) = \langle 3\cos t, 3\sin t, t\rangle$.

Solution: From Example 20.3, we have

$$\mathbf{T}(t) = \left\langle -\frac{3\sin t}{\sqrt{10}}, \frac{3\cos t}{\sqrt{10}}, \frac{1}{\sqrt{10}}\right\rangle.$$

We find $\mathbf{T}'(t)$:

$$\mathbf{T}'(t) = \left\langle -\frac{3\cos t}{\sqrt{10}}, -\frac{3\sin t}{\sqrt{10}}, 0\right\rangle.$$

Note that

$$|\mathbf{T}'(t)| = \sqrt{\left(-\frac{3\cos t}{\sqrt{10}}\right)^2 + \left(-\frac{3\sin t}{\sqrt{10}}\right)^2} = \frac{3}{\sqrt{10}}.$$

Thus,

$$\mathbf{N}(t) = \frac{\mathbf{T}'(t)}{|\mathbf{T}'(t)|} = \frac{\left\langle -\frac{3\cos t}{\sqrt{10}}, -\frac{3\sin t}{\sqrt{10}}, 0\right\rangle}{\frac{3}{\sqrt{10}}} = \langle -\cos t, -\sin t, 0\rangle.$$

Observe that $|\mathbf{N}(t)| = 1$ and that $\mathbf{N}\cdot\mathbf{T} = 0$.

A vector valued function $\mathbf{r}(t)$ with a constant magnitude, $|\mathbf{r}(t)| = c$, traces a path along a circle (or portion thereof) of radius c in R^2, or a path on a sphere of radius c in R^3. In such a case, $\mathbf{r}(t)\cdot\mathbf{r}'(t) = 0$ for all t in the relevant domain. This was mentioned in Example 17.6 and appears here in Examples 20.3 and 20.4.

See an error? Have a suggestion?
Please see www.surgent.net/vcbook

21. Curvature

Let C be a continuous path in R^n. The expression $\left|\frac{d\mathbf{T}}{ds}\right|$ represents the rate of change in a unit tangent vector per unit segment along the path—roughly, how "fast" $\mathbf{T}$ turns per unit segment s. This is called **curvature**, and it offers a way to quantify the "curviness" of a path. Intuitively, we would expect that a line has no (or zero) curvature, while a path with high curvature would turn quickly, in the extreme case appearing as a corner in the sense that the turn happened instantaneously.

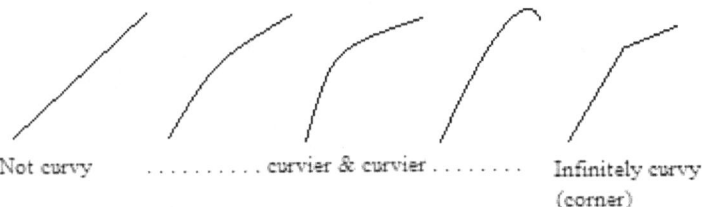

Not curvy .... curvier & curvier .... Infinitely curvy (corner)

Also, it is plausible (and as it turns out, it is correct) to assume that a circle has constant curvature. Small circles have high curvature since the turning is happening faster, while big circles have small curvature. It is then plausible to assume there is an inverse relationship between the radius of a circle and its curvature. That is, for a circle of radius r, its curvature, denoted κ (lower-case letter *kappa*), should be $\frac{1}{r}$. As we will see in an example, this is true: $\kappa = \frac{1}{r}$.

Since $\mathbf{T}(t) = \frac{\mathbf{r}'(t)}{|\mathbf{r}'(t)|}$ and $\frac{ds}{dt} = |\mathbf{r}'(t)|$, we have

$$\kappa(t) = \left|\frac{d\mathbf{T}}{ds}\right| = \left|\frac{d\mathbf{T}/dt}{ds/dt}\right| = \frac{\left|\frac{d}{dt}\left(\frac{\mathbf{r}'(t)}{|\mathbf{r}'(t)|}\right)\right|}{|\mathbf{r}'(t)|}.$$

♦ • ♦ • ♦ • ♦ • ♦

Example 21.1: Find the curvature of a circle with radius r.

Solution: Parameterize the circle: $\mathbf{r}(t) = \langle r\cos t, r\sin t\rangle$, with $0 \leq t \leq 2\pi$. Differentiating, we have

$$\mathbf{r}'(t) = \langle -r\sin t, r\cos t\rangle$$

and

$$|\mathbf{r}'(t)| = \sqrt{(-r\sin t)^2 + (r\cos t)^2} = \sqrt{r^2(\cos^2 t + \sin^2 t)} = r.$$

Thus,
$$T(t) = \frac{r'(t)}{|r'(t)|} = \frac{\langle -r\sin t, r\cos t\rangle}{r} = \langle -\sin t, \cos t\rangle.$$

Differentiating $T(t)$, we have
$$\frac{d}{dt}T(t) = \frac{d}{dt}\left(\frac{r'(t)}{|r'(t)|}\right) = \langle -\cos t, -\sin t\rangle.$$

Its magnitude is
$$\left|\frac{d}{dt}\left(\frac{r'(t)}{|r'(t)|}\right)\right| = \sqrt{(-\cos t)^2 + (-\sin t)^2} = 1.$$

Recalling that $|r'(t)| = r$, the curvature of a circle of radius r is
$$\kappa(t) = \left|\frac{dT}{ds}\right| = \left|\frac{dT/dt}{ds/dt}\right| = \frac{\left|\frac{d}{dt}\left(\frac{r'(t)}{|r'(t)|}\right)\right|}{|r'(t)|} = \frac{1}{r}.$$

The implications of this result allow us to infer that a large circle has small curvature, so that zero curvature would be appropriate for a line (roughly speaking, a circle with a very large radius would have sides of imperceptible curvature, so allowing the radius to tend to infinity as a limit, the arc of such a circle would be a line in the limiting sense). Similarly, a very small circle has a very large curvature, and taking this to its logical extreme, a circle of zero radius would have infinite curvature, and such a concept would be appropriate for a path with a corner.

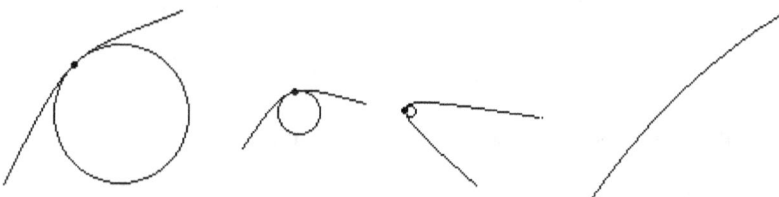

Circles of osculation: as the circles decrease in radius (left three images), they "fit" paths of greater curvature. On the right is a portion of a circle with a very large radius. It appears to be nearly linear. Note that curvature can change in value for different points along a path.

Example 21.2: Find the curvature of a line, $y = mx + b$.

Solution: Letting $x = t$, the parameterization of the line is $\mathbf{r}(t) = \langle t, mt + b \rangle$. Thus, $\mathbf{r}'(t) = \langle 1, m \rangle$ and $|\mathbf{r}'(t)| = \sqrt{1 + m^2}$. The unit tangent vector is

$$\mathbf{T}(t) = \frac{\mathbf{r}'(t)}{|\mathbf{r}'(t)|} = \frac{\langle 1, m \rangle}{\sqrt{1 + m^2}}.$$

Note that this is a constant vector. Therefore, $\frac{d}{dt}\mathbf{T}(t) = 0$ and as a result, the curvature of a line is $\kappa(t) = 0$, as expected.

◆ • ◆ • ◆ • ◆ • ◆

Example 21.3: Find the curvature of the parabola $y = x^2$.

Solution: We have $\mathbf{r}(t) = \langle t, t^2 \rangle$, so that $\mathbf{r}'(t) = \langle 1, 2t \rangle$ and $|\mathbf{r}'(t)| = \sqrt{1 + 4t^2}$. Thus,

$$\mathbf{T}(t) = \frac{\mathbf{r}'(t)}{|\mathbf{r}'(t)|} = \frac{\langle 1, 2t \rangle}{\sqrt{1 + 4t^2}} = \left\langle \frac{1}{\sqrt{1 + 4t^2}}, \frac{2t}{\sqrt{1 + 4t^2}} \right\rangle.$$

Differentiating, we have

$$\frac{d}{dt}\mathbf{T}(t) = \left\langle -\frac{4t}{(1 + 4t^2)^{3/2}}, \frac{2}{(1 + 4t^2)^{3/2}} \right\rangle.$$

Its magnitude is

$$\left| \frac{d}{dt}\mathbf{T}(t) \right| = \sqrt{\left(-\frac{4t}{(1 + 4t^2)^{3/2}} \right)^2 + \left(\frac{2}{(1 + 4t^2)^{3/2}} \right)^2}$$

$$= \sqrt{\frac{16t^2}{(1 + 4t^2)^3} + \frac{4}{(1 + 4t^2)^3}}$$

$$= \sqrt{\frac{4 + 16t^2}{(1 + 4t^2)^3}}$$

$$= \sqrt{\frac{4(1 + 4t^2)}{(1 + 4t^2)^3}}$$

$$= \frac{2}{1 + 4t^2}.$$

Thus, the curvature is

$$\kappa(t) = \left|\frac{dT}{ds}\right| = \left|\frac{dT/dt}{ds/dt}\right| = \frac{\left(\frac{2}{1+4t^2}\right)}{\sqrt{1+4t^2}} = \frac{2}{(1+4t^2)^{3/2}}.$$

Since a parabola does not have constant curvature, it is no surprise that the curvature can vary as a function of t. At the origin, where $t = 0$, the curvature is $\kappa(0) = 2$. Thus, a circle of osculation at the origin would have a radius of $\frac{1}{2}$. As t trends away from 0, the curvature decreases to 0; equivalently, the circles of osculation become larger. This should be plausible, as the parabola is "curviest" at its vertex, and less "curvy" farther away.

◆ • ◆ • ◆ • ◆ • ◆

Curvature should not be confused with concavity.

Finding the curvature of a curve defined parametrically can involve many steps. An equivalent formula for determining curvature of a path in R^2 given by $y = f(t)$ is

$$\kappa(t) = \frac{y''}{(1+(y')^2)^{3/2}},$$

assuming that its first and second derivative exist.

Notation can vary: κ (kappa) is often used in situations where the sign of the curvature is ignored (i.e. always taken as a non-negative value). Since it is possible for $y'' < 0$, then curvature may take on negative values in which case lower-case k is used to represent signed curvature. However, using κ in these situations is usually acceptable.

◆ • ◆ • ◆ • ◆ • ◆

Example 21.4: Find the curvature of the parabola $y = t^2$ using the alternative formula given above.

Solution: We have $y' = 2t$ and $y'' = 2$. Thus,

$$\kappa(t) = \frac{y''}{(1+(y')^2)^{3/2}} = \frac{2}{(1+(2t)^2)^{3/2}} = \frac{2}{(1+4t^2)^{3/2}}.$$

22. Projectile Motion

On Earth, the gravitational acceleration constant is -9.8 meters per second per second (or m/s^2). If we superimpose an xy-axis system with the positive y axis being "up", then **acceleration** can be written as a vector-valued function,

$$\mathbf{a}(t) = \langle 0, -9.8 \rangle.$$

The x-component of acceleration is 0, since falling bodies will not accelerate horizontally due to gravity. Note that both components are constants.

Integrating acceleration, we get **velocity**, which is also a vector-valued function:

$$\begin{aligned}\mathbf{v}(t) &= \int \mathbf{a}(t) \, dt \\ &= \int \langle 0, -9.8 \rangle \, dt \\ &= \langle 0, -9.8t \rangle + \langle v_x, v_y \rangle \\ &= \langle v_x, -9.8t + v_y \rangle.\end{aligned}$$

Here, $\langle v_x, v_y \rangle$ are constants of integration and represents the initial velocity in the x-direction and in the y-direction, respectively. **Speed** is the magnitude of velocity. Speed is a scalar value.

Integrating velocity, we get **displacement** (or **position**), also a vector-valued function:

$$\begin{aligned}\mathbf{r}(t) &= \int \mathbf{v}(t) \, dt \\ &= \int \langle v_x, -9.8t + v_y \rangle \, dt \\ &= \langle v_x t + r_x, -4.9t^2 + v_y t + r_y \rangle.\end{aligned}$$

Similar to above, $\langle r_x, r_y \rangle$ represent the initial position of the object in the x-direction and y-direction respectively. The placement of the origin is arbitrary but is usually done so that $r_x = 0$ since it is almost always practical to assume no initial horizontal distance.

A typical object that is released and then allowed to return to earth under gravity alone will follow a parabolic arc. Its position vector $\mathbf{r}(t)$, velocity vector $\mathbf{v}(t)$ and acceleration vector $\mathbf{a}(t)$ are a "tool kit" of equations that completely describes the motion of the object. In most projectile motion situations, we assume that air friction is ignored, any spin on the object is not considered, and that time t is in seconds and is usually reckoned from $t = 0$, representing when the object was released.

Here are some common questions that may be posed regarding a falling object:

- What is the maximum height of the object? When does it achieve its maximum height?
- How far away does the object travel in the horizontal direction? (Its **range**).
- How fast is the object travelling when it impacts the ground for the first time?

Certainly, there are more.

> Build the complete set of displacement, velocity and acceleration vectors first, before solving any subsequent questions.

◆ • ◆ • ◆ • ◆ • ◆ • ◆

Example 22.1: A ball is propelled off a 100 m tall cliff at an initial speed of 50 meters per second at an angle of 30 degrees above the horizontal.

a) Find the maximum height of the ball.
b) Find the range of the ball when hits the ground for the first time.
c) How fast is the ball travelling when it impacts the ground for the first time?
d) At what angle does the ball impact the ground for the first time?

Solution: Starting with acceleration, $\mathbf{a}(t) = \langle 0, -9.8 \rangle$, integrate to obtain

$$\mathbf{v}(t) = \langle v_x, -9.8t + v_y \rangle.$$

To find v_x and v_y, note that its initial speed is $|\mathbf{v}(0)| = 50$ at an angle of 30 degrees. This suggests a right triangle, in which $|\mathbf{v}(0)| = 50$ is the hypotenuse, and v_x and v_y are the horizontal and vertical legs, respectively.

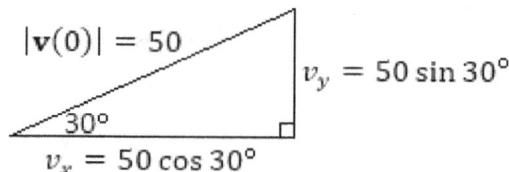

Thus, $v_x = 50 \cos 30 = 25\sqrt{3} \approx 43.301$, and $v_y = 50 \sin 30 = 25$, and the velocity vector is

$$\mathbf{v}(t) = \langle 43.301, -9.8t + 25 \rangle.$$

Integrating **v**, we obtain the displacement vector

$$\mathbf{r}(t) = \langle 43.301t + r_x, -4.9t^2 + 25t + r_y \rangle.$$

Since the ball was thrown off the top of a cliff, set the origin at the base of the cliff, so that $r_x = 0$ and $r_y = 100$. Therefore, the displacement vector is

$$\mathbf{r}(t) = \langle 43.301t, -4.9t^2 + 25t + 100 \rangle.$$

We now have sufficient information to answer the posed questions. In all cases, assume $t \geq 0$.

a) The ball reaches its maximum height when the *y*-component of velocity is 0 since the ball's vertical velocity is 0 (momentarily stops) when it reaches it maximum height. Thus, we solve $-9.8t + 25 = 0$ to find *when* the ball has reached its maximum height. This happens at $t = 25/9.8 = 2.551$ seconds. Substituting this into the *y*-component of displacement, we now determine the height of the ball at this time:

Maximum height $= -4.9(2.551)^2 + 25(2.551) + 100 = 131.888$ meters.

b) The ball impacts the ground when the *y*-component of displacement is 0. We use the quadratic formula to find the roots of $-4.9t^2 + 25t + 100 = 0$:

$$t = \frac{-25 \pm \sqrt{25^2 - 4(-4.9)(100)}}{2(-4.9)} \quad \rightarrow \quad t \approx -2.637 \text{ or } 7.739.$$

The negative result is ignored. The ball lands at $t \approx 7.739$ seconds. The range in which the ball travelled from the base is found by evaluating the *x*-component of displacement by this *t*:

Range $= 43.301(7.739) = 335.106$ m.

c) We find the velocity at $t = 7.739$ seconds:

$$\mathbf{v}(7.739) = \langle 43.301, -9.8(7.739) + 25 \rangle = \langle 43.301, -50.842 \rangle.$$

The impact speed of the ball is the magnitude of this vector:

$$\text{Speed} = |\langle 43.301, -50.842\rangle| = \sqrt{43.301^2 + (-50.842)^2}$$
$$\approx 66.702 \text{ m/s}.$$

d) To find the angle at which the ball impacts the ground, sketch a diagram to be sure that the components of velocity are properly in place:

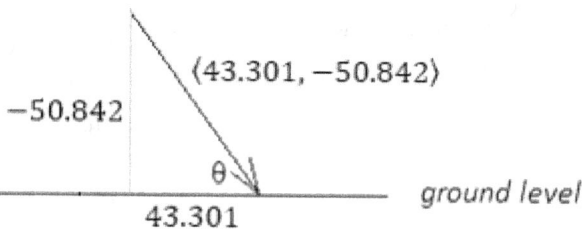

The angle of impact is $\theta = \tan^{-1}\frac{-50.842}{43.301} = -49.6$ degrees. The negatives can be ignored, so that the impact angle is 49.6 degrees.

Example 22.2: An airplane is flying horizontally 5,000 meters above flat ground at a velocity of 300 kilometers per hour. An object is released below the airplane and allowed to fall to earth.

a) How long is the object airborne?
b) How far downrange does the object impact the ground?

Solution: We need to convert 300 kilometers per hour into meters per second:

$$\frac{300 \text{ km}}{1 \text{ hr}} \times \frac{1 \text{ hr}}{3600 \text{ sec}} \times \frac{1000 \text{ m}}{1 \text{ km}} = \frac{(300)(1000)}{3600} = \frac{250}{3} \frac{\text{meters}}{\text{second}}.$$

The initial velocity is therefore $\mathbf{v}(0) = \langle v_x, v_y\rangle = \langle\frac{250}{3}, 0\rangle$, where the 0 in the y-component suggests that the object was merely released with no initial push in the vertical direction. Integrating the acceleration vector $\mathbf{a}(t) = \langle 0, -9.8\rangle$, the velocity vector is

$$\mathbf{v}(t) = \langle v_x, -9.8t + v_y\rangle = \langle\frac{250}{3}, -9.8t\rangle, \qquad t \geq 0.$$

For initial position, the object was released 5000 meters above the ground suggests that $r_y = 5000$, while $r_x = 0$. Therefore, the displacement function is

$$\mathbf{r}(t) = \left\langle \frac{250}{3}t, -4.9t^2 + 5000 \right\rangle, \qquad t \geq 0.$$

a) The object is airborne for the time needed for the y-component of displacement to equal 0 meters, the height of the ground. Thus, we have

$$-4.9t^2 + 5000 = 0.$$

Solving, we get $t = \sqrt{\frac{5000}{4.9}} \approx 31.944$ seconds.

b) The range is found by evaluating the x-component of displacement at $t = 31.944$ seconds. We find that the object travelled $\frac{250}{3}(31.944) = 2{,}662$ meters.

◆ ◆ ◆ ◆ ◆ ◆ ◆ ◆ ◆

Example 22.3: A cannon is angled at 25 degrees above the flat ground. A cannonball is shot from this cannon, and it lands 50 meters downrange. Assume the cannonball left the cannon 1 meter above the ground.

a) What was the initial speed of the cannonball?
b) How many seconds was the cannonball in the air?
c) What was the maximum height of the cannonball?

Solution: Let v_0 represent the initial speed. Using trigonometry, the x and y components of the initial velocity are $v_x = v_0 \cos 25°$ and $v_y = v_0 \sin 25°$. Therefore, the velocity vector is

$$\mathbf{v}(t) = \langle v_0 \cos 25°, -9.8t + v_0 \sin 25° \rangle, \qquad t \geq 0.$$

By placing the origin directly below where the cannonball began its trajectory, we have $r_x = 0$ and $r_y = 1$. The displacement vector is

$$\mathbf{r}(t) = \langle (v_0 \cos 25°)t, -4.9t^2 + (v_0 \sin 25°)t + 1 \rangle, \qquad t \geq 0.$$

a) The cannonball landed 50 meters downrange, so the x-component of displacement is set equal to 50, and we can find an expression for t in terms of v_0:

$$(v_0 \cos 25°)t = 50; \quad \text{so that} \quad t = \frac{50}{v_0 \cos 25°}.$$

When the cannonball landed, the y-component of displacement is 0 meters, so we have $-4.9t^2 + (v_0 \sin 25°)t + 1 = 0$. Mow substitute $t = \frac{50}{v_0 \cos 25°}$ into this equation, simplify, and solve for v_0:

$$-4.9 \left(\frac{50}{v_0 \cos 25°}\right)^2 + (v_0 \sin 25°)\left(\frac{50}{v_0 \cos 25°}\right) = -1$$

$$\frac{-12{,}250}{v_0^2 \cos^2 25°} + \frac{50 \sin 25°}{\cos 25°} = -1$$

Multiply by the common denominator, $v_0^2 \cos^2 25°$ to clear fractions:

$$-12{,}250 + 50 v_0^2 \sin 25° \cos 25° = -v_0^2 \cos^2 25°.$$

Collect terms:

$$v_0^2 (50 \sin 25° \cos 25° + \cos^2 25°) = 12{,}250.$$

Isolate v_0:

$$v_0 = \left(\frac{12{,}250}{50 \sin 25° \cos 25° + \cos^2 25°}\right)^{1/2} \approx 24.766 \; \frac{\text{meters}}{\text{second}}.$$

The cannonball left the cannon at a speed of about 24.766 meters per second.

b) Since $t = \frac{50}{v_0 \cos 25°}$ and $v_0 = 24.766$, we have $t = \frac{50}{(24.766) \cos 25°} \approx$ 2.228 seconds. This is the time the cannonball was in the air.

c) The maximum height occurs when the y-component of velocity is 0:

$$-9.8t + 24.766 \sin 25° = 0 \quad \to \quad t = \frac{24.766 \sin 25°}{9.8} \approx 1.068 \text{ seconds.}$$

This is substituted into the y-component of displacement:

$$-4.9(1.068)^2 + (24.766 \sin 25°)(1.068) + 1 = 5.59 \text{ meters.}$$

♦ • ♦ • • ♦ • • ♦

Example 22.4: An object is projected from the top of a 50-meter-tall cliff. It remains in the air for 8 seconds, landing 250 meters downrange. Find the following:

a) The maximum height above ground that the object achieved;
b) The initial speed and angle above the horizontal that the object was released.

Solution: Starting with $\mathbf{a}(t) = \langle 0, -9.8 \rangle$, we develop vector-valued functions for $\mathbf{v}(t)$ and $\mathbf{r}(t)$:

$$\mathbf{v}(t) = \langle v_x, -9.8t + v_y \rangle \ \& \ \mathbf{r}(t) = \langle v_x t + r_x, -4.9t^2 + v_y t + r_y \rangle, \qquad t \geq 0.$$

where $\langle v_x, v_y \rangle$ are the initial components of the velocity when the object is released, and $\langle r_x, r_y \rangle$ is its initial position. By setting the origin at the base of the cliff directly below the point where the object was released, we have $r_x = 0$ and $r_y = 50$. Thus, we have

$$\mathbf{r}(t) = \langle v_x t, -4.9t^2 + v_y t + 50 \rangle.$$

At $t = 8$ seconds, the object lands 250 meters downrange, at which time its vertical component of position will be 0. We examine the components of $\mathbf{r}(t)$:

$$v_x(8) = 250$$
$$-4.9(8)^2 + v_y(8) + 50 = 0.$$

The first equation gives $v_x = \frac{250}{8} = 31.25$ meters per second, and the second equation gives $v_y = \frac{(4.9)(64)-50}{8} = 32.95$ meters per second. Thus,

$$\mathbf{v}(t) = \langle 31.25, -9.8t + 32.95 \rangle; \quad \text{and} \quad \mathbf{r}(t) = \langle 31.25t, -4.9t^2 + 32.95t + 50 \rangle.$$

With the velocity and position vectors established, we address the questions:

a) The object reaches its maximum height when the y-component of velocity is 0, or when $-9.8t + 32.95 = 0$, giving $t = \frac{32.95}{9.8} \approx 3.362$ seconds. This is then evaluated into the y-component of position. The object's maximum height is

$$-4.9(3.362)^2 + 32.95(3.362) + 50 \approx 105.393 \text{ meters.}$$

b) The initial velocity is $\mathbf{v}(0) = \langle 31.25, 32.95 \rangle$. Thus, the initial speed of the object is

$$\sqrt{31.25^2 + 32.95^2} \approx 45.4 \text{ meters per second,}$$

and its initial angle above the horizontal is

$$\theta = \arctan\left(\frac{v_y}{v_x}\right) = \arctan\left(\frac{32.95}{31.25}\right) \approx 46.52°.$$

♦ ♦ ♦ ♦ ♦ ♦ ♦ ♦ ♦

Example 22.5: An object is propelled off a building that is 75 m high, the object propelled at an initial angle of 40° to the horizontal. It reaches a maximum height after 1.75 seconds. How far downrange does the object land, assuming the ground below to be flat?

Solution: We start with velocity $\mathbf{v}(t) = \langle v_x, -9.8t + v_y \rangle$ and position $\mathbf{r}(t) = \langle v_x t + r_x, -4.9t^2 + v_y t + r_y \rangle$. Assuming a starting position of $\langle v_x, v_y \rangle = \langle 0, 75 \rangle$, we can further refine the position vector function as $\mathbf{r}(t) = \langle v_x t, -4.9t^2 + v_y t + 75 \rangle$.

From the velocity vector, the object reaches its maximum height when its y-component is 0. Thus, we have $-9.8(1.75) + v_y = 0$, which gives $v_y = 17.15$ meters per second.

From this, we can reconstruct the initial velocity components. We have a right triangle with an angle of 40° and an opposite leg of 17.15 meters per second. Thus, the adjacent leg is given by $\tan 40° = \frac{\text{opposite}}{\text{adjacent}} = \frac{17.15}{v_x}$, so that $v_x = \frac{17.15}{\tan 40°} = 20.439$ meters per second.

This allows us to completely define the position vector, $\mathbf{r}(t) = \langle 20.439t, -4.9t^2 + 17.15t + 75 \rangle, t \geq 0$. The object impacts the ground when the y-component of $\mathbf{r}(t)$ is 0:

$$-4.9t^2 + 17.15t + 75 = 0.$$

Using the quadratic formula, we get two results:

$$t = \frac{-(17.15) \pm \sqrt{(17.15)^2 - 4(-4.9)(75)}}{2(-4.9)} \quad \rightarrow \quad t \approx -2.536, t \approx 6.036.$$

The negative result is ignored. The positive result, 6.036 seconds, is the time after release that the object impacts the ground. Its distance from the base of the building is given by the x-component of position:

$$20.439(6.036) = 123.37.$$

Thus, the object travelled a total of about 123.37 meters in the horizontal direction before impacting the ground.

◆ ◆ ◆ ◆ ◆ ◆ ◆ ◆ ◆

Example 22.6: A golf ball is hit from the ground by an astronaut on a distant planet. It reaches a maximum height of 50 meters after 4 seconds of flight. What is the gravitational constant on this planet?

Solution: We start with $\mathbf{a}(t) = \langle 0, -a \rangle$, then develop $\mathbf{v}(t)$ and $\mathbf{r}(t)$:

$$\mathbf{v}(t) = \langle v_x, -at + v_y \rangle \quad \text{and} \quad \mathbf{r}(t) = \langle v_x t + r_x, -\frac{a}{2}t^2 + v_y t + r_y \rangle.$$

Assume that the initial position is $\langle r_x, r_y \rangle = \langle 0, 0 \rangle$. Thus, position is given by $\mathbf{r}(t) = \langle v_x t, -\frac{a}{2}t^2 + v_y t \rangle, \ t \geq 0$.

When the ball reaches the maximum height, its vertical component of velocity is 0, while the vertical component of position is 50. This creates a pair of equations:

$$-at + v_y = 0$$
$$-\frac{a}{2}t^2 + v_y t = 50.$$

When $t = 4$, we obtain $-4a + v_y = 0$ and $-\frac{a}{2}(4)^2 + 4v_y = 50$. From the first equation, we have

$$v_y = 4a,$$

and from the second equation, we have

$$-8a + 4(4a) = 50.$$

Solving for a, we obtain

$$-8a + 16a = 50$$
$$8a = 50$$
$$a = \frac{50}{8} = 6.25.$$

Evidently, objects on this planet fall at a rate of 6.25 meters per second².

♦ ♦ ♦ ♦ ♦ ♦ ♦ ♦ ♦ ♦

23. Level Curves and Contour Maps

Let $z = f(x, y)$ be a function whose graph is a surface embedded in R^3. Suppose this graph is intersected by a plane $z = k$, parallel to the xy-plane. This is equivalent to holding z constant and reducing the equation into an implicit function of x and y only, written $f(x, y) = k$.

The intersection of $z = f(x, y)$ with $z = k$ forms a **level curve** on the surface of f. It is a curve where x and y can vary, but z does not change. Imagine standing on a hill and taking a step such that you neither go uphill nor downhill. If you do this repeatedly, you will (theoretically) walk along a path that is level, and end at the same point from which you started.

A level curve projected onto the xy-plane is called a **contour**.

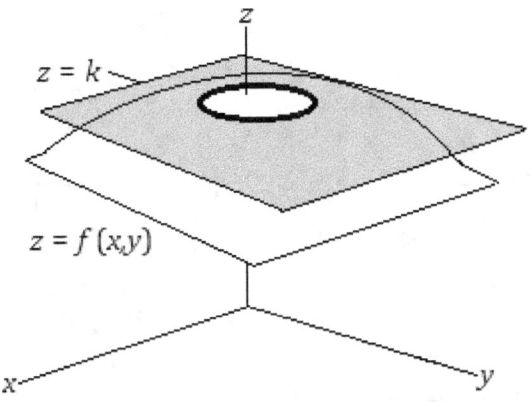

A surface representing $z = f(x, y)$ is intersected by a plane $z = k$.
The bold path is "level" in that the z-values on it do not change.

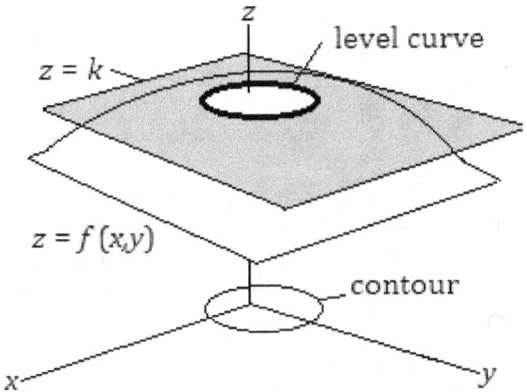

A contour is the projection of a level curve onto the domain plane.

For any surface, sketch a number of level curves (for different values of z), whose contours then form a contour map of the surface. With practice, one can locate minimum points, maximum points, saddle points, ridges, valleys, and other "terrain" forms. These maps are usually rendered using software, and they are an effective way to represent a surface in two dimensions.

The usual rules for drawing and reading a contour map include:

- The contours are marked by the z-values for reference. For complicated surfaces, the z-values may differ by some set value, *e.g.* by multiples of 10. Too many contours, and the map is cluttered, while too few contours, important details may be lost.

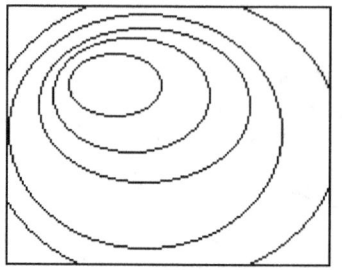 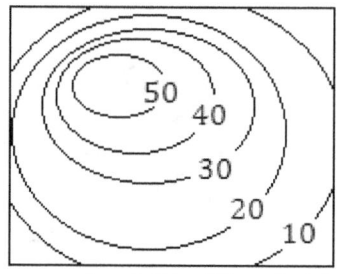

Left: This is a vague contour map. There are no z-values. Is this a hill or a basin?
Right: With z-values now in place, we see that this represents a hill.

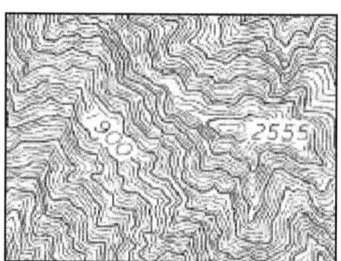 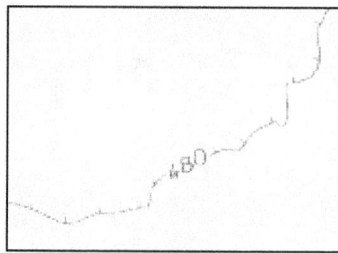

Left: The map is too cluttered. **Right:** Now it's too vague.

- Contour lines for different z-values never touch. This would violate the vertical-line rule for functions in R^3. Contours that are close together indicate a large change in z, or "steepness", while contours spaced apart indicate flatter terrain. Some topographical maps may show what appears to be contours that "touch". These represent cliffs.

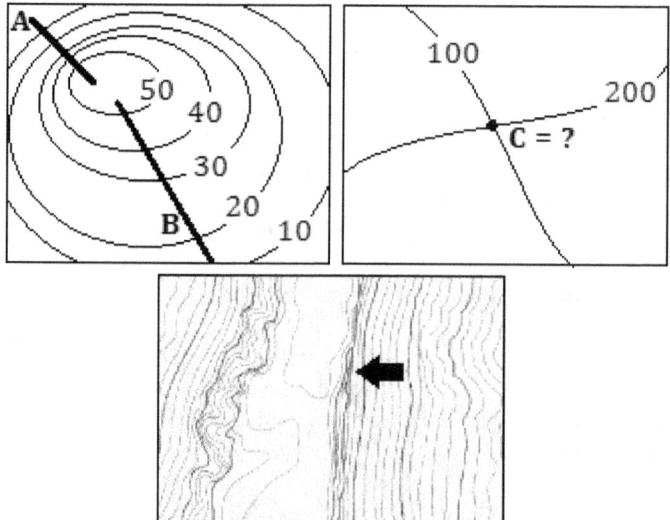

Top Left: Paths A and B go to the top of the hill, but A is steeper than B. **Top Right:** Contours cannot intersect. What would C's z-value be? **Above:** These contours are so close they seem to touch. This is a cliff beside a bluff in the Grand Canyon of Arizona.

- Contours can form closed loops and nest (and or may be nested within) other contours. Hills are found where the nested contours increase in values of z. Basins are found where the nested contours decrease in values of z.

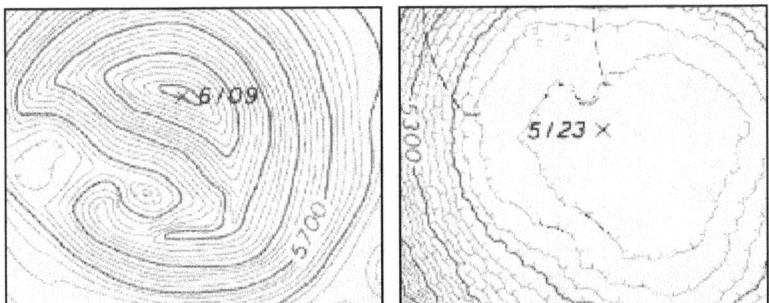

Left: A hill. Note the darker contours increase by 100, the lighter contours by 20.
Right: A basin. The contour's z-values decrease toward the center.
(This is the contour map of Meteor Crater near Winslow, Arizona)

- Saddle points are locations where the point may be a minimum along some paths crossing through the point, and a maximum for other paths passing through the point. On most contour maps, a saddle is usually inferred where a contour "pinches" in, then widens again.

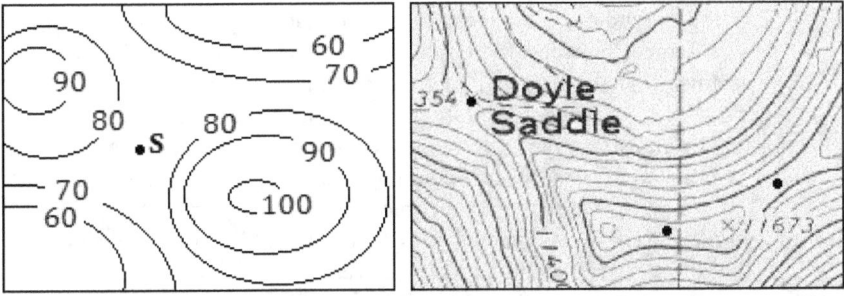

Left: A simplified contour map showing two hills (upper left and lower right). A person walking from one hill to the other would pass through S, and this would be the person's lowest point on his walk. Another person walking from the lower left to the upper right would pass through S, and this would be the highest point for this other person's walk. Thus, S is a saddle point. **Right:** Saddles are noted by dots. Note that the contours pinch in near these points.

All Maps: © USGS

Example 23.1: Given the contour map below for $z = f(x,y)$ where $0 \leq x \leq 5$ and $0 \leq y \leq 5$. Find the following: (a) $f(2,4)$; (b) $f(3,2)$; (c) $f(3,4)$; (d) $f(2,y) = 70$.

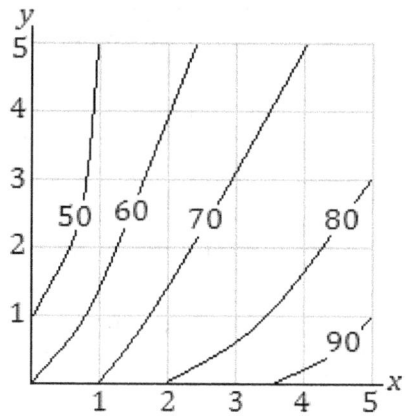

Solution:

a) Locate $x = 2$ and $y = 4$ and follow the grid lines to where they meet. At this location, it appears that $z = 60$, so we conclude that $f(2,4) = 60$.
b) Using a technique similar to (a), we see that the gridlines for $x = 3$ and $y = 2$ meet between the contours for $z = 70$ and $z = 80$. Here, we surmise that the value of $f(3,2)$ is probably closer to 70 than to 80 since it is situated closer to the contour for $z = 70$. However, to determine z exactly is impossible without a finer contour map. We can say that $f(3,2) \approx 73$, for example. However, any z value such that $70 < z < 80$ would be a plausible (acceptable) answer.
c) Similar to (b), we estimate that $f(3,4) \approx 67$, although any z value such that $60 < z < 70$ is a plausible answer.
d) Following the $z = 70$ contour over the grid for $x = 2$, we see that the corresponding y value is about 1.5. Thus, $f(2, 1.5) = 70$.

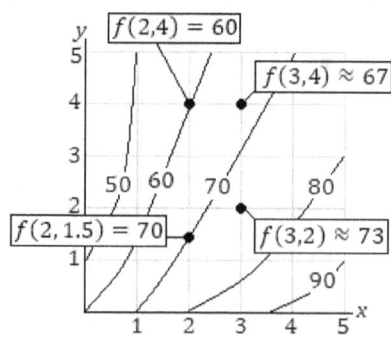

Typical Contour Maps for Planes and Paraboloids

Planes: A plane of the form $z = ax + by + c$ has a contour map of equally-spaced lines.

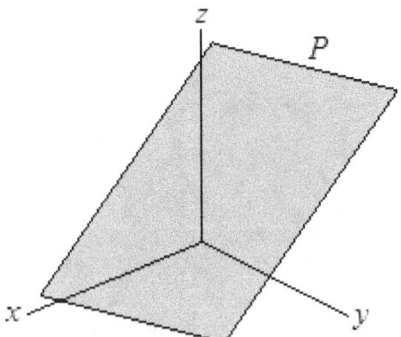

Plane P exists (is embedded) in R^3.

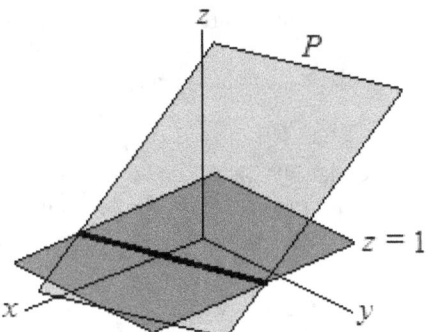

Plane $z = 1$ is parallel to the xy-plane.
The bold line is a level curve on P cut by $z = 1$.
All points on this bold line have a z value of 1.

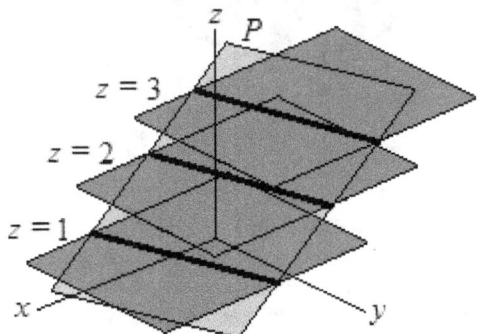

More planes of the form $z = k$ are drawn,
Intersecting plane P, and more level curves result.

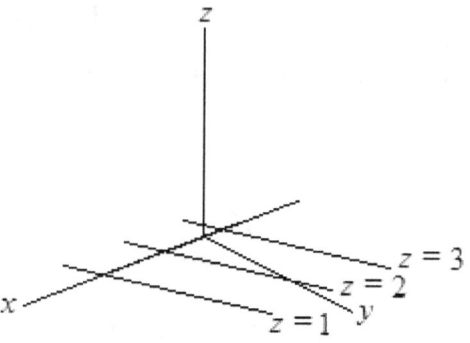

Plane P and the planes $z = k$ are removed.
The level curves are projected onto the xy-plane.
They appear to be a sequence of parallel lines.

Paraboloids: A paraboloid of the form $z = ax^2 + by^2 + cx + dy + e$, where a and b are the same sign, has concentric contours emanating outward from the vertex. If $a = b$, then the contours are circles. Otherwise, the contours are ellipses. If the function contains an xy term, then the contours may be angled so that their axes are not parallel to the x and y axes.

In the simplest case, $z = x^2 + y^2$, an intersection of this surface with a plane $z = k$ forms a level curve that is a circle, and a contour on the xy-plane that is also a circle. This process is repeated. Thus, the contour map of this paraboloid will be concentric circles centered at the origin.

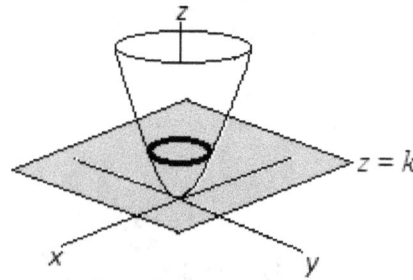

The paraboloid $z = x^2 + y^2$ intersected by a plane $z = k$ forms a circle.

The following contour map represents the paraboloid $z = x^2 + y^2 + 2x - 4y$, whose vertex is $(-1, 2)$. Since the coefficients of the quadratic terms are the same, the contours are circles. Note that the farther away from the origin, the contours are spaced closer. This is a trick of perspective. In reality, the contours are spaced evenly when looking along the z-axis. However, as x and y are chosen farther awar from the vertex, the change in z is greater. The close-in contours indicate a steeper rate of change.

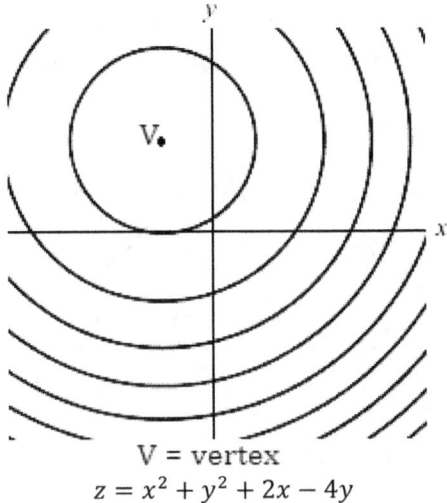

V = vertex
$$z = x^2 + y^2 + 2x - 4y$$

The contour map below is for the paraboloid $z = x^2 + 2y^2 + xy + 2x - 4y$, whose vertex is approximately $(-1.71, 1.43)$. Note that the coefficients of the quadratic terms are not equal, so that the contours are ellipses. Also, the xy term causes the orientation of the major and minor elliptic axes to rotate slightly.

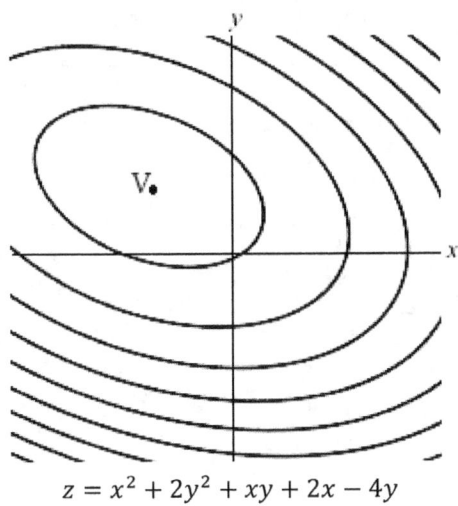

$$z = x^2 + 2y^2 + xy + 2x - 4y$$

For the surface $z = x^2 + y^2 + 4xy + x$, the vertex is $\left(\frac{1}{6}, -\frac{1}{3}\right)$, but the *xy* term has "deformed" the paraboloid into a **saddle** surface. Note the "pinching" effect of the contour lines.

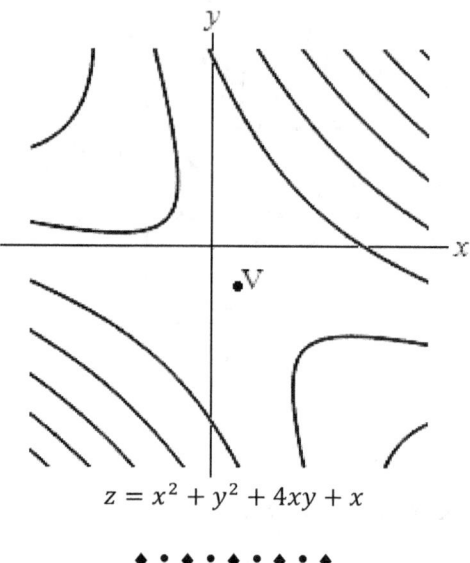

$z = x^2 + y^2 + 4xy + x$

♦ · ♦ · ♦ · ♦ · ♦ · ♦

Example 23.2: The image below is a contour map indicating the temperature for certain locations within a rectangular county.

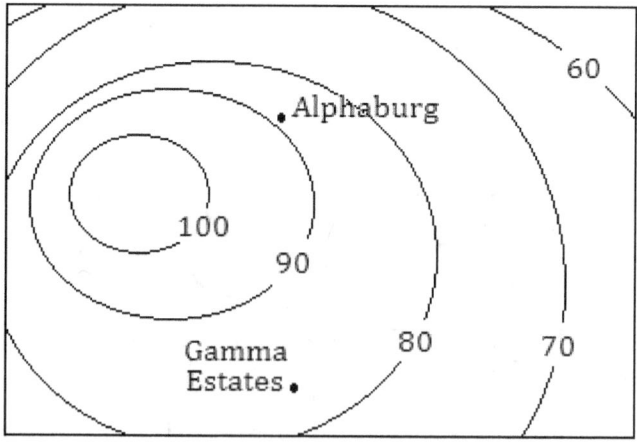

Based on this image, answer the following questions.

    a) What is the approximate temperature in Alphaburg?
    b) What is the approximate temperature in Gamma Estates?
    c) What is the approximate hottest temperature within the county?

Solution:

a) Since the location of Alphaburg lies in the area between the contours representing z = 80 degrees and z = 90 degrees, its temperature is 80 < z < 90. However, since the location is close to the z = 90 contour, it's reasonable to assume that Alphaburg's temperature is in the high 80s, perhaps 88 or 89 degrees. It is *not* reasonable to assume that Alphaburg's temperature is 90 degrees since the location is not on the 90-degree contour.

b) Gamma Estates' temperature would be within the 80 < z < 90 range. Since it is closer to the z = 80 contour, it is reasonable to assume the temperature is approximately 82 or 83 degrees.

c) The hottest temperature is within the z = 100 contour. We can assume that it is in the range 100 < z < 110. It is not reasonable to assume that the temperature is 110 degrees or higher. An approximate figure of z = 105 degrees would be reasonable.

◆ ◆ ◆ ◆ ◆ ◆ ◆ ◆ ◆

24. Partial Differentiation

The derivative of a single variable function, $\frac{d}{dx}f(x)$, always assumes that the independent variable is increasing in the usual manner. Visually, the derivative's value at a point $x = a$ is the slope of the tangent line of $y = f(x)$ at $x = a$, and the slope's value only "makes sense" if x increases to the right, as viewed on a standard *xy*-axis system. This is the *direction* of $\frac{d}{dx}f(x)$. On the real line representing the independent variable x, there are just two directions in which x can vary: to the right or to the left. However, with a multivariable function $z = f(x, y)$, the number of possible directions in which the independent variables can vary (together) is infinite.

Thus, when finding the instantaneous rate of change between the dependent variable and one of the independent variables of a multivariable function $z = f(x, y)$, we must clearly specify a direction in which we are comparing this rate of change.

For example, if given a function $z = f(x, y)$, and assuming for now that its graph is continuous everywhere and smooth, in that it lacks corners and folds, then there are two possible "convenient" directions in which to calculate an instantaneous rate of change: the positive *x* direction, or the positive *y* direction. (There are infinitely-many directions. This is discussed in Section 26.)

The instantaneous rate of change of z with respect to x is called **the partial derivative of z with respect to x**, and is written

$$\frac{\partial z}{\partial x} \quad \text{or} \quad \frac{\partial f}{\partial x}, \quad \text{or informally as } z_x \text{ or } f_x.$$

In this case, the variable y is held constant. It does not vary.

Similarly, the instantaneous rate of change of z with respect to y is called **the partial derivative of z with respect to y**, and is written

$$\frac{\partial z}{\partial y} \quad \text{or} \quad \frac{\partial f}{\partial y}, \quad \text{or informally as } z_y \text{ or } f_y.$$

Here, x is now held constant.

• • ◆ • • ◆ • • ◆

Example 24.1: Use the contour map below, representing $z = f(x, y)$, to answer the questions that follow. Assume that f is smooth and continuous.

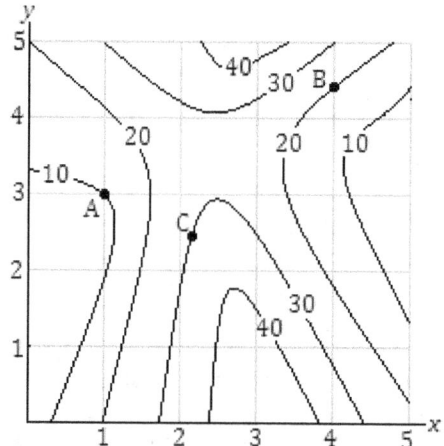

a) Use a convenient nearby point to estimate the slope of a tangent line drawn at A, in the positive x direction, then estimate the slope of a tangent line at this same point, now in the positive y direction.
b) Use a convenient nearby point to estimate the slope of a tangent line drawn at B, in the positive x direction, then estimate the slope of a tangent line at this same point, now in the positive y direction.
c) Estimate the sign of the slope of a tangent line drawn at C, in the positive x direction, then estimate the sign of the slope of a tangent line at this same point, now in the positive y direction.

Solution:

a) Observe that A is given by the ordered triple $(1, 3, 10)$. A convenient nearby point in the positive x direction is $(1.6, 3, 20)$. When moving in the positive x direction, variable y remains constant. Thus, a reasonable estimation of the slope of a tangent line drawn at A in the positive x direction is $\frac{\partial}{\partial x} f(A) \approx \frac{20-10}{1.6-1} = \frac{10}{0.6} = 16.7$. Note that $\frac{\partial}{\partial x} f(A)$ is positive. When at A, and moving in the positive x direction, we see that we would be walking toward higher ground, as shown by the $z = 20$ contour.

In a similar way, we use the ordered triple $(1, 4.2, 20)$ as a convenient point in the positive y direction. Thus, a reasonable estimation of the slope of a tangent line drawn at A in the positive y direction is $\frac{\partial}{\partial y} f(A) \approx \frac{20-10}{4.2-3} = \frac{10}{1.2} = 8.3$. Note that when moving in the positive y direction, variable x remains constant.

b) Point B is $(4, 4.5, 20)$. A convenient point in the positive x direction is $(5, 4.5, 10)$. Thus, a reasonable estimation of the slope of a tangent line drawn at B in the positive x direction is $\frac{\partial}{\partial x} f(B) \approx \frac{10-20}{5-4} = -10$. Here, moving in the positive x direction means a negative (downward) change in z.

In the y direction, we use $(4, 5, 30)$, and a reasonable estimation of the slope of a tangent line drawn at B in the positive y direction is $\frac{\partial}{\partial y} f(B) \approx \frac{30-20}{5-4.5} = \frac{10}{0.5} = 20$. Moving in the positive y directions means a positive (upward) change in z.

c) Rather than choosing points nearby to C, we will study the contour map and make a judgement of the signs of $\frac{\partial}{\partial x} f(C)$ and $\frac{\partial}{\partial y} f(C)$.

The z-value at C is 30. Moving a small distance in the positive x direction would place us in a region where $30 < z < 40$. Thus, this means that any immediate movement off of C in the positive x direction would result in an increase in the z value. Thus, we can conclude that $\frac{\partial}{\partial x} f(C)$ is positive.

Moving a small distance in the positive y direction puts us in a region where $20 < z < 30$. This means that any immediate movement off of C in the positive y direction would result in a decrease in the z value. Thus, we can conclude that $\frac{\partial}{\partial y} f(C)$ is negative.

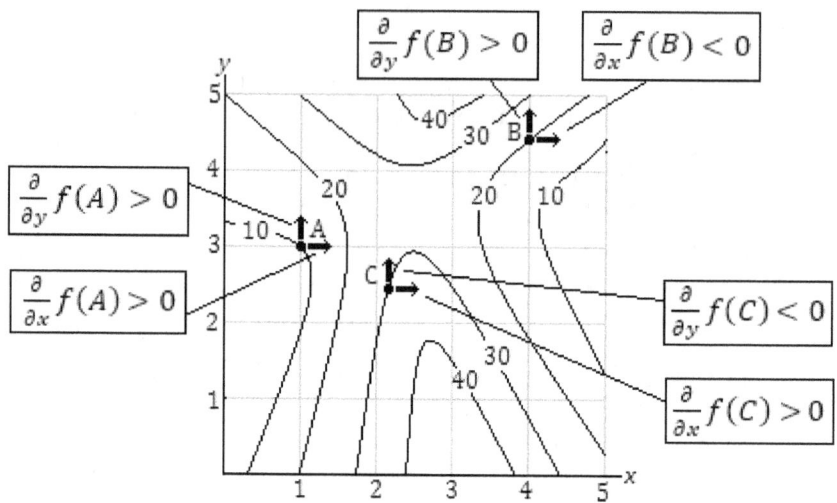

Small arrows are drawn in the positive x and y directions to help suggest the signs of the partial derivatives in these directions, at points A, B and C.

◆ • ◆ • ◆ • ◆ • ◆

Example 24.2: Use the contour map below, representing a paraboloid $z = f(x, y)$ that opens in the positive z direction, to answer the questions that follow. Assume that f is smooth and continuous, and that the vertex V is at the origin and is the minimum point. Determine signs of the partial derivatives of f with respect to x and with respect to y at points A, B, C and V.

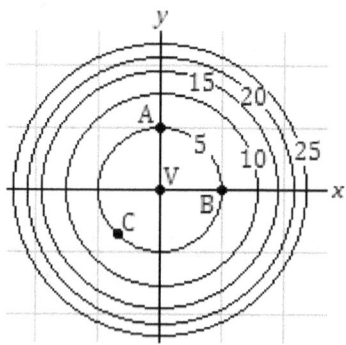

Solution: As in the previous example, we can sketch in small arrows at each point to help suggest the sign of the partial derivative in a particular direction. *But this method must be used carefully!*

We assume that if a surface representing f is smooth and continuous, then the partial derivatives of f with respect to x and with respect to y are 0 at all minimum, maximum and saddle points.

Thus, we can immediately identify the signs of the partial derivatives of f with respect to x and with respect to y at the vertex V. Since V is a minimum, then $\frac{\partial}{\partial x}f(V) = 0$ and $\frac{\partial}{\partial y}f(V) = 0$.

For A, we note that any movement in the positive y direction will mean a positive change in z. Thus, $\frac{\partial}{\partial y}f(A) > 0$. However, note that movement in the positive x direction is tangential to the level curve. In such cases, the change in z with respect to x is 0. Thus, $\frac{\partial}{\partial x}f(A) = 0$.

For B, we have $\frac{\partial}{\partial x}f(B) > 0$ and $\frac{\partial}{\partial y}f(B) = 0$, since movement in the y direction is tangential to the level curve.

For C, movements in the x or the y direction are not tangential to the level curve. In both cases, z will decrease in value, so that $\frac{\partial}{\partial x}f(C) < 0$ and $\frac{\partial}{\partial y}f(C) < 0$.

♦ • ♦ • ♦ • ♦ • ♦

> On a contour map representing the surface of a smooth and continuous function f, the values of the partial derivatives of f with respect to x and with respect to y are 0 at all minimum, maximum and saddle points. If movement in the x or y direction happens to be tangential to the contour at a point, then the value of the partial derivative of f with respect to the x or y direction is 0. That is, tangential movement along a level curve always means no change in z.

♦ • ♦ • ♦ • ♦ • ♦

Example 24.3: The contour map at right represents the surface of a smooth and continuous function $z = g(x, y)$. Assume that points B, C, D and G are minimum, maximum or saddle points. State the sign (positive, negative or zero) of the partial derivative of g with respect to x and with respect to y, at each of the points A through G.

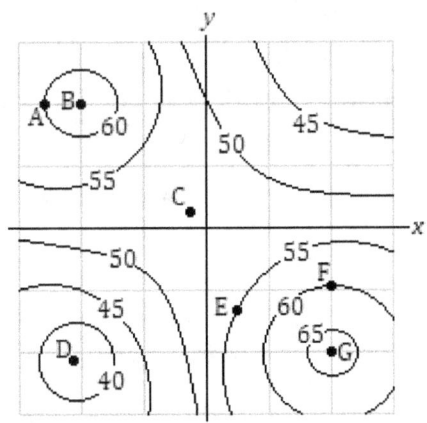

Solution:

For point A, we have $\frac{\partial}{\partial x}g(A) > 0$ and $\frac{\partial}{\partial y}g(A) = 0$.

For point B, we have $\frac{\partial}{\partial x}g(B) = 0$ and $\frac{\partial}{\partial y}g(B) = 0$. B is a local maximum.

For point C, we have $\frac{\partial}{\partial x}g(C) = 0$ and $\frac{\partial}{\partial y}g(C) = 0$. C is a saddle point.

For point D, we have $\frac{\partial}{\partial x}g(D) = 0$ and $\frac{\partial}{\partial y}g(D) = 0$. D is a local minimum.

For point E, we have $\frac{\partial}{\partial x}g(E) > 0$ and $\frac{\partial}{\partial y}g(E) < 0$.

For point F, we have $\frac{\partial}{\partial x}g(F) = 0$ and $\frac{\partial}{\partial y}g(F) < 0$.

For point G, we have $\frac{\partial}{\partial x}g(G) = 0$ and $\frac{\partial}{\partial y}g(G) = 0$. G is a local maximum.

Note that A is tangent to the contour in the y direction, and that F is tangent to the contour in the x direction.

The rules for partial differentiation are identical to single variable integration. The Product Rule, Quotient Rule and Chain Rule are all used as necessary.

Example 24.4: Let $z = f(x, y) = x^2y + 3x^3y^4 + 2x - 4y$. Find $\frac{\partial z}{\partial x}$ and $\frac{\partial z}{\partial y}$.

Solution: When finding $\frac{\partial z}{\partial x}$, treat the y as a constant. If it is in a term by itself, then the whole term is treated as a constant. If it is connected to x through multiplication, then it is treated as a coefficient. Thus, we have

$$\frac{\partial z}{\partial x} = \frac{\partial}{\partial x}(x^2y + 3x^3y^4 + 2x - 4y)$$
$$= \frac{\partial}{\partial x}(x^2y) + \frac{\partial}{\partial x}(3x^3y^4) + \frac{\partial}{\partial x}(2x) - \frac{\partial}{\partial x}(4y)$$
$$= (2x)y + 3(3x^2)y^4 + 2(1) - 0$$
$$= 2xy + 9x^2y^4 + 2.$$

Similarly, to find $\frac{\partial z}{\partial y}$, we treat x as a constant or a coefficient:

$$\frac{\partial z}{\partial y} = \frac{\partial}{\partial y}(x^2y + 3x^3y^4 + 2x - 4y)$$
$$= \frac{\partial}{\partial y}(x^2y) + \frac{\partial}{\partial y}(3x^3y^4) + \frac{\partial}{\partial y}(2x) - \frac{\partial}{\partial y}(4y)$$
$$= x^2(1) + 3x^3(4y^3) + 0 - 4(1)$$
$$= x^2 + 12x^3y^3 - 4.$$

Example 24.5: Let $z = g(x, y) = xye^y$. Find $\frac{\partial g}{\partial x}$ and $\frac{\partial g}{\partial y}$.

Solution: For $\frac{\partial g}{\partial x}$, the factors ye^y are attached to x by multiplication and are treated as a coefficient of x. Thus,

$$\frac{\partial g}{\partial x} = \frac{\partial}{\partial x}(xye^y) = (1)ye^y = ye^y, \qquad \text{where } \frac{\partial}{\partial x}x = 1.$$

For $\frac{\partial g}{\partial y}$, the x is now treated as a coefficient, and the Product Rule of differentiation is used:

$$\frac{\partial g}{\partial y} = \frac{\partial}{\partial y}(xye^y) = xy\frac{\partial}{\partial y}(e^y) + e^y\frac{\partial}{\partial y}(xy) = xye^y + xe^y.$$

Example 24.6: Let $z = x^3 \sin(x^2y^3)$. Find z_x and z_y.

Solution: For z_x, note that x is present in two factors attached by multiplication. Thus, we use the Product Rule of differentiation and the Chain Rule:

$$z_x = x^3(\cos(x^2y^3)\, 2xy^3) + 3x^2 \sin(x^2y^3)$$
$$= 2x^4y^3 \cos(x^2y^3) + 3x^2 \sin(x^2y^3).$$

For z_y, we do not need the Product Rule, treating the x^3 as a coefficient of the sine operator. However, we do need the Chain Rule:

$$z_y = x^3 \cos(x^2y^3)\, x^2(3y^2)$$
$$= 3x^5y^2 \cos(x^2y^3).$$

◆ • ◆ • ◆ • • ◆ • ◆

Partial differentiation can be used for functions with more than two variables.

Example 24.7: The function $A(p, r, t) = p(1 + r)^t$ gives the future value A of p dollars invested at an annual percentage rate r, compounded annually, after t years. Find A_p, A_t and A_r.

Solution: To find A_p, note that $(1 + r)^t$ is treated as a constant multiplier to p. Since $\frac{\partial}{\partial p}(p) = 1$, we have

$$A_p = \frac{\partial}{\partial p}(p(1+r)^t) = (1)(1+r)^t = (1+r)^t.$$

To find A_t, we use the differentiation rule for exponentials, $\frac{d}{dx}(a^x) = a^x \ln a$. Thus, we have

$$A_t = \frac{\partial}{\partial t}(p(1+r)^t) = p(1+r)^t \ln(1+r).$$

To find A_r, note that p and t are constants. Thus, we can use the Power Rule of differentiation:

$$A_r = \frac{\partial}{\partial r}(p(1+r)^t) = pt(1+r)^{t-1}.$$

◆ • ◆ • ◆ • ◆ • ◆

Higher-Order Partial Derivatives & Clairaut's Theorem

Partial differentiation can also be used to find second-order derivatives, and so on. Suppose $z = f(x,y)$ is given. There are two first partial derivatives,

$$f_x = \frac{\partial f}{\partial x} \quad \text{and} \quad f_y = \frac{\partial f}{\partial y}.$$

Each partial derivative is itself a function of two variables. Thus, each has two partial derivatives of its own. For example, $f_x(x,y)$ has two partial derivatives:

$$(f_x)_x = \frac{\partial}{\partial x}\left(\frac{\partial f}{\partial x}\right) = \frac{\partial^2 f}{\partial x^2} \quad \text{and} \quad (f_x)_y = \frac{\partial}{\partial y}\left(\frac{\partial f}{\partial x}\right) = \frac{\partial^2 f}{\partial y\, \partial x}.$$

Similarly, $f_y(x,y)$ has two partial derivatives:

$$(f_y)_y = \frac{\partial}{\partial y}\left(\frac{\partial f}{\partial y}\right) = \frac{\partial^2 f}{\partial y^2} \quad \text{and} \quad (f_y)_x = \frac{\partial}{\partial x}\left(\frac{\partial f}{\partial y}\right) = \frac{\partial^2 f}{\partial x\, \partial y}.$$

Usually, second derivatives are noted by using subscripts without parentheses. Thus,

$$f_{xx} = (f_x)_x, \quad f_{yy} = (f_y)_y, \quad f_{xy} = (f_x)_y \quad \text{and} \quad f_{yx} = (f_y)_x.$$

Second derivatives such as f_{xx} and f_{yy} are informally called *homogeneous* second derivatives, while f_{xy} and f_{yx} are called *mixed* second derivatives.

Under "typical" circumstances, e.g. the function f being smooth and continuous, and twice-differentiable over its relevant domain, the mixed second derivatives will be equal:

$$f_{xy} = f_{yx} \quad \text{(Clairaut's Theorem)}.$$

As one might expect, second derivatives of a smooth and continuous function offer insight to the concavity of the function.

Higher-order derivatives are found in a similar manner. For example,

$$f_{xxy} = ((f_x)_x)_y = \frac{\partial}{\partial y}\left(\frac{\partial}{\partial x}\left(\frac{\partial}{\partial x}f(x,y)\right)\right) = \frac{\partial^3 f}{\partial y\, \partial x^2}.$$

♦ • • ♦ • • ♦ • • ♦

Example 24.8: Given $z = f(x,y) = x^2 y + 3x^3 y^4 + 2x - 4y$. Find f_{xx}, f_{yy}, f_{xy} and f_{yx}.

Solution: From a previous example, we found the two first partial derivatives:

$$f_x(x,y) = 2xy + 9x^2 y^4 + 2 \quad \text{and} \quad f_y(x,y) = x^2 + 12x^3 y^3 - 4.$$

Thus, we have

$$f_{xx} = \frac{\partial}{\partial x} f_x(x,y) = \frac{\partial}{\partial x}(2xy + 9x^2 y^4 + 2) = 2y + 18xy^4$$

and

$$f_{yy} = \frac{\partial}{\partial y} f_y(x,y) = \frac{\partial}{\partial y}(x^2 + 12x^3 y^3 - 4) = 36x^3 y^2.$$

Furthermore, we have

$$f_{xy} = \frac{\partial}{\partial y} f_x(x,y) = \frac{\partial}{\partial y}(2xy + 9x^2 y^4 + 2) = 2x + 36x^2 y^3$$

and

$$f_{yx} = \frac{\partial}{\partial x} f_y(x,y) = \frac{\partial}{\partial x}(x^2 + 12x^3 y^3 - 4) = 2x + 36x^2 y^3.$$

Note that $f_{xy} = f_{yx}$.

Example 24.9: Given $a = b^3 c^4 d^5$. Show that $a_{bcd} = a_{dbc}$.

Solution: We find successive partial derivatives by reading the subscripts left to right. For example, $a_{bcd} = ((a_b)_c)_d$. We have

$$a_b = 3b^2 c^4 d^5,$$
$$a_{bc} = 3b^2(4c^3)d^5 = 12b^2 c^3 d^5,$$
$$a_{bcd} = 12b^2 c^3 (5d^4) = 60b^2 c^3 d^4.$$

Similarly,

$$a_d = 5b^3 c^4 d^4,$$
$$a_{db} = 5(3b^2)c^4 d^4 = 15b^2 c^4 d^4,$$
$$a_{dbc} = 15b^2(4c^3)d^4 = 60b^2 c^3 d^4.$$

There are six ways in which to take the derivative of a with respect to b, c and d in any order. We have found two. You should find the other four and verify all are equal.

◆ • ◆ • ◆ • ◆ • ◆

Example 24.10: Find $\frac{\partial f}{\partial x}$ and $\frac{\partial f}{\partial y}$, where $f(x, y) = \int_x^y 3t^2 \, dt$.

Solution: Defining functions as integrals is not uncommon. In this case, we can antidifferentiate the integrand, and evaluate at the limits of integration:

$$f(x, y) = \int_x^y 3t^2 \, dt = [t^3]_x^y = y^3 - x^3.$$

Taking partial derivatives, we have,

$$\frac{\partial f}{\partial x} = \frac{\partial}{\partial x}(y^3 - x^3) = -3x^2 \quad \text{and} \quad \frac{\partial f}{\partial y} = \frac{\partial}{\partial y}(y^3 - x^3) = 3y^2.$$

Note that the results look similar to the original integrand. Was it necessary to do the antidifferentiation step? See the next example.

Example 24.11: Find $\frac{\partial f}{\partial x}$ and $\frac{\partial f}{\partial y}$, where $f(x, y) = \int_x^y \sqrt{t^4 - 2t + 7}\, dt$.

Solution: Repeating the steps of the previous example leads to what appears to be an impossible step: the integrand does not antidifferentiate "conveniently" into common elementary functions. For now, define $H(t)$ to be the antiderivative of $\sqrt{t^4 - 2t + 7}$. We cannot determine $H(t)$, but we know its derivative is $\sqrt{t^4 - 2t + 7}$. Thus, we have

$$f(x, y) = \int_x^y \sqrt{t^4 - 2t + 7}\, dt = [H(t)]_x^y = H(y) - H(x).$$

Thus,

$$\frac{\partial f}{\partial x} = \frac{\partial}{\partial x}(H(y) - H(x)) = -\sqrt{x^4 - 2x + 7},$$

where $\frac{\partial}{\partial x}(H(y)) = 0$ and where $\frac{\partial}{\partial x}(H(x))$ is the derivative of $H(t)$ with x in place of t.

In a similar manner,

$$\frac{\partial f}{\partial y} = \frac{\partial}{\partial y}(H(y) - H(x)) = \sqrt{y^4 - 2y + 7}.$$

♦ • ♦ • ♦ • ♦ • ♦

See an error? Have a suggestion?
Please see www.surgent.net/vcbook

25. Chain Rule

The Chain Rule is present in all differentiation. If $z = f(x, y)$ represents a two-variable function, then it is plausible to consider the cases when x and y may be functions of other variable(s). For example, consider the function $f(x, y) = x^2 + y^3$, where $x(t) = 2t + 1$ and $y(t) = 3t^2 + 4t$. In such a case, we can find the derivative of f with respect to t by direct substitution, so that f is written as a function of t only, or we may use a form of the Chain Rule for multi-variable functions to find this derivative.

♦ • ♦ • • ♦ • ♦ • ♦

Example 25.1: Given $f(x, y) = x^2 + y^3$, where $x(t) = 2t + 1$ and $y(t) = 3t^2 + 4t$. Find $\frac{df}{dt}$.

Solution: Substitute $x(t) = 2t + 1$ and $y(t) = 3t^2 + 4t$ into the function f:

$$f(x(t), y(t)) = (2t + 1)^2 + (3t^2 + 4t)^3$$

Now, f is a function of t only. Expand by multiplication:

$$f(t) = 4t^2 + 4t + 1 + 27t^6 + 108t^5 + 144t^4 + 64t^3$$

Thus, $f(t) = 27t^6 + 108t^5 + 144t^4 + 64t^3 + 4t^2 + 4t + 1$. Its derivative is found by applying the Power Rule to each term:

$$\frac{df}{dt} = f'(t) = 162t^5 + 540t^4 + 576t^3 + 192t^2 + 8t + 4.$$

Now, let's try a different approach. Keeping the x and y variables present, write the derivative of f using the Chain Rule:

$$\frac{df}{dt} = \left(\frac{\partial f}{\partial x}\right)\left(\frac{dx}{dt}\right) + \left(\frac{\partial f}{\partial y}\right)\left(\frac{dy}{dt}\right) = (2x)(2) + (3y^2)(6t + 4).$$

Now we substitute $x(t) = 2t + 1$ and $y(t) = 3t^2 + 4t$ into the expression, and simplify:

$$\begin{aligned}\frac{df}{dt} &= (2x)(2) + (3y^2)(6t + 4) \\ &= (2(2t + 1))(2) + (3(3t^2 + 4t)^2)(6t + 4) \\ &= 8t + 4 + 3(9t^4 + 24t^3 + 16t^2)(6t + 4) \\ &= 8t + 4 + 3(54t^5 + 180t^4 + 192t^3 + 64t^2) \\ &= 162t^5 + 540t^4 + 576t^3 + 192t^2 + 8t + 4.\end{aligned}$$

Both methods work, but the second method, by writing out all derivatives using all variables present, is more general, and also allows us to see patterns in how these derivatives are written.

A useful ways to visualize the form of the Chain Rule is to sketch a derivative tree. In the previous example, we had f as a function of x and y, and then x and y as functions of t. Thus, we would write the tree as shown below. Then, the derivative form is found by multiplying along paths, and summing the separate paths:

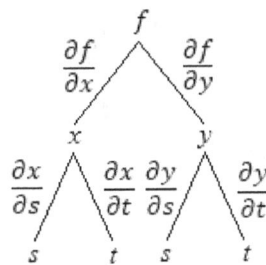

$$\frac{df}{dt} = \left(\frac{\partial f}{\partial x}\right)\left(\frac{dx}{dt}\right) + \left(\frac{\partial f}{\partial y}\right)\left(\frac{dy}{dt}\right)$$

♦ • ♦ • ♦ • ♦ • ♦

Example 25.2: Suppose $f(x,y) = 2xy^2$ and $x(s,t) = 3s - 2t$ and $y(s,t) = s^2 + 4t$. Find $\frac{\partial f}{\partial s}$ and $\frac{\partial f}{\partial t}$.

Solution: Note that f is a function of x and y, and that x and y are both functions of s and t. The derivative tree is shown below, with partial derivative notation attached to the "limbs":

For example, the form of the partial derivative of f with respect to s is

$$\frac{\partial f}{\partial s} = \left(\frac{\partial f}{\partial x}\right)\left(\frac{\partial x}{\partial s}\right) + \left(\frac{\partial f}{\partial y}\right)\left(\frac{\partial y}{\partial s}\right).$$

In a similar way, the form of the partial derivative of f with respect to t is

$$\frac{\partial f}{\partial t} = \left(\frac{\partial f}{\partial x}\right)\left(\frac{\partial x}{\partial t}\right) + \left(\frac{\partial f}{\partial y}\right)\left(\frac{\partial y}{\partial t}\right).$$

The tree helps us visualize the form of the derivatives:

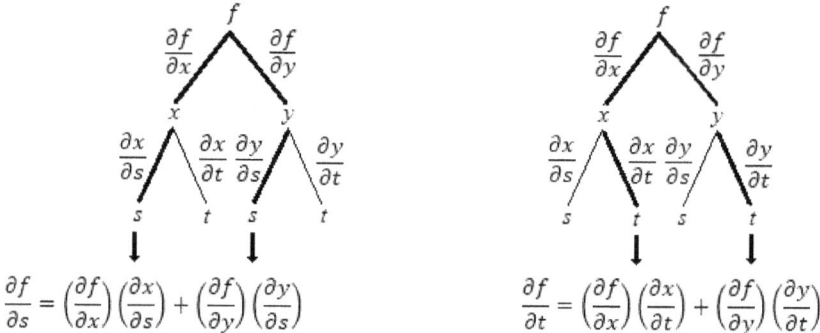

Thus, to find $\frac{\partial f}{\partial s}$, we have

$$\frac{\partial f}{\partial s} = \left(\frac{\partial f}{\partial x}\right)\left(\frac{\partial x}{\partial s}\right) + \left(\frac{\partial f}{\partial y}\right)\left(\frac{\partial y}{\partial s}\right)$$
$$= (2y^2)(3) + (4xy)(2s)$$
$$= 6y^2 + 8xys$$
$$= 6(s^2 + 4t)^2 + 8(3s - 2t)(s^2 + 4t)s \quad \begin{cases} y = s^2 + 4t \\ x = 3s - 2t \end{cases}$$

This is simplified to

$$\frac{\partial f}{\partial s} = 30s^4 - 16s^3 t + 144s^2 t - 64st^2 + 96t^2.$$

To find $\frac{\partial f}{\partial t}$, we have

$$\frac{\partial f}{\partial t} = \left(\frac{\partial f}{\partial x}\right)\left(\frac{\partial x}{\partial t}\right) + \left(\frac{\partial f}{\partial y}\right)\left(\frac{\partial y}{\partial t}\right)$$
$$= (2y^2)(-2) + (4xy)(4)$$
$$= -4y^2 + 16xy$$
$$= -4(s^2 + 4t)^2 + 16(3s - 2t)(s^2 + 4t) \quad \begin{cases} y = s^2 + 4t \\ x = 3s - 2t \end{cases}$$

This simplifies to

$$\frac{\partial f}{\partial t} = -4s^4 + 48s^3 - 64s^2 t + 192st - 192t^2.$$

◆ • ◆ • ◆ • ◆ • ◆

Example 25.3: Let $y = g(u, v, w)$ and let u, v and w be functions of m and n. Suppose that $\frac{\partial g}{\partial m} = 19, \frac{\partial g}{\partial u} = 4, \frac{\partial g}{\partial v} = 2, \frac{\partial g}{\partial w} = 3, \frac{\partial u}{\partial m} = 5, \frac{\partial u}{\partial n} = 11, \frac{\partial v}{\partial n} = 1, \frac{\partial w}{\partial m} = -5$ and $\frac{\partial w}{\partial n} = 12$. Find $\frac{\partial v}{\partial m}$.

Solution: Using a derivative tree (or recognizing the pattern of the Chain Rule), we have

$$\frac{\partial g}{\partial m} = \left(\frac{\partial g}{\partial u}\right)\left(\frac{\partial u}{\partial m}\right) + \left(\frac{\partial g}{\partial v}\right)\left(\frac{\partial v}{\partial m}\right) + \left(\frac{\partial g}{\partial w}\right)\left(\frac{\partial w}{\partial m}\right).$$

By substitution, we have

$$19 = (4)(5) + (2)\left(\frac{\partial v}{\partial m}\right) + (3)(-5)$$
$$19 = 20 + 2\left(\frac{\partial v}{\partial m}\right) - 15$$
$$\frac{19 - 20 + 15}{2} = \frac{\partial v}{\partial m}$$
$$\frac{\partial v}{\partial m} = 7.$$

Note that we did not need to use the information provided for $\frac{\partial u}{\partial n}, \frac{\partial v}{\partial n}$ and $\frac{\partial w}{\partial n}$.

Implicit Differentiation can be performed by employing the chain rule of a multivariable function. Often, this technique is much faster than the "traditional" direct method seen in single-variable calculus can be applied to functions of many variables with ease.

Example 25.4: Use implicit differentiation to find $\frac{dy}{dx}$ where $x^2y + y^3 = x^4$.

Solution: The "traditional" method is to differentiate each term in place, with respect to x. Note that the product rule is used on the first term, where $\frac{d}{dx}(x^2y) = x^2\frac{dy}{dx} + 2xy$:

$$\frac{d}{dx}(x^2y) + \frac{d}{dx}(y^3) = \frac{d}{dx}(x^4),$$

which gives $\quad x^2\frac{dy}{dx} + 2xy + 3y^2\frac{dy}{dx} = 4x^3.$

Then algebraically isolate $\frac{dy}{dx}$:

$$\frac{dy}{dx}(x^2 + 3y^2) = 4x^3 - 2xy, \quad \text{so that} \quad \frac{dy}{dx} = \frac{4x^3 - 2xy}{x^2 + 3y^2}.$$

To use the Chain Rule, rewrite the equation $x^2y + y^3 = x^4$ with all terms to one side:

$$x^2y + y^3 - x^4 = 0.$$

Call the left side $F(x, y) = x^2y + y^3 - x^4$. We seek $\frac{dy}{dx}$, so differentiate both sides with respect to x. Using the Chain Rule, the derivative of F with respect to x is written $\left(\frac{\partial F}{\partial x}\right)\left(\frac{dx}{dx}\right) + \left(\frac{\partial F}{\partial y}\right)\left(\frac{dy}{dx}\right)$. Note that the right side gives us $\frac{d}{dx}0 = 0$. We have

$$\left(\frac{\partial F}{\partial x}\right)\left(\frac{dx}{dx}\right) + \left(\frac{\partial F}{\partial y}\right)\left(\frac{dy}{dx}\right) = 0.$$

Now, $\frac{\partial F}{\partial x} = 2xy - 4x^3$ and $\frac{\partial F}{\partial y} = x^2 + 3y^2$. Furthermore, $\frac{dx}{dx} = 1$, and $\frac{dy}{dx}$ is the unknown. Make the substitutions and solve for the unknown:

$$(2xy - 4x^3)(1) + (x^2 + 3y^2)\frac{dy}{dx} = 0$$
$$(x^2 + 3y^2)\frac{dy}{dx} = 4x^3 - 2xy$$
$$\frac{dy}{dx} = \frac{4x^3 - 2xy}{x^2 + 3y^2}.$$

◆ ◆ ◆ ◆ ◆ ◆ ◆ ◆ ◆

In general, if x and y are implicitly related, collect all terms to one side and call the collected expression $F(x, y)$. Thus,

$$\frac{dy}{dx} = -\frac{F_x}{F_y} \quad \text{and} \quad \frac{dx}{dy} = -\frac{F_y}{F_x}.$$

This is true for implicit functions of three or more variable, too.

◆ ◆ ◆ ◆ ◆ ◆ ◆ ◆ ◆

Example 25.5: Given $xy^2z + 3xz^3 = yz^6$, find $\frac{dy}{dz}$.

Solution: Call $F(x, y, z) = xy^2z + 3xz^3 - yz^6$. Using the formula $\frac{dy}{dz} = -\frac{F_z}{F_y}$, we have

$$F_z = xy^2 + 9xz^2 - 6yz^5 \quad \text{and} \quad F_y = 2xyz - z^6.$$

Thus,

$$\frac{dy}{dz} = -\frac{F_z}{F_y} = -\left(\frac{xy^2 + 9xz^2 - 6yz^5}{2xyz - z^6}\right) = \frac{6yz^5 - xy^2 - 9xz^2}{2xyz - z^6}.$$

26. Directional Derivatives & The Gradient

Given a multivariable function $z = f(x, y)$ and a point on the xy-plane $P_0 = (x_0, y_0)$ at which f is differentiable (it is smooth with no discontinuities, folds or corners), there are infinitely many directions (relative to the xy-plane) in which to sketch a tangent line to f at P_0. A **directional derivative** is the slope of a tangent line to f at P_0 in which a *unit* direction vector $\mathbf{u} = \langle u_1, u_2 \rangle$ has been specified, and is given by the formula

$$D_{\mathbf{u}} f(x_0, y_0) = f_x(x_0, y_0) u_1 + f_y(x_0, y_0) u_2.$$

The right side of the equation can be viewed as the result of a dot product:

$$D_{\mathbf{u}} f(x_0, y_0) = \langle f_x(x_0, y_0), f_y(x_0, y_0) \rangle \cdot \langle u_1, u_2 \rangle.$$

The vector-valued function $\langle f_x(x_0, y_0), f_y(x_0, y_0) \rangle$ is called the **gradient** of f at $x = x_0$ and $y = y_0$, and is written $\nabla f(x_0, y_0)$. Thus, the directional derivative of f at P_0 in the direction of $\mathbf{u}$ is written in the shortened form

$$D_{\mathbf{u}} f(x_0, y_0) = \nabla f(x_0, y_0) \cdot \mathbf{u}.$$

♦ • ♦ • ♦ • ♦ • ♦

Example 26.1: Find $\nabla f(x, y)$, where $f(x, y) = x^2 y + 2xy^3$.

Solution: Since $\nabla f(x, y) = \langle f_x(x, y), f_y(x, y) \rangle$, we have

$$\nabla f(x, y) = \langle 2xy + 2y^3, x^2 + 6xy^2 \rangle.$$

♦ • ♦ • ♦ • ♦ • ♦

Example 26.2: Find the slope of the tangent line of $f(x, y) = x^2 y + 2xy^3$ at $x_0 = -1$, $y_0 = 2$ in the direction of $\mathbf{u} = \langle 4, 3 \rangle$.

Solution: From the previous example, $\nabla f(x, y) = \langle 2xy + 2y^3, x^2 + 6xy^2 \rangle$. When evaluated at $x_0 = -1$ and $y_0 = 2$, we have

$$\nabla f(-1, 2) = \langle 2(-1)(2) + 2(2)^3, (-1)^2 + 6(-1)(2)^2 \rangle = \langle 12, -23 \rangle.$$

The direction $\mathbf{u}$ is not a unit vector. Since $|\mathbf{u}| = \sqrt{4^2 + 3^2} = \sqrt{25} = 5$, the unit vector in the direction of $\mathbf{u}$ is $\left\langle \frac{4}{5}, \frac{3}{5} \right\rangle$. Thus,

$$D_{\mathbf{u}} f(-1, 2) = \langle 12, -23 \rangle \cdot \left\langle \frac{4}{5}, \frac{3}{5} \right\rangle = 12 \left(\frac{4}{5} \right) - 23 \left(\frac{3}{5} \right) = -\frac{21}{5}.$$

Example 26.3: Find the slope of the tangent line of $g(x,y) = \frac{x}{y^2}$ at $x_0 = 3$ and $y_0 = 5$, in the direction of the origin.

Solution: The vector from $(3,5)$ to $(0,0)$ is given by $\langle 0-3, 0-5 \rangle = \langle -3, -5 \rangle$. Its magnitude is $\sqrt{(-3)^2 + (-5)^2} = \sqrt{34}$. Thus, the unit direction vector is

$$\mathbf{u} = \left\langle -\frac{3}{\sqrt{34}}, -\frac{5}{\sqrt{34}} \right\rangle.$$

The gradient of g is

$$\nabla g(x,y) = \left\langle \frac{1}{y^2}, -\frac{2x}{y^3} \right\rangle.$$

Therefore,

$$\nabla g(3,5) = \left\langle \frac{1}{(5)^2}, -\frac{2(3)}{(5)^3} \right\rangle = \left\langle \frac{1}{25}, -\frac{6}{125} \right\rangle.$$

The slope of the tangent line of g at $x_0 = 3$ and $y_0 = 5$ in the direction of $\mathbf{u}$ is

$$D_{\mathbf{u}}g(3,5) = \left\langle \frac{1}{25}, -\frac{6}{125} \right\rangle \cdot \left\langle -\frac{3}{\sqrt{34}}, -\frac{5}{\sqrt{34}} \right\rangle$$

$$= \left(\frac{1}{25}\right)\left(-\frac{3}{\sqrt{34}}\right) + \left(-\frac{6}{125}\right)\left(-\frac{5}{\sqrt{34}}\right)$$

$$= -\frac{15}{125\sqrt{34}} + \frac{30}{125\sqrt{34}} = \frac{15}{125\sqrt{34}} \approx 0.0206.$$

◆ ◆ ◆ ◆ ◆ ◆ ◆ ◆

Example 26.4: Find the slope of the tangent line of $h(x,y) = \sqrt{1 + x^2 + y^2}$ where $P_0 = (1,2)$ and the direction is given by a ray from P_0 oriented at $\theta = \frac{\pi}{6}$ radians, relative to the positive x-direction.

Solution: The direction vector is given by $\mathbf{u} = \left\langle \cos\left(\frac{\pi}{6}\right), \sin\left(\frac{\pi}{6}\right) \right\rangle = \left\langle \frac{\sqrt{3}}{2}, \frac{1}{2} \right\rangle$. It is a unit vector. The gradient of h is

$$\nabla h(x,y) = \left\langle \frac{x}{\sqrt{1+x^2+y^2}}, \frac{y}{\sqrt{1+x^2+y^2}} \right\rangle.$$

Upon substitution,

$$\nabla h(1,2) = \left\langle \frac{(1)}{\sqrt{1+(1)^2+(2)^2}}, \frac{(2)}{\sqrt{1+(1)^2+(2)^2}} \right\rangle = \left\langle \frac{1}{\sqrt{6}}, \frac{2}{\sqrt{6}} \right\rangle.$$

The directional derivative of h at $(1,2)$ in the direction of **u** is

$$D_{\mathbf{u}} h(1,2) = \left\langle \frac{\sqrt{3}}{2}, \frac{1}{2} \right\rangle \cdot \left\langle \frac{1}{\sqrt{6}}, \frac{2}{\sqrt{6}} \right\rangle$$

$$= \left(\frac{\sqrt{3}}{2}\right)\left(\frac{1}{\sqrt{6}}\right) + \left(\frac{1}{2}\right)\left(\frac{2}{\sqrt{6}}\right)$$

$$= \frac{\sqrt{3}+2}{2\sqrt{6}} \approx 0.762.$$

◆ • ◆ • ◆ • ◆ • ◆

Directional derivatives can be extended into higher dimensions.

Example 26.5: Find the slope of the tangent line of $f(x,y,z) = xy^2z^3$ when $x_0 = 2, y_0 = 1$ and $z_0 = 3$ in the direction of $\langle 2, 4, -5 \rangle$.

Solution: The gradient of f is

$$\nabla f(x,y,z) = \langle f_x, f_y, f_z \rangle = \langle y^2 z^3, 2xyz^3, 3xy^2 z^2 \rangle.$$

At $(2,1,3)$, we have

$$\nabla f(2,1,3) = \langle 27, 108, 54 \rangle.$$

The unit direction vector is $\mathbf{u} = \left\langle \frac{2}{\sqrt{45}}, \frac{4}{\sqrt{45}}, -\frac{5}{\sqrt{45}} \right\rangle$. The slope of the tangent line of f at $(2,1,3)$ in the direction of **u** is

$$D_{\mathbf{u}} f(2,1,3) = \nabla f(2,1,3) \cdot \mathbf{u}$$

$$= \langle 27, 108, 54 \rangle \cdot \left\langle \frac{2}{\sqrt{45}}, \frac{4}{\sqrt{45}}, -\frac{5}{\sqrt{45}} \right\rangle$$

$$= \frac{54}{\sqrt{45}} + \frac{432}{\sqrt{45}} - \frac{270}{\sqrt{45}} \approx 32.2.$$

Using the cosine form of the formula for the dot product of two vectors, $\mathbf{u} \cdot \mathbf{v} = |\mathbf{u}||\mathbf{v}| \cos \theta$, we can rewrite $D_{\mathbf{u}} f(x_0, y_0) = \nabla f(x_0, y_0) \cdot \mathbf{u}$ as

$$D_{\mathbf{u}} f(x_0, y_0) = |\nabla f(x_0, y_0)||\mathbf{u}| \cos \theta.$$

Since $\mathbf{u}$ is a unit vector, then $|\mathbf{u}| = 1$, so that

$$|\nabla f(x_0, y_0)||\mathbf{u}| \cos \theta = |\nabla f(x_0, y_0)| \cos \theta,$$

where θ is the angle between the gradient vector at (x_0, y_0), and the direction vector $\mathbf{u}$. From this, we can infer that $|\nabla f(x_0, y_0)| \cos \theta$ is maximized when $\nabla f(x_0, y_0)$ and $\mathbf{u}$ are parallel, or when $\theta = 0$ (so that $\cos \theta = 1$). This leads to a significant result in directional derivatives.

Given a function $z = f(x, y)$ and a point $P_0 = (x_0, y_0, z_0)$:

- The **direction of steepest ascent** at P_0 is given by $\nabla f(x_0, y_0) = \langle f_x(x_0, y_0), f_y(x_0, y_0) \rangle$. In this case, it is permissible to state the direction as a non-unit vector.

- The **slope of steepest ascent** at P_0 is given by $|\nabla f(x_0, y_0)|$.

- The **direction of steepest descent** at P_0 is opposite the direction of steepest ascent, and is given by $-\nabla f(x_0, y_0) = \langle -f_x(x_0, y_0), -f_y(x_0, y_0) \rangle$.

- The **slope of steepest descent** at P_0 is $-|\nabla f(x_0, y_0)|$.

A path that follows the directions of steepest ascent is called a **gradient path** and is always orthogonal to the contours of the surface.

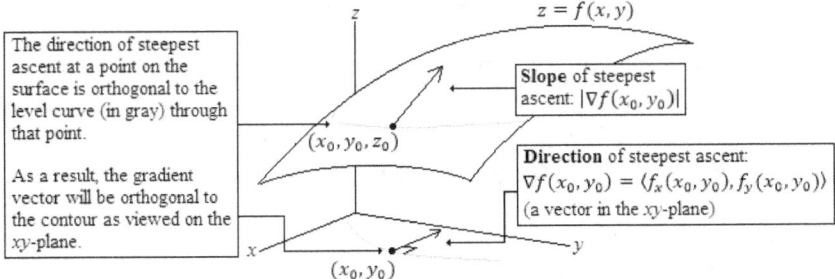

139

Example 26.6: Let $f(x,y) = x^2 + 2xy^2$. State the direction(s) in which the slope of the tangent line at $x_0 = 2$ and $y_0 = 1$ is 0.

Solution: We have $\nabla f(x,y) = \langle 2x + 2y^2, 4xy \rangle$. Let $\mathbf{u} = \langle u_1, u_2 \rangle$. We have

$$D_\mathbf{u} f(2,1) = \nabla f(2,1) \cdot \mathbf{u}$$
$$= \langle 6,8 \rangle \cdot \langle u_1, u_2 \rangle$$
$$= 6u_1 + 8u_2.$$

If the slope is to be 0, we set $6u_1 + 8u_2 = 0$. Thus, whenever $u_2 = -\frac{3}{4}u_1$, then the slope of the tangent line at $x_0 = 2$ and $y_0 = 1$ will be 0.

◆ • ◆ • ◆ • ◆ • ◆

Example 26.7: Find the direction of steepest ascent of $f(x,y) = x^2y + 2xy^3$ at $x_0 = -1$ and $y_0 = 2$, then find the slope of steepest ascent.

Solution: From Example 26.2, we have $\nabla f(x,y) = \langle 2xy + 2y^3, x^2 + 6xy^2 \rangle$ so that $\nabla f(-1,2) = \langle 12, -23 \rangle$. This is the *direction* of steepest ascent. The *slope* of steepest ascent is $|\langle 12, -23 \rangle| = \sqrt{12^2 + (-23)^2} \approx 25.94$.

> When finding a directional derivative where the direction is stated or to be determined, you *must* be sure that it is stated as a unit vector. However, when asked to find a direction of steepest ascent, it is permissible to leave it as a non-unit vector since you will likely be calculating the slope as well. While it is not incorrect to state the direction of steepest ascent as a unit vector, a common error is to then use that unit vector to find the slope, in which case the answer will be 1, which is likely incorrect.

◆ • ◆ • ◆ • ◆ • ◆

Example 26.8: Suppose the slope of the tangent line of $z = f(x,y)$ at $P_0 = (x_0, y_0)$ in the direction of $\langle 3,1 \rangle$ is $\sqrt{10}$, and that the slope of the tangent line at the same point in the direction of $\langle 1,4 \rangle$ is $\frac{18}{\sqrt{17}}$. What is the direction of steepest ascent of f at P_0, and what is the slope in this direction?

Solution: We don't know f, but we can treat the components in its gradient, $\nabla f(x_0, y_0) = \langle f_x(x_0, y_0), f_y(x_0, y_0) \rangle$, as a pair of unknowns. In the direction of $\langle 3,1 \rangle$, the slope of the tangent line is $\sqrt{10}$. Considering the unit direction vector $\mathbf{u} = \left\langle \frac{3}{\sqrt{10}}, \frac{1}{\sqrt{10}} \right\rangle$, we have $D_\mathbf{u} f(x_0, y_0) = \nabla f(x_0, y_0) \cdot \mathbf{u} = \sqrt{10}$. Thus, we have

$$\langle f_x(x_0, y_0), f_y(x_0, y_0) \rangle \cdot \left\langle \frac{3}{\sqrt{10}}, \frac{1}{\sqrt{10}} \right\rangle = \sqrt{10},$$

which gives

$$f_x(x_0, y_0) \frac{3}{\sqrt{10}} + f_y(x_0, y_0) \frac{1}{\sqrt{10}} = \sqrt{10}. \quad (1)$$

In a similar way, we consider the unit direction vector in the direction of $\langle 1, 4 \rangle$, which is $\left\langle \frac{1}{\sqrt{17}}, \frac{4}{\sqrt{17}} \right\rangle$. The slope in this direction is $\frac{18}{\sqrt{17}}$. We have

$$\langle f_x(x_0, y_0), f_y(x_0, y_0) \rangle \cdot \left\langle \frac{1}{\sqrt{17}}, \frac{4}{\sqrt{17}} \right\rangle = \frac{18}{\sqrt{17}},$$

which gives

$$f_x(x_0, y_0) \frac{1}{\sqrt{17}} + f_y(x_0, y_0) \frac{4}{\sqrt{17}} = \frac{18}{\sqrt{17}}. \quad (2)$$

Taking equations **(1)** and **(2)** together, we have a system of two unknowns in two equations:

$$f_x(x_0, y_0) \frac{3}{\sqrt{10}} + f_y(x_0, y_0) \frac{1}{\sqrt{10}} = \sqrt{10}$$
$$f_x(x_0, y_0) \frac{1}{\sqrt{17}} + f_y(x_0, y_0) \frac{4}{\sqrt{17}} = \frac{18}{\sqrt{17}}.$$

The first equation is multiplied by $\sqrt{10}$, and the second by $\sqrt{17}$ to clear fractions:

$$f_x(x_0, y_0)(3) + f_y(x_0, y_0)(1) = 10$$
$$f_x(x_0, y_0)(1) + f_y(x_0, y_0)(4) = 18.$$

The bottom equation is multiplied by -3:

$$f_x(x_0, y_0)(3) + f_y(x_0, y_0)(1) = 10$$
$$f_x(x_0, y_0)(-3) + f_y(x_0, y_0)(-12) = -54.$$

Adding the second equation to the first, we have $-11 f_y(x_0, y_0) = -44$. Thus, $f_y(x_0, y_0) = 4$. Substituting this into either of the equations **(1)** or **(2)**, we find that $f_x(x_0, y_0) = 2$. Therefore, we now know $\nabla f(x_0, y_0)$, which is $\langle 2, 4 \rangle$. This is the direction of steepest ascent of f. The slope at P_0 in this direction is $\sqrt{2^2 + 4^2} = \sqrt{20} = 2\sqrt{5} \approx 4.47$.

Example 26.9: A plane tilts to the north at a 6% grade – that is, for every 100 feet one moves horizontally north, he or she will gain 6 feet vertically. Find the slope and the grade if someone walks to the northeast.

Solution: Assume the plane passes through the origin, assuming also that the y-axis is north and south, and the x-axis is east and west, in the usual map orientation. When $y = 100$, we have $z = 6$, so that another ordered triple on the plane is $(0,100,6)$. Thus, we can write $z = \frac{6}{100}y = 0.06y$ as the equation of the plane. The gradient of f is $\nabla f(x,y) = \langle 0, 0.06 \rangle$. Note that x is an independent variable but has no effect on the values of z. If it helps, write the plane as $z = 0x + 0.06y$.

Furthermore, at the origin, we still have $\nabla f(0,0) = \langle 0, 0.06 \rangle$. Meanwhile, movement to the northeast can be modeled by the vector $\langle 1,1 \rangle$, or as a unit vector, $\mathbf{u} = \left\langle \frac{1}{\sqrt{2}}, \frac{1}{\sqrt{2}} \right\rangle$.

The slope at the origin in the direction of northeast is given by

$$D_\mathbf{u} f(0,0) = \nabla f(0,0) \cdot \mathbf{u}$$

$$= \langle 0, 0.06 \rangle \cdot \left\langle \frac{1}{\sqrt{2}}, \frac{1}{\sqrt{2}} \right\rangle$$

$$= \frac{0.06}{\sqrt{2}} \approx 0.0424.$$

The grade can be inferred by the fact that 1 foot of movement in the northeast direction results in a rise of 0.0424 feet vertically. Thus, the grade is about 4.24%.

Note that a movement east or west would result in no change in z. The directional derivative in either direction (the positive or negative x direction) is 0. Let $\mathbf{u} = \langle 1, 0 \rangle$ or $\langle -1, 0 \rangle$ and verify that the directional derivative would be 0.

27. Tangent Planes & Approximations

If $z = f(x, y)$ is a differentiable surface in R^3 and (x_0, y_0, z_0) is a point on this surface, then it is possible to construct a plane passing through this point, tangent to the surface of f.

Recall that a plane is constructed by determining a vector $\mathbf{n} = \langle a, b, c \rangle$ normal to the plane and identifying a point (x_0, y_0, z_0) on the plane. With this information, the equation of the plane is given by

$$a(x - x_0) + b(y - y_0) + c(z - z_0) = 0.$$

To find this desired normal vector $\mathbf{n}$, we temporarily write $z = f(x, y)$ as a function of three variables, $F(x, y, z) = f(x, y) - z$, noting that the equation $f(x, y) - z = 0$ is now a level curve of the graph of F. Recall from Section 26 that the vectors in the gradient of F will be orthogonal to all level curves of the graph of F. We start with $\nabla F = \langle F_x, F_y, F_z \rangle$, and observe that $F_x = f_x$ and that $F_y = f_y$, and since we wrote $F(x, y, z) = f(x, y) - z$, that $F_z = -1$. Therefore,

$$\mathbf{n} = \langle a, b, c \rangle = \langle f_x(x_0, y_0), f_y(x_0, y_0), -1 \rangle.$$

Tying this all together, the **equation of the tangent plane** to a point (x_0, y_0, z_0) on the surface of $z = f(x, y)$ is given by

$$f_x(x_0, y_0)(x - x_0) + f_y(x_0, y_0)(y - y_0) - (z - z_0) = 0.$$

◆ • • • ◆ • • • ◆

Example 27.1: Let $z = f(x, y) = x^2 + 2xy^3$. Find the equation of the tangent plane to f when $x_0 = 1$ and $y_0 = 2$.

Solution: When $x_0 = 1$ and $y_0 = 2$, then $z_0 = f(x_0, y_0) = f(1,2) = (1)^2 + 2(1)(2)^3 = 17$. Thus, the point of tangency is $(x_0, y_0, z_0) = (1,2,17)$.

The partial derivatives are $f_x(x, y) = 2x + 2y^3$ and $f_y(x, y) = 6xy^2$. Evaluated at $x_0 = 1$ and $y_0 = 2$, we have $f_x(1,2) = 18$ and $f_y(1,2) = 24$. Thus, the plane of tangency is

$$18(x - 1) + 24(y - 2) - (z - 17) = 0.$$

Simplified, the plane is $18x + 24y - z = 49$, or with z isolated, we obtain $z = 18x + 24y - 49$.

Example 27.2: Find the equation of the tangent plane to $z = g(x,y) = \frac{2x+y}{3y^2}$ when $x_0 = -2$ and $y_0 = 3$.

Solution: The point of tangency is $(x_0, y_0, z_0) = \left(-2, 3, -\frac{1}{27}\right)$, where $z_0 = \frac{2(-2)+(3)}{3(3)^2} = -\frac{1}{27}$.

The partial derivatives are

$$g_x(x,y) = \frac{2}{3y^2} \quad \text{and} \quad g_y(x,y) = -\frac{y+4x}{3y^3}.$$

Evaluated at $x_0 = -2$ and $y_0 = 3$, we have $g_x(-2,3) = \frac{2}{27}$ and $g_y(-2,3) = \frac{5}{81}$. Thus, the equation of the plane of tangency is

$$\frac{2}{27}(x-(-2)) + \frac{5}{81}(y-3) - \left(z - \left(-\frac{1}{27}\right)\right) = 0.$$

Multiplying by 81 to clear fractions and then distributing to clear parentheses, the equation is simplified to $6x + 5y - 81z = 6$.

This process can be extended to surfaces in higher dimensions.

Example 27.3: Find the equation of the tangent plane to $w = f(x,y,z) = x^2y^3z^4$ at $(2, 1, -2, 64)$.

Solution: The partial derivatives are evaluated at $x_0 = 2$, $y_0 = 1$ and $z_0 = -2$:

$$f_x(x,y,z) = 2xy^3z^4 \rightarrow f_x(2,1,-2) = 64,$$
$$f_y(x,y,z) = 3x^2y^2z^4 \rightarrow f_y(2,1,-2) = 192,$$
$$f_z(x,y,z) = 4x^2y^3z^3 \rightarrow f_z(2,1,-2) = -128.$$

Thus, the plane of tangency is

$$64(x-2) + 192(y-1) - 128(z-(-2)) - 1(w-64) = 0.$$

Solving for w, we have

$$w = 64(x-2) + 192(y-1) - 128(z+2) + 64.$$

Simplified, we have $w = 64x + 192y - 128z - 512$.

Tangent planes can be used to estimate values on the surface of a multi-variable function f.

Example 27.4: Given that $z = 18x + 24y - 49$ is the equation of the plane tangent to the surface $f(x, y) = x^2 + 2xy^3$ when $x_0 = 1$ and $y_0 = 2$, estimate the value of $f(1.1, 1.9)$.

Solution: Since planes consist only of linear and constant terms, it is usually easier to evaluate points on a plane rather than points on a surface. In this case, we have

$$z = 18(1.1) + 24(1.9) - 49 = 16.4.$$

Observe that the point $(1.1, 1.9, 16.4)$ lies on the tangent plane, not on the surface of f. However, if we were to evaluate f at $x = 1.1$ and $y = 1.9$, we obtain

$$f(1.1, 1.9) = (1.1)^2 + 2(1.1)(1.9)^3 = 16.2998.$$

The estimated value of 16.4 is an excellent approximation of the actual value of 16.2998. Using planes to estimate values on a surface requires that the point of evaluation be "close" to the point of tangency. In this example, 1.1 is close to 1, and 1.9 is close to 2. However, suppose that we wanted to use the tangent plane to estimate $f(1.5, 2.4)$. We get

$$z = 18(1.5) + 24(2.4) - 49 = 35.6,$$

The actual point on the surface is $f(1.5, 2.4) = 43.722$. We see that the estimated value of z is not close to the actual value of z.

◆ • ◆ • ◆ • • ◆ • ◆

Example 27.5: Given the surface $w = f(x, y, z) = x^2 y^3 z^4$ at $(2, 1, -2, 64)$ in Example 27.3, estimate the value of $w = f(2.01, 0.99, -1.98)$.

Solution: We use the equation $w = 64(x - 2) + 192(y - 1) - 128(z + 2) + 64$ from Example 27.3. We then substitute $x = 2.01$, $y = 0.99$ and $z = -1.98$:

$$\begin{aligned} w &= 64(2.01 - 2) + 192(0.99 - 1) - 128(-1.98 + 2) + 64 \\ &= 64(0.01) + 192(-0.01) - 128(0.02) + 64 \\ &= 0.64 - 1.92 - 2.56 + 64 \\ &= -3.84 + 64 \\ &= 60.16. \end{aligned}$$

The actual w-value is $w = f(2.01, 0.99, -1.98) = (2.01)^2(0.99)^3(-1.98)^4 = 60.25$. The estimation is very close to the actual value.

♦ • ♦ • ♦ • ♦ • ♦

Example 27.6: Let $f(x, y) = 3x^2 - 2y$. Find the acute angle that the tangent plane of f, when $x_0 = -2$ and $y_0 = 3$, makes with the xy-plane.

Solution: The partial derivatives are $f_x(x, y) = 6x$ and $f_y(x, y) = -2$. Thus, the normal vector **n** is

$$\mathbf{n} = \langle f_x(-2,3), f_y(-2,3), -1 \rangle$$
$$= \langle 6(-2), -2, -1 \rangle$$
$$= \langle -12, -2, -1 \rangle.$$

The xy-plane has two "convenient" normal vectors, the positive z-axis represented by the vector $\mathbf{z}^+ = \langle 0,0,1 \rangle$ and the negative z-axis represented by the vector $\mathbf{z}^- = \langle 0,0,-1 \rangle$. Since **n** points in the direction of the negative z-axis, we will compare **n** to $\mathbf{z}^-$.

Recall that the angle between two planes is the same as the angle between its normal vectors, and that two planes always meet acutely (except when they are orthogonal). Thus, to find the angle between the xy-plane and the plane of tangency, it is sufficient to determine the angle between the two normal vectors.

The angle between the two vectors is

$$\theta = \cos^{-1}\left(\frac{\mathbf{n} \cdot \mathbf{z}^-}{|\mathbf{n}||\mathbf{z}^-|}\right) = \cos^{-1}\left(\frac{1}{\sqrt{149}}\right) \approx 85.3°.$$

Therefore, the angle that the tangent plane of $f(x, y) = 3x^2 - 2y$, when $x_0 = -2$ and $y_0 = 3$, makes with the xy-plane, is 85.3°.

♦ • ♦ • ♦ • ♦ • ♦

Example 27.7: A surface is defined parametrically by $\mathbf{r}(u, v) = \langle 2u + v, v - 3u, uv \rangle$. Find the equation of the tangent plane at the point $(4, -11, -6)$.

Solution: Observe that $x(u, v) = 2u + v$, $y(u, v) = v - 3u$ and $z(u, v) = uv$. From the point $(4, -11, -6)$, we can infer that $x = 2u + v = 4$ and $y = v - 3u = -11$. This is a system:

$$2u + v = 4$$
$$-3u + v = -11.$$

Solving the system, we find that $u = 3$ and $v = -2$. Note that this also checks for $z = uv = (2)(-3) = -6$.

We now need to find a vector **n** normal to the surface. Taking partial derivatives of **r**, we have

$$\mathbf{r}_u(u,v) = \langle 2, -3, v \rangle \quad \text{and} \quad \mathbf{r}_v(u,v) = \langle 1, 1, u \rangle.$$

Evaluating at $u = 3$ and $v = -2$, we have

$$\mathbf{r}_u(3, -2) = \langle 2, -3, -2 \rangle \quad \text{and} \quad \mathbf{r}_v(3, -2) = \langle 1, 1, 3 \rangle.$$

Thus, the normal vector **n** is

$$\mathbf{n} = \mathbf{r}_u \times \mathbf{r}_v = \langle -7, -8, 5 \rangle.$$

The plane tangent to the surface $\mathbf{r}(u,v) = \langle 2u + v, v - 3u, uv \rangle$ at the point $(4, -11, -6)$ is

$$-7(x - 4) - 8(y - (-11)) + 5(z - (-6)) = 0.$$

Simplifying, we have

$$-7(x - 4) - 8(y + 11) + 5(z + 6) = 0$$
$$-7x + 28 - 8y - 88 + 5z + 30 = 0$$
$$-7x - 8y + 5z = 30.$$

◆ • ◆ • • ◆ • • ◆

See an error? Have a suggestion?
Please see www.surgent.net/vcbook

28. Differentials

The equation of the tangent plane to a point (x_0, y_0, z_0) on the surface of $z = f(x, y)$ is given by

$$f_x(x_0, y_0)(x - x_0) + f_y(x_0, y_0)(y - y_0) - (z - z_0) = 0.$$

Add $(z - z_0)$ to both sides:

$$(z - z_0) = f_x(x_0, y_0)(x - x_0) + f_y(x_0, y_0)(y - y_0).$$

Now, view the expression $z - z_0$ as a change in z, written Δz. Do the same for $(x - x_0)$ and $(y - y_0)$. We have

$$\Delta z = f_x(x_0, y_0)\Delta x + f_y(x_0, y_0)\Delta y.$$

For sufficiently small changes in the variables, we can assume that $dx \approx \Delta x$, and so on. Thus, the above equation can be written using differentials:

$$dz = f_x(x_0, y_0)dx + f_y(x_0, y_0)dy.$$

We use this formula to study the effect that small changes in x and y have on z.

◆ ◆ ◆ ◆ ◆ ◆ ◆ ◆ ◆

Example 28.1: The exterior of a circular cylindrical tank is measured to be 4 meters in radius and 5 meters high. Assume that the measurements have a tolerance of 0.02 meters for the radius and 0.03 meters for the height. What effect do the possible variances in radius or height have on the volume of the tank?

Solution: The volume is given by $V(r, h) = \pi r^2 h$, where r is the radius of the base, and h is the vertical height. The differentials dV, dr and dh are related by the formula

$$\begin{aligned} dV &= V_r(r_0, h_0)dr + V_h(r_0, h_0)dh. \\ &= 2\pi r_0 h_0 \, dr + \pi r_0^2 \, dh \\ &= 2\pi(4)(5)(0.02) + \pi(4)^2(0.03) \\ &= 40\pi(0.02) + 16\pi(0.03). \\ &= 4.02 \text{ cubic meters.} \end{aligned}$$

It might be surprising that being off by just 0.02 meters (2 cm) when measuring the radius and 0.03 meters (3 cm) when measuring the height would translate into a change of approximately 4 cubic meters for the volume.

However, we can calculate exact volumes for these measurements and compare. If the radius is exactly 4 meters and the height exactly 5 meters, the presumptive volume is

$$V(4,5) = \pi(4)^2(5) = 251.327 \text{ m}^3.$$

If both measures are "low", that is, $r = 3.98$ meters and $h = 4.97$ meters, then the volume is

$$V(3.98, 4.97) = \pi(3.98)^2(4.97) = 247.327 \text{ m}^3.$$

The difference between the two volume figures is $247.327 - 251.327 = -4 \text{ m}^3$. Thus, measuring low results in approximately 4 fewer cubic meters of volume.

If both measures are "high", $r = 4.02$ and $h = 5.03$, then the volume is

$$V(4.02, 5.03) = \pi(4.02)^2(5.03) = 255.37 \text{ m}^3.$$

The difference between this higher figure and the presumed volume figure is $255.37 - 251.327 = 4.043 \text{ m}^3$. Again, the change in volume is roughly 4 cubic meters.

◆ • ◆ • ◆ • ◆ • ◆

Example 28.2: The surface area of a rectangular box of length l, width w and height h is given by $A(l, w, h) = 2(wl + wh + lh)$. Suppose workers measure the length to be 20 feet, the width 8 feet and the height 5 feet. If the tolerance of the surface area is to be no more than 6 square feet (low or high), what should the tolerances on the length, width and height be, assuming all to be the same?

Solution: Written in differential form, dA is related to dl, dw and dh by

$$dA = A_l dl + A_w dw + A_h dh$$
$$= 2(w + h)dl + 2(l + h)dw + 2(l + w)dh.$$

We assume that $dl = dw = dh$. Substituting, we have

$$6 = 2(20 + 8)dl + 2(20 + 5)dw + 2(8 + 5)dh$$
$$6 = 56dl + 50dw + 26dh$$
$$6 = 132dl \quad (\text{since } dl = dw = dh)$$

Thus, $dl = \frac{6}{132} \approx 0.045$ feet, or slightly over half an inch, in allowable tolerance. If the workers can keep their measurements for the length, width and height within this small tolerance, the actual surface area should not vary by more than 6 square feet from the presumptive surface area.

Example 28.3: A conical pyramid of sand has a circular base with radius $r = 6$ meters and a height $h = 4$ meters. If sand is added to the pile in such a way that the change in radius and the change in height are the same, what will have more of an effect on the volume, a change in the radius or a change in the height?

Solution: The volume of a conical pyramid is given by $V(r, h) = \frac{1}{3}\pi r^2 h$, where r is the radius of the base, and h is the vertical height. In differential form, we have

$$dV = \left(\frac{2}{3}\pi r h\right) dr + \left(\frac{1}{3}\pi r^2\right) dh.$$

Evaluated at $r = 6$ meters and a height $h = 4$ meters, we have

$$dV = \left(\frac{2}{3}\pi(6)(4)\right) dr + \left(\frac{1}{3}\pi(6)^2\right) dh = 16\pi\, dr + 12\pi\, dh.$$

Assuming that $dr = dh$, then since $16\pi > 12\pi$, a change in the radius will have a greater effect on the volume than an equal change in height would.

◆ • ◆ • ◆ • ◆ • ◆

Taking the generic differential form for $z = f(x, y)$, which is $dz = f_x(x_0, y_0)dx + f_y(x_0, y_0)dy$, we can divide both sides by dt, in effect forming a **related rate** in which x, y and z are functions of a parameter variable t. We use the Chain Rule and obtain:

$$\frac{dz}{dt} = f_x(x_0, y_0)\frac{dx}{dt} + f_y(x_0, y_0)\frac{dy}{dt}.$$

Example 28.4: A circular cylinder is being heated in such a way that its radius is increasing at the rate of 0.05 feet/minute and the height is shrinking at the rate of 0.02 feet/minute. Find the rate at which the surface area is changing when its base radius is 3 feet and the height is 7 feet.

Solution: Using the formula for surface area of a circular cylinder, $A(r, h) = 2\pi r h + 2\pi r^2$, we differentiate each term with respect to t:

$$\frac{dA}{dt} = A_r \frac{dr}{dt} + A_h \frac{dh}{dt} = (2\pi h + 4\pi r)\frac{dr}{dt} + (2\pi r)\frac{dh}{dt}.$$

Substituting, we have

$$\frac{dA}{dt} = (2\pi(7) + 4\pi(3))(0.05) + (2\pi(3))(-0.02) \approx 3.71 \; \frac{\text{feet}^2}{\text{minute}}.$$

29. Unconstrained Optimization

Optimization is the process of determining the highest (maximum) and lowest (minimum) points on a graph. Maximum and minimum points are collectively called **extreme points**, or **extrema**.

Let $z = f(x, y)$ be a function in R^3, and assume that f exists and is continuous over the entire xy-plane. That is, its domain is R^2, there being no restrictions on variables x and y.

A **critical point** (x_c, y_c, z_c), where $z_c = f(x_c, y_c)$, is a point where $f_x(x_c, y_c) = 0$ or does not exist, *and* where $f_y(x_c, y_c) = 0$ or does not exist. These are the possible extreme points. All minimum and maximum points are **local** (or **relative**), meaning that the point is the lowest or highest point within some open region in R^2 that includes the point. If it is the lowest or highest point over the entire domain, then the point is an **absolute** minimum or maximum.

The second derivative test for R^2 is one way to determine if a critical point (x_c, y_c, z_c) is a minimum, a maximum, or neither. The formula is

$$D = f_{xx}(x_c, y_c) f_{yy}(x_c, y_c) - \left(f_{xy}(x_c, y_c)\right)^2.$$

- If $D > 0$ and if $f_{xx}(x_c, y_c) > 0$, then the graph of f is concave upward, and (x_c, y_c, z_c) is a relative minimum.
- If $D > 0$ and if $f_{xx}(x_c, y_c) < 0$, then the graph of f is concave downward, and (x_c, y_c, z_c) is a relative maximum.
- If $D < 0$, then (x_c, y_c, z_c) is not a minimum nor a maximum. It is a saddle point.
- If $D = 0$, then no conclusion about (x_c, y_c, z_c) can be inferred. Other methods need to be used to classify the critical point.

When $D > 0$, the signs of $f_{xx}(x_c, y_c)$ and $f_{yy}(x_c, y_c)$ will be the same. Thus, when $D > 0$, it is sufficient to note the sign of one since the sign of the other will be identical.

When there are no restrictions on the domain, this process is called **unconstrained optimization**.

Example 29.1: Let $z = f(x,y) = x^2 + y^2 + 6x - 4y + 2$. Find its critical points and classify these points as minima, maxima or saddle.

Solution: Find the first partial derivatives:

$$f_x(x,y) = 2x + 6$$
$$f_y(x,y) = 2y - 4.$$

Note that the derivatives (as well as the function itself) are defined for all x and all y in R^2. Thus, there are no possible locations where the derivatives "do not exist". Then set the partial derivatives to 0, and solve:

$$\begin{matrix} 2x + 6 = 0 \\ 2y - 4 = 0 \end{matrix} \quad \text{which gives} \quad \begin{matrix} x = -3 \\ y = 2. \end{matrix}$$

Thus, we have one critical point, $(-3, 2, f(-3,2))$, where $f(-3,2) = -11$. To classify this critical point, we use the second derivative test. The second derivatives are found first (recall that $f_{xy}(x,y) = f_{yx}(x,y)$):

$$f_{xx}(x,y) = 2, \qquad f_{yy}(x,y) = 2 \quad \text{and} \quad f_{xy}(x,y) = 0.$$

By the second derivative test, we have

$$D = f_{xx}(-3,2)f_{yy}(-3,2) - \left(f_{xy}(-3,2)\right)^2$$
$$= (2)(2) - 0^2$$
$$= 4.$$

Note that $D > 0$ and that $f_{xx} > 0$. Therefore, $(-3, 2, -11)$ is a local minimum point. The graph of $z = f(x,y) = x^2 + y^2 + 6x - 4y + 2$ is a paraboloid that opens upward (in the direction of positive z). Its vertex is $(-3, 2, -11)$. This point is also the absolute minimum point over the entire domain.

◆ • ◆ • ◆ • ◆ • ◆

Example 29.2: Find the critical points of $z = g(x,y) = x^4 + y^4$, and classify these points as minima, maxima or saddle.

Solution: The first partial derivatives are $g_x(x,y) = 4x^3$ and $g_y(x,y) = 4y^3$. Setting each to 0, we get $x = 0$ and $y = 0$. Note that $z = g(0,0) = 0$, so that $(0,0,0)$ is the lone critical point.

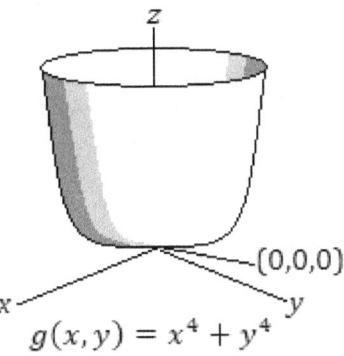

$$g(x,y) = x^4 + y^4$$

The second derivatives are $g_{xx}(x,y) = 12x^2$, $g_{yy}(x,y) = 12y^2$ and $g_{xy}(x,y) = 0$. Using the second derivative test, we have

$$D = g_{xx}(0,0)g_{yy}(0,0) - \left(g_{xy}(0,0)\right)^2$$
$$= [12(0)^2][12(0)^2] - 0$$
$$= 0.$$

The second derivative test yields no useful information. However, note that the cross sections of this surface are $z = x^4$ (when $y = 0$) and $z = y^4$ (when $x = 0$). In each case, the point (0,0) is a minimum, so we can infer that (0,0) is a local minimum point on the surface of $z = x^4 + y^4$. The surface is bowl-shaped, with a flattened bottom, where (0,0,0) is its vertex. Viewing its graph suggests that the point is the absolute minimum, too.

♦ • ♦ • ♦ • ♦ • ♦

Example 29.3: Find the critical points of $z = h(x,y) = |x| + |y|$, and classify these points as minima, maxima or saddle.

Solution: Since $|x| = \begin{cases} -x, & x < 0 \\ x, & x \geq 0 \end{cases}$, then by inspection, $\frac{d}{dx}|x| = \begin{cases} -1, & x < 0 \\ 1, & x > 0 \end{cases}$, where the derivative is not defined (does not exist) at $x = 0$. A similar argument shows that for $|y|$, the derivative does not exist at $y = 0$. Therefore, the point (0,0,0) is a critical point. However, the second derivative test is not applicable. Instead, we can classify the critical point by observing the graph of h, where we see that (0,0,0) is a local and absolute minimum point.

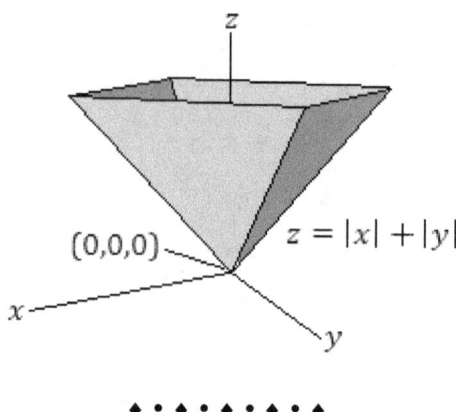

♦ • ♦ • ♦ • ♦ • ♦

Example 29.4: Let $z = f(x, y) = x^3 + y^3 - 3x - 27y + 7$. Find its critical points and classify these points as minima, maxima or neither.

Solution: We find the partial derivatives:

$$f_x(x, y) = 3x^2 - 3$$
$$f_y(x, y) = 3y^2 - 27.$$

These are set equal to 0 and solved for the variable:

$$\begin{matrix} 3x^2 - 3 = 0 \\ 3y^2 - 27 = 0 \end{matrix}, \quad \text{which simplifies as} \quad \begin{matrix} 3(x^2 - 1) = 0 \\ 3(y^2 - 9) = 0. \end{matrix}$$

From the first equation, $x^2 - 1 = 0$, we get $x = 1$ and $x = -1$. From the second equation, $y^2 - 9 = 0$, we get $y = 3$ and $y = -3$. We combine these solutions in all possible ways, and we have four critical points:

$$(1, 3, f(1,3)),$$
$$(1, -3, f(1,-3)),$$
$$(-1, 3, f(-1,3)),$$
$$(-1, -3, f(-1,-3)).$$

The z values are $f(1,3) = -49$, $f(1,-3) = 59$, $f(-1,3) = -45$ and $f(-1,-3) = 63$.

To classify these critical points, use the second derivative test. The second derivatives are

$$f_{xx}(x,y) = 6x, \quad f_{yy}(x,y) = 6y \quad \text{and} \quad f_{xy}(x,y) = 0.$$

Thus, using the formula, we have

$$D = f_{xx}(x,y)f_{yy}(x,y) - \left(f_{xy}(x,y)\right)^2 = (6x)(6y).$$

- When $x=1$ and $y=3$, we have $D = (6)(18)$, which is a positive number. Note that $f_{xx}(1,3)$ is also positive. Thus, the critical point $(1, 3, -49)$ is a local minimum.

- When $x=1$ and $y=-3$, we have $D = (6)(-18)$, which is a negative number. Thus, the critical point $(1, -3, 59)$ is a saddle point.

- When $x=-1$ and $y=3$, we have $D = (-6)(18)$, which is a negative number. Thus, the critical point $(-1, 3, -45)$ is a saddle point.

- When $x = -1$ and $y = -3$, we have $D = (-6)(-18)$, which is a positive number. Note that $f_{xx}(-1,-3)$ is negative. Thus, the critical point $(-1, -3, 63)$ is a local maximum.

When finding D, it's not important to determine its actual value. It's more important to determine its sign. Thus, calculating $(6)(18)$ is not as important as observing that the product of two positive values is positive. Furthermore, by leaving the expression as $(6)(18)$ rather than simplifying it, we can also quickly see that the value 6, representing $f_{xx}(1,3)$, is positive.

The graph of $z = f(x,y) = x^3 + y^3 - 3x - 27y + 7$ is below:

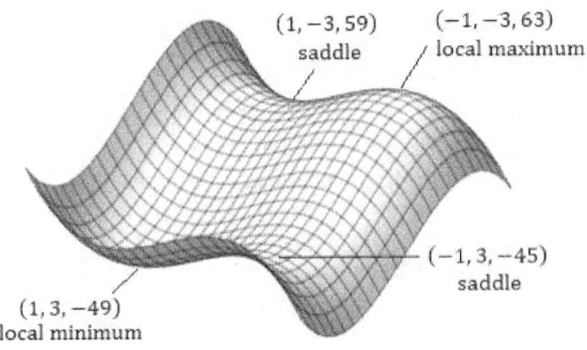

Often, a system must be solved to determine critical points.

Example 29.5: Let $z = g(x,y) = x^2 + y^2 - 4x + y + xy + 3$. Find its critical points and classify these points as minima, maxima or saddle.

Solution: Find the partial derivatives:

$$g_x(x,y) = 2x - 4 + y$$
$$g_y(x,y) = 2y + 1 + x.$$

These are set equal to 0 and a linear system in two variables results:

$$\begin{array}{c} 2x - 4 + y = 0 \\ 2y + 1 + x = 0 \end{array}, \text{ which simplifies as } \begin{array}{c} 2x + y = 4 \\ x + 2y = -1. \end{array}$$

Multiply the second equation by –2:

$$2x + y = 4$$
$$-2x - 4y = 2.$$

Summing, we have $-3y = 6$, so that $y = -2$. Back substituting, we find that $x = 3$. Thus, $(3, -2, g(3, -2))$ is the critical point. The z value is $g(3, -2) = -4$.

The second derivatives are

$$g_{xx}(x,y) = 2, \quad g_{yy}(x,y) = 2 \text{ and } g_{xy}(x,y) = 1.$$

Using the second derivative test, we have

$$D = g_{xx}(3,-2)g_{yy}(3,-2) - \left(g_{xy}(3,-2)\right)^2$$
$$= (2)(2) - 1^2$$
$$= 3.$$

Since $D > 0$ and $g_{xx}(3,-2) > 0$, this point is a local minimum. The graph is a paraboloid with vertex $(3, -2, -4)$ opening upward. The domain is the entire xy-plane. Thus, the point is also an absolute minimum.

Example 29.6: Let $z = f(x, y) = x^3 - y^3 - 2x^2 + xy + 3y$. Find its critical points and classify these points as minima, maxima or saddle.

Solution: The partial derivatives are

$$f_x(x, y) = 3x^2 - 4x + y$$
$$f_y(x, y) = -3y^2 + x + 3.$$

Setting these to zero, develop a non-linear system:

$$3x^2 - 4x + y = 0$$
$$-3y^2 + x + 3 = 0.$$

Unlike the previous example, we cannot use the elimination method. Instead, we use substitution. In the first equation, solve for y:

$$y = 4x - 3x^2.$$

This is substituted into the second equation, then simplified:

$$-3(4x - 3x^2)^2 + x + 3 = 0$$
$$-3(16x^2 - 24x^3 + 9x^4) + x + 3 = 0$$
$$-27x^4 + 72x^3 - 48x^2 + x + 3 = 0.$$

Using a graphing calculator, we find four roots to this quartic equation. They are

$$x \approx -0.21, \quad x \approx 0.36, \quad x \approx 0.92 \quad \text{and} \quad x \approx 1.59.$$

For each x value above, use the equation $y = 4x - 3x^2$ to find the corresponding y value. The z-values are then found by evaluating f at each x and y value. There are four critical points:

$$(-0.21, -0.97, -1.89), \quad (0.364, 1.06, 2.19),$$
$$(0.92, 1.14, 2.073) \quad \text{and} \quad (1.59, -1.24, -4.82).$$

Note that the z-values alone do not provide enough information to classify these points as minimum, maximum or neither. We must use the second derivative test. The second derivatives are

$$f_{xx}(x, y) = 6x - 4, \quad f_{yy}(x, y) = -6y, \quad f_{xy}(x, y) = 1.$$

Thus, we have

$$D = f_{xx}(x_c, y_c) f_{yy}(x_c, y_c) - \left(f_{xy}(x_c, y_x)\right)^2$$
$$= (6x_c - 4)(-6y_c) - (1)^2,$$

where x_c and y_c are the input values of a critical point.

Each critical point is evaluated into the second derivative test formula. Only x and y are used, z is not:

- At $(-0.21, -0.97, -1.89)$, we get $D = (6(-0.21) - 4)(-6(-0.97)) - 1 = -31.48$. Since D is negative, the point $(-0.21, -0.97, -1.89)$ is a saddle point.

- At $(0.36, 1.06, 2.19)$, we get $D = (6(0.36) - 4)(-6(1.06)) - 1 = 10.54$. Since D is positive and since $f_{xx}(0.36, 1.06)$ is negative (as is $f_{yy}(0.36, 1.06)$), the point $(0.36, 1.06, 2.19)$ is a local maximum.

- At $(0.92, 1.14, 2.07)$, we get $D = (6(0.92) - 4)(-6(1.14)) - 1 = -11.37$. Since D is negative, the point $(0.92, 1.14, 2.07)$ is a saddle point.

- At $(1.59, -1.24, -4.82)$, we get $D = (6(1.59) - 4)(-6(-1.24)) - 1 = 40.14$. Since D is positive and since $f_{xx}(1.59, -1.24)$ is positive (as is $f_{yy}(1.59, -1.24)$), the point $(1.59, -1.24, -4.82)$ is a local minimum.

Note that the z-values play no role in classifying these points.

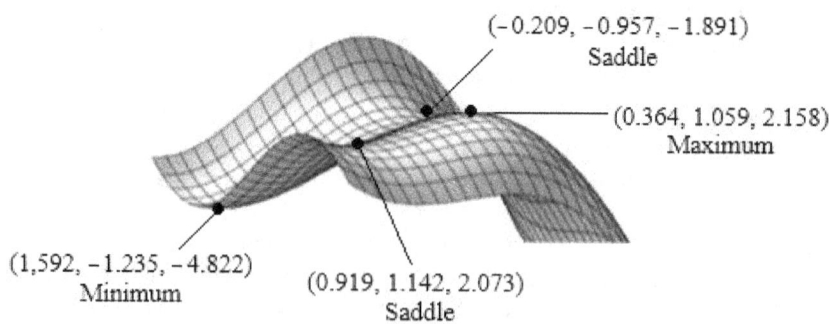

30. Constrained Optimization

The graph of $z = f(x, y)$ is represented by a surface in R^3. Normally, x and y are chosen independently of one another so that one may "roam" over the entire surface of f (within any domain restrictions on x and y). Determining minimum or maximum points on f under this circumstance is called unconstrained optimization.

If x and y are related to one another by an equation, then only one of the variables can be independent. In such a case, we may determine a minimum or maximum point on the surface of f subject to the constraint placed on x and y. This is called **constrained optimization**. Such constraints are usually written where x and y are combined implicitly, $g(x, y) = c$.

Suppose you are hiking on a hill. If there is no restriction on where you may walk, then you are "unconstrained" and you may seek the hill's maximum point. However, if you are constrained to a hiking path, then it is possible to determine a maximum point on the hill, but only that part along the hiking path.

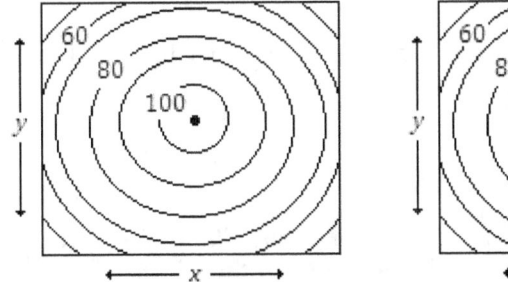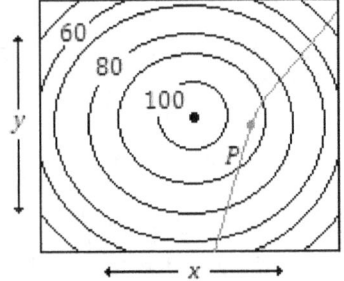

(**Left**) **Unconstrained optimization:** The maximum point of this hill is marked by a black dot, and is roughly $z = 105$. (**Right**) **Constrained optimization:** The highest point on the hill, subject to the constraint of staying on path P, is marked by a gray dot, and is roughly $z = 93$.

The two common ways of solving constrained optimization problems is through substitution, or a process called The Method of Lagrange Multipliers (which is discussed in a later section). Using substitution, the biggest challenge is the amount of algebra that may occur.

◆ • ◆ • • ◆ • • ◆

Example 30.1: Find the minimum or maximum point on the surface of $z = f(x, y) = x^2 + y^2$ subject to the constraint $-3x + y = 2$.

Solution: The surface of f is a paraboloid with its vertex $(0,0,0)$ at the origin, opening in the positive z direction (or "up"). Its unconstrained minimum point is $(0,0,0)$. There is no maximum point on this surface.

Now, note that with the constraint $y = 3x + 2$ in place, x and y are no longer independent variables. Once a value for x is chosen, then y is determined. We are now restricted to this "path" on the surface of f.

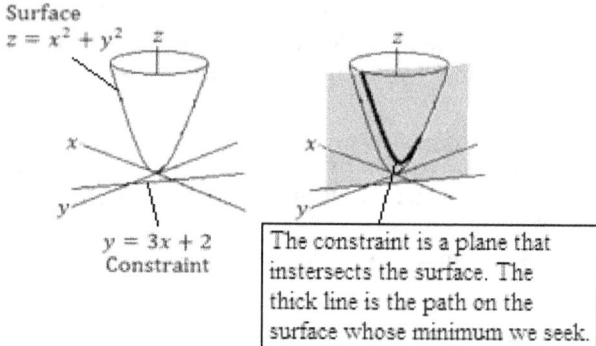

| $y = 3x + 2$ Constraint | The constraint is a plane that instersects the surface. The thick line is the path on the surface whose minimum we seek. |

To find the minimum or maximum point on this paraboloid subject to the constraint $y = 3x + 2$, substitute the constraint into the function f and simplify:

$$f(x, 3x + 2) = x^2 + (3x + 2)^2 = 10x^2 + 12x + 4.$$

Differentiating, we have $f'(x) = 20x + 12$, and to find the critical value of x, we set $f'(x) = 0$:

$$20x + 12 = 0 \quad \text{which gives} \quad x = -\frac{3}{5}.$$

Observe that $f(x) = 10x^2 + 12x + 4$ is a parabola in R^2 that opens upward. Thus, the critical value for x will correspond to the minimum point on this parabola. Find y by substitution into the constraint:

$$y = 3\left(-\frac{3}{5}\right) + 2 = \frac{1}{5}.$$

Lastly, we find z:

$$z = f\left(-\frac{3}{5}, \frac{1}{5}\right) = \left(-\frac{3}{5}\right)^2 + \left(\frac{1}{5}\right)^2 = \frac{10}{25} = \frac{2}{5}.$$

Thus, the minimum *point* on the surface of f subject to the constraint $y = 3x + 2$ is $\left(-\frac{3}{5}, \frac{1}{5}, \frac{2}{5}\right)$. The minimum *value* of z on the surface of f subject to the constraint $y = 3x + 2$ is $= \frac{2}{5}$.

Example 30.2: Find the minimum and maximum points on the surface of $z = f(x, y) = xy$ subject to the constraint $x^2 + y^2 = 4$.

Solution: Solving for one of the variables in the constraint, we have $y = \pm\sqrt{4 - x^2}$, which implies that $-2 \leq x \leq 2$. This is substituted into the function $f(x, y) = xy$:

$$f\left(x, \sqrt{4 - x^2}\right) = x\sqrt{4 - x^2} \quad \text{or} \quad f\left(x, -\sqrt{4 - x^2}\right) = -x\sqrt{4 - x^2}.$$

Differentiate. We will start with $f(x) = x\sqrt{4 - x^2}$, using the product rule and combining into a single expression, noting any restrictions:

$$f'(x) = x\left(\frac{-2x}{2\sqrt{4 - x^2}}\right) + \sqrt{4 - x^2} = \frac{4 - 2x^2}{\sqrt{4 - x^2}}, \quad (x \neq \pm 2).$$

Setting this equal to 0, we can find critical values for x. Note that this reduces to showing where the numerator, $4 - 2x^2$, is 0:

$$4 - 2x^2 = 0 \quad \text{gives} \quad x = \pm\sqrt{2}.$$

Substituting this back into the constraint, each x value results in two y values:

$$x = \sqrt{2}: \ \left(\sqrt{2}\right)^2 + y^2 = 4 \quad \text{gives} \quad y = \pm\sqrt{2},$$
$$x = -\sqrt{2}: \ \left(-\sqrt{2}\right)^2 + y^2 = 4 \quad \text{gives} \quad y = \pm\sqrt{2}.$$

In this example, repeating the above steps with $f(x) = -x\sqrt{4 - x^2}$ results in the same critical values for x and y (you verify). Thus, we have four critical points, where the z-value is found by evaluating at the given x and y values:

$$\left(\sqrt{2}, \sqrt{2}, 2\right), \quad \left(\sqrt{2}, -\sqrt{2}, -2\right), \quad \left(-\sqrt{2}, \sqrt{2}, -2\right), \quad \left(-\sqrt{2}, -\sqrt{2}, 2\right).$$

By inspection, the minimum points on $f(x, y) = xy$ subject to the constraint $x^2 + y^2 = 4$ are $\left(\sqrt{2}, -\sqrt{2}, -2\right)$ and $\left(-\sqrt{2}, \sqrt{2}, -2\right)$, and the maximum points are $\left(\sqrt{2}, \sqrt{2}, 2\right)$ and $\left(-\sqrt{2}, -\sqrt{2}, 2\right)$.

This is plausible: the surface f is positive in quadrants 1 and 3, negative in quadrants 2 and 4, and symmetric across the origin. The constraint is also symmetric across the origin. Because the constraint is a closed loop, it must achieve both minimum and maximum values.

The values for which the derivative is not defined, $x = \pm 2$, both imply that $y = 0$, and that $f(2,0) = 0$ and $f(-2,0) = 0$. These are clearly neither minimum nor maximum points, and thus can be ignored.

Example 30.3: Find the extreme points of $f(x, y) = x^3 + y^2 + 2xy$ subject to the constraint $x = y^2 - 1$.

Solution: Since x is already isolated in the constraint, substitute it into the function:

$$f(y^2 - 1, y) = (y^2 - 1)^3 + y^2 + 2(y^2 - 1)y.$$

Simplifying, we have a function in one variable:

$$f(y) = y^6 - 3y^4 + 2y^3 + 4y^2 - 2y - 1.$$

Differentiating, we have

$$f'(y) = 6y^5 - 12y^3 + 6y^2 + 8y - 2.$$

We set this to 0 to determine critical values for y. The challenge is that we have to somehow glean solutions from this 5th-degree polynomial. In this case, we graph $f'(y)$ and note where it crosses the input axis (if using a graphing calculator, the y variable will be renamed x in the calculator. Just keep track of this). The critical y values are

$$y \approx -1.377, \quad y \approx -0.831 \quad \text{and} \quad y \approx 0.228.$$

Recall that x and y are related by the equation $x = y^2 - 1$. Thus, the corresponding x values are:

$$y = -1.377: \ x = (-1.377)^2 - 1 = 0.896,$$
$$y = -0.831: \ x = (-0.831)^2 - 1 = -0.309,$$
$$y = 0.228: \quad x = (0.228)^2 - 1 = -0.948.$$

(From this point forward, it's accepted that all x, y and z values will be approximated values.)

The full coordinates for the critical points are

$$(0.896, -1.377, 0.148), \ (-0.309, -0.831, 1.175)$$
$$\text{and } (-0.948, 0.228, -1.232).$$

By inspection, the minimum point on the surface $f(x, y) = x^3 + y^2 + 2xy$ subject to the constraint $x = y^2 - 1$ is $(-0.948, 0.228, -1.232)$. However, the other two points are neither minimum nor maximum points. Note that as y increases in value, so does x, and so will $z = f(x, y)$. Thus, an ant walking on this parabolic path on the surface of f can get no lower than $z = -1.232$, but can achieve as high a z value as it desires, assuming it walks far enough, all the while staying on the path.

Example 30.4: Consider the portion of the plane $2x + 4y + 5z = 20$ in the first octant. Find the point on the plane closest to the origin.

Solution: The point on the plane closest to the origin will lie on a line orthogonal to the plane. Let (x, y, z) be a point on the plane, so the distance d between this point and the origin $(0,0,0)$ is

$$d(x, y, z) = \sqrt{(x - 0)^2 + (y - 0)^2 + (z - 0)^2} = \sqrt{x^2 + y^2 + z^2}.$$

However, note that not all variables are independent—they are constrained to one another by the plane's equation. We can isolate one of the variables in the plane. For example, $z = 4 - \frac{2}{5}x - \frac{4}{5}y$. Thus, d can be written as a function of x and y only, and the radicand is expanded:

$$d(x, y) = \sqrt{x^2 + y^2 + \left(4 - \frac{2}{5}x - \frac{4}{5}y\right)^2}$$

$$= \sqrt{\frac{29}{25}x^2 + \frac{41}{25}y^2 + \frac{16}{25}xy - \frac{16}{5}x - \frac{32}{5}y + 16}.$$

Variables x and y also obey another constraint: both must be non-negative. This will ensure that z is also non-negative.

Taking partial derivatives and simplifying, we have

$$d_x = \frac{\frac{29}{25}x + \frac{8}{25}y - \frac{8}{5}}{\sqrt{\frac{29}{25}x^2 + \frac{41}{25}y^2 + \frac{16}{25}xy - \frac{16}{5}x - \frac{32}{5}y + 16}},$$

$$d_y = \frac{\frac{8}{25}x + \frac{41}{25}y - \frac{16}{5}}{\sqrt{\frac{29}{25}x^2 + \frac{41}{25}y^2 + \frac{16}{25}xy - \frac{16}{5}x - \frac{32}{5}y + 16}}.$$

When set to 0, the denominators can be ignored. Thus, only the numerators are considered, and we have

$$\frac{29}{25}x + \frac{8}{25}y - \frac{8}{5} = 0 \quad \text{and} \quad \frac{8}{25}x + \frac{41}{25}y - \frac{16}{5} = 0.$$

Placing the constants to the right of the equality and multiplying by 25 to clear fractions, we then have a simplified system in two variables:

$$29x + 8y = 40$$
$$8x + 41y = 80.$$

Using any method (such as elimination) to solve this system, we find that $x = \frac{8}{9}$ and $y = \frac{16}{9}$, and substituting these into the plane's equation, we have $z = \frac{20}{9}$. Thus, the point $\left(\frac{8}{9}, \frac{16}{9}, \frac{20}{9}\right)$ is the point on the plane closest to the origin.

We can check this by using normal vectors and techniques used when discussing planes and lines:

Note that a vector orthogonal to the plane is $\mathbf{v} = \langle 2,4,5 \rangle$ and that a line passing through the origin and parallel to this vector (thus, the line is orthogonal to the plane) has the equation

$$\langle 0,0,0 \rangle + t\langle 2,4,5 \rangle = \langle 2t, 4t, 5t \rangle.$$

We determine where this line intersects the plane (see Example 13.4):

$$2(2t) + 4(4t) + 5(5t) = 20$$
$$4t + 16t + 25t = 20$$
$$45t = 20$$
$$t = \frac{4}{9}.$$

Substituting $t = \frac{4}{9}$ into $\langle 2t, 4t, 5t \rangle$ gives the vector $\langle 2\left(\frac{4}{9}\right), 4\left(\frac{4}{9}\right), 5\left(\frac{4}{9}\right)\rangle = \langle \frac{8}{9}, \frac{16}{9}, \frac{20}{9} \rangle$. Recall that this vector's foot is at the origin, so its head corresponds to the actual point, $\left(\frac{8}{9}, \frac{16}{9}, \frac{20}{9}\right)$.

An alternative method that involves less algebra is shown in Example 32.5.

Example 30.5: Consider the portion of the plane $2x + 4y + 5z = 20$ in the first octant. A rectangular box is situated with one corner at the origin and its opposite corner on the plane so that the box's edges lie along (or are parallel to) the x-axis, y-axis or z-axis. Find the largest possible volume of such a box, keeping the box to within the first octant.

Solution: The volume of a box with edges of length x, y and z is $V(x, y, z) = xyz$. However, as in the previous example, z is dependent on x and y by the equation $z = 4 - \frac{2}{5}x - \frac{4}{5}y$, so that we have

$$V(x, y) = xy\left(4 - \frac{2}{5}x - \frac{4}{5}y\right) = 4xy - \frac{2}{5}x^2y - \frac{4}{5}xy^2.$$

The partial derivatives are

$$V_x = 4y - \frac{4}{5}xy - \frac{4}{5}y^2 \quad \text{and} \quad V_y = 4x - \frac{2}{5}x^2 - \frac{8}{5}xy.$$

These are set equal to zero and multiplied by 5 to clear fractions. Note that we can factor y from V_x and x from V_y:

$$V_x = 0: \quad y(20 - 4x - 4y) = 0$$
$$V_y = 0: \quad x(20 - 2x - 8y) = 0.$$

One solution to this system is $x = 0$ and $y = 0$, but this can be dismissed since it would result in a box with volume 0, a minimum volume, but not a maximum as we seek. Thus, we examine the other factors:

$$20 - 4x - 4y = 0$$
$$20 - 2x - 8y = 0.$$

This system is simplified slightly:

$$x + y = 5$$
$$x + 4y = 10.$$

The solution is $x = \frac{10}{3}$ and $y = \frac{5}{3}$, so that $z = \frac{4}{3}$. If we so desire, we can use the second derivative test for two-variable functions to show that this gives a maximum volume.

Thus, the largest possible box will have a volume of $\left(\frac{10}{3}\right)\left(\frac{5}{3}\right)\left(\frac{4}{3}\right) = \frac{200}{27} \approx 7.407$ cubic units.

An alternative method is shown in Example 32.6.

31. Constrained Optimization: The Extreme Value Theorem

Suppose a continuous function $z = f(x, y)$ in R^3 has constraints on the independent variables x and y in such a way that the constraint region (when viewed as a region in the xy-plane) is a closed and bounded subset of the plane. By *closed*, we mean that the region includes its boundaries, and by *bounded*, the region is of finite area with no asymptotic "tails" trending to infinity.

Under these conditions, it is guaranteed that both absolute minimum and absolute maximum points must exist on the surface representing f subject to the constraints. This is called the **extreme value theorem** (EVT).

◆ • ◆ • ◆ • ◆ • ◆

Example 31.1: Let $z = f(x, y) = x^2 + y^2 - 8x - 7y + 3xy$. Find its extreme values such that $x \geq 0$, $y \geq 0$ and $x + 2y \leq 6$.

Solution: First, sketch the constraint region in the xy-plane. Recall that $x = 0$ represents the y-axis and that $x \geq 0$ suggests shading to the right of the y-axis. Use similar logic when sketching the other two boundary lines. Note that we have a triangle. It is closed (includes its boundaries) and bounded (of finite area). We are not graphing f itself. We will handle f purely analytically.

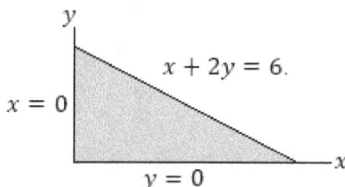

We check for possible critical points contained within the region. Using routine optimization techniques, we differentiate with respect to x and with respect to y:

$$f_x(x, y) = 2x - 8 + 3y$$
$$f_y(x, y) = 2y - 7 + 3x.$$

Setting these equations equal to 0, we solve a system:

$$\begin{array}{l} 2x - 8 + 3y = 0 \\ 2y - 7 + 3x = 0, \end{array} \text{ which simplifies to } \begin{array}{l} 2x + 3y = 8 \\ 3x + 2y = 7. \end{array}$$

The solution of this system is $x = 1$ and $y = 2$. Note that this point satisfies all three constraints simultaneously. This is a valid critical point.

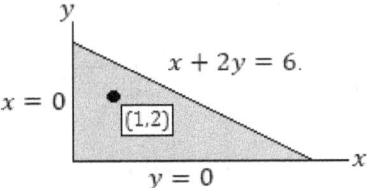

Next, we identify all vertices (corners) of the region. These will be critical points too.

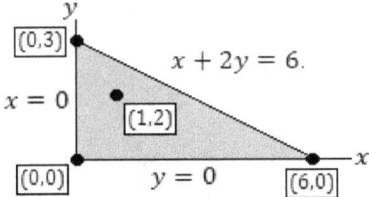

Then we check for critical points along each boundary, one at a time.

- For $x = 0$ (the y-axis), where $0 \leq y \leq 3$, we substitute into the function:

$$f(0, y) = (0)^2 + y^2 - 8(0) - 7y + 3(0)y$$
$$f(y) = y^2 - 7y. \quad \text{(simplified)}$$

Differentiating, we have $f'(y) = 2y - 7$, and when set equal to 0, we find that $y = \frac{7}{2}$. However, this value is outside the range $0 \leq y \leq 3$, so it is ignored.

- For $y = 0$ (the x-axis), where $0 \leq x \leq 6$, we substitute into the function:

$$f(x, 0) = x^2 + (0)^2 - 8x - 7(0) + 3x(0)$$
$$f(x) = x^2 - 8x. \quad \text{(simplified)}$$

Differentiating, we have $f'(x) = 2x - 8$, and when set equal to 0, we find that $x = 4$ is a critical value. This is inside the range $0 \leq x \leq 6$, so it is included.

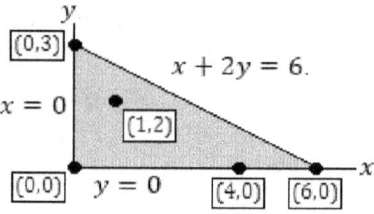

- For $x + 2y = 6$, we solve for a convenient variable and substitute. Let's use $x = 6 - 2y$.

$$f(6 - 2y, y) = (6 - 2y)^2 + y^2 - 8(6 - 2y) - 7y - 3(6 - 2y)y$$
$$f(y) = 36 - 24y + 4y^2 + y^2 - 48 + 16y - 7y - 18y + 6y^2$$
$$f(y) = 11y^2 - 33y - 12. \text{ (simplified)}$$

Differentiating, we have $f'(y) = 22y - 33$, and when set equal to 0, we find that $y = \frac{3}{2}$ is a critical value. Since $x = 6 - 2y$, we have $x = 6 - 2\left(\frac{3}{2}\right) = 3$. These values are within the respective ranges for the x and y variables, so this is also a critical point.

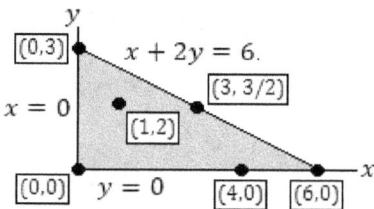

We have six critical points. These are evaluated into the function:

$$z = f(0,0) = (0)^2 + (0)^2 - 8(0) - 7(0) + 3(0)(0) = 0,$$
$$z = f(4,0) = (4)^2 + (0)^2 - 8(4) - 7(0) + 3(4)(0) = -16,$$
$$z = f(6,0) = (6)^2 + (0)^2 - 8(6) - 7(0) + 3(6)(0) = -12,$$
$$z = f(0,3) = (0)^2 + (3)^2 - 8(0) - 7(3) + 3(0)(3) = -12,$$
$$z = f(1,2) = (1)^2 + (2)^2 - 8(1) - 7(2) + 3(1)(2) = -11,$$
$$z = f(3,3/2) = (3)^2 + (3/2)^2 - 8(3) - 7(3/2) + 3(3)(3/2) = -39/4.$$

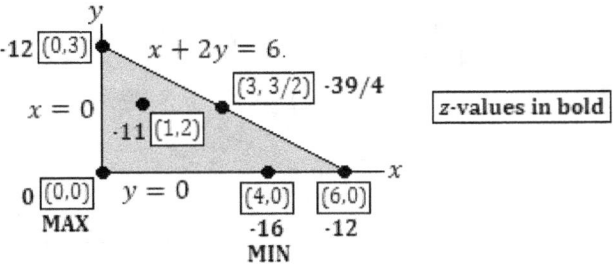

By inspection, the absolute maximum value on the surface of f subject to the constraints is $z = 0$, and it occurs at $(0,0,0)$, the absolute maximum point. The absolute minimum value is $z = -16$ and occurs at the absolute minimum point $(4,0,-16)$. The other points are then ignored.

♦ • ♦ • ♦ • ♦ • ♦

In the previous example and the one that follows, it is important to remember that we are searching for highest and lowest points on a surface. In the images in each example, we tend to concentrate on the projection of the region onto the xy-plane and track our calculations on this projection. However, if these regions are projected back to the surface, it will conform to the surface. In a sense, it is similar to fencing that encloses a yard on hilly terrain. The extreme points may lie within the yard, along one of the fences, or at a corner of fences.

♦ • ♦ • ♦ • ♦ • ♦

Example 31.2: Let $z = f(x,y) = (x-1)^2 + (y-1)^2$. Find its extreme values such that $y \geq 0$ and $x^2 + y^2 \leq 9$.

Solution: Sketch the constraint region in the xy-plane:

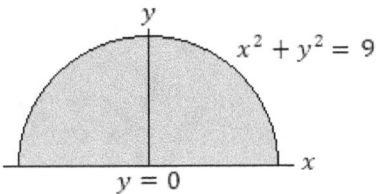

Note that $f(x,y) = (x-1)^2 + (y-1)^2$ is the paraboloid $z = x^2 + y^2$ that has been shifted one unit in the positive x direction and one unit in the positive y direction (it is not graphed). Its vertex is at $(1,1,0)$. When $x = 1$ and $y = 1$, it is contained within the region as defined by the constraints. This is a critical point.

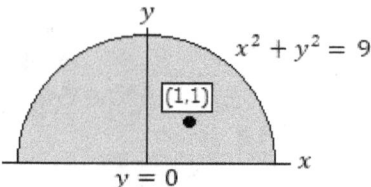

The points at which the circle meets the x-axis are critical points too.

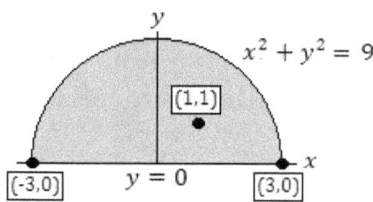

Along the x-axis, we let $y = 0$, noting that $-3 \leq x \leq 3$, and substitute into the function f:

$$f(x, 0) = (x - 1)^2 + (0 - 1)^2$$
$$f(x) = x^2 - 2x + 2. \text{ (simplified)}$$

Differentiating, we have $f'(x) = 2x - 2$, and setting this equal to 0, we have that $x = 1$. This is within the bounds of x, so it is a critical value, while (1,0) is a critical point.

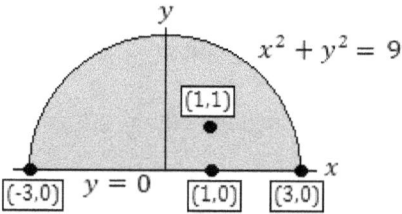

For the circular boundary, we rewrite it as $y = \sqrt{9 - x^2}$, with $-3 \leq x \leq 3$, noting that we need only the positive root. This is substituted into f:

$$f\left(x, \sqrt{9 - x^2}\right) = (x - 1)^2 + \left(\sqrt{9 - x^2} - 1\right)^2$$
$$f(x) = x^2 - 2x + 1 + 9 - x^2 - 2\sqrt{9 - x^2} + 1$$
$$f(x) = -2x + 11 - 2\sqrt{9 - x^2}.$$

Differentiating, we have

$$f'(x) = -2 + \frac{2x}{\sqrt{9-x^2}}, \quad \text{for } -3 < x < 3.$$

This is set equal to 0, and simplified:

$$-2 + \frac{2x}{\sqrt{9-x^2}} = 0$$
$$\frac{2x}{\sqrt{9-x^2}} = 2$$
$$x = \sqrt{9-x^2}$$
$$x^2 = \left(\sqrt{9-x^2}\right)^2$$
$$x^2 = (9-x^2)$$
$$2x^2 = 9$$
$$x^2 = \frac{9}{2}.$$

(Note that the values $x = \pm 3$ have already been accounted for previously.)

Thus, we have $x = 3/\sqrt{2}$ and $x = -3/\sqrt{2}$. These are approximately $x = \pm 2.12$. They fall within the interval $-3 < x < 3$. Since $y = \sqrt{9-x^2}$ along the boundary, when $x = 3/\sqrt{2}$, we have $y = \sqrt{9 - (3/\sqrt{2})^2} = \sqrt{9 - (9/2)} = \sqrt{9/2} = 3/\sqrt{2}$. Similarly, when $x = -3/\sqrt{2}$, then $y = 3/\sqrt{2}$. These are also critical points.

We also consider the point (0,3), since the variable y will be constrained within the interval $0 \leq y \leq 3$. Thus, we have a total of seven critical points.

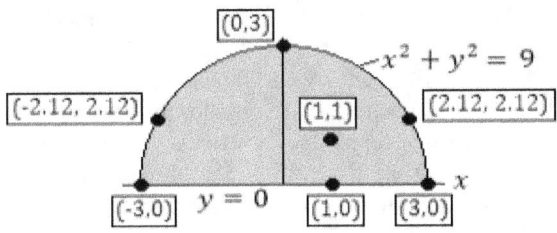

These are evaluated into the function:

$$z = f(1,0) = ((1) - 1)^2 + ((0) - 1)^2 = 1,$$
$$z = f(3,0) = ((3) - 1)^2 + ((0) - 1)^2 = 5,$$
$$z = f(-3,0) = ((-3) - 1)^2 + ((0) - 1)^2 = 17,$$
$$z = f(1,1) = ((1) - 1)^2 + ((1) - 1)^2 = 0,$$
$$z = f(-3/\sqrt{2}, 3/\sqrt{2}) = ((-3/\sqrt{2}) - 1)^2 + ((3/\sqrt{2}) - 1)^2 \approx 11,$$
$$z = f(3/\sqrt{2}, 3/\sqrt{2}) = ((3/\sqrt{2}) - 1)^2 + ((3/\sqrt{2}) - 1)^2 \approx 2.515,$$
$$z = f(0,3) = ((0) - 1)^2 + ((3) - 1)^2 = 5.$$

The absolute maximum point is $(-3, 0, 17)$, and the absolute minimum point is $(1, 1, 0)$. The rest are ignored.

◆ ◆ ◆ ◆ ◆ ◆ ◆ ◆ ◆

32. Method of Lagrange Multipliers

The Method of Lagrange Multipliers is a generalized approach to solving constrained optimization problems. Assume that we are seeking to optimize a function $z = f(x, y)$ subject to a "path" constraint defined implicitly by $g(x, y) = c$. The process usually follows these steps:

1. Define a function $L(x, y, \lambda) = f(x, y) - \lambda(g(x, y) - c)$.
2. Find the partial derivatives L_x, L_y and L_λ. Note that $L_\lambda = -g(x, y) + c$.
3. Set these partial derivatives to 0. Note that $L_\lambda = 0$ is the same as $g(x, y) = c$, the original path constraint. Also note any restrictions on x and y, as it may be necessary to consider locations where the derivative fails to exist.
4. Isolate the λ in the equations $L_x = 0$ and $L_y = 0$, then equate the two expressions. This will "drop out" the λ, leaving an equation in x and y only. If possible, isolate x or y.
5. Substitute the result from step 4 into the equation $g(x, y) = c$, which will now be a single-variable equation. Solve for the remaining variable.
6. Back substitute to find corresponding values for the other variables.
7. Compare z values. The smallest will be a minimum, the largest a maximum. If there is just one z value, then other observations, such as cross-sections, may be needed to determine whether the point is a minimum or maximum.

Reminder: the z values by themselves do not provide any indication whether the point is a minimum, maximum or saddle.

The following examples illustrate possible situations that may occur.

Example 32.1: Find the minimum value of $z = f(x,y) = x^2 + y^2 - 2x - 2y$ subject to the constraint $x + 2y = 4$.

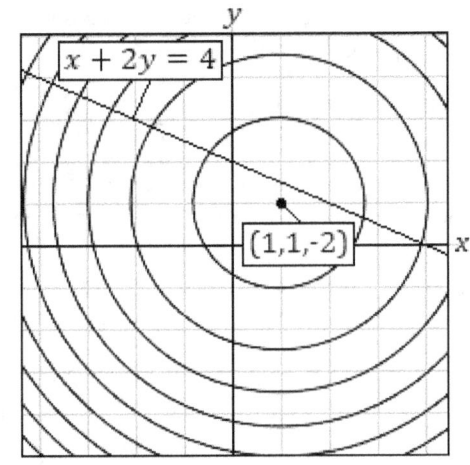

Solution: To the right is a contour map in R^2 of the surface defined by f, and the constraint $x + 2y = 4$ shown as a line. The actual surface is a paraboloid that opens up and has a minimum point at $(1,1,-2)$, its vertex. The path, when conformed to the surface, is a cross-section of the paraboloid, itself a parabola. Thus, by inspecting the geometry of the problem, the extreme point on this path/parabola will be a minimum point.

Note that the path is slightly off-set from the vertex. It is reasonable to assume that the lowest point on the constraint path will be near the vertex, but clearly cannot be at the paraboloid's vertex.

First, create a new function L, clearing parentheses at the end:

$$L(x,y,\lambda) = f(x,y) - \lambda(g(x,y) - c)$$
$$L(x,y,\lambda) = x^2 + y^2 - 2x - 2y - \lambda(x + 2y - 4)$$
$$L(x,y,\lambda) = x^2 + y^2 - 2x - 2y - \lambda x - 2\lambda y + 4\lambda.$$

Next, find its partial derivatives:

$$L_x = 2x - 2 - \lambda$$
$$L_y = 2y - 2 - 2\lambda$$
$$L_\lambda = -x - 2y + 4.$$

Now, set these to 0:

$$2x - 2 - \lambda = 0 \quad (1)$$
$$2y - 2 - 2\lambda = 0 \quad (2)$$
$$-x - 2y + 4 = 0. \quad (3)$$

In equations **(1)** and **(2)**, isolate the λ:

$$\lambda = 2x - 2 \quad \text{and} \quad \lambda = y - 1.$$

There are no restrictions on x or y. Now, equate and simplify. Note that λ is no longer present.

$$2x - 2 = y - 1, \quad \text{which gives} \quad y = 2x - 1.$$

Note that equation **(3)** from above is the same as the constraint $x + 2y = 4$. Substitute the equation $y = 2x - 1$ into the simplified form of equation **(3)**, and solve for x:

$$x + 2(2x - 1) = 4$$
$$5x - 2 = 4$$
$$5x = 6$$
$$x = \frac{6}{5}.$$

Find y by substituting $x = \frac{6}{5}$ into the equation $y = 2x - 1$:

$$y = 2\left(\frac{6}{5}\right) - 1 = \frac{7}{5}.$$

Lastly, find z using the original function f:

$$f\left(\frac{6}{5},\frac{7}{5}\right) = \left(\frac{6}{5}\right)^2 + \left(\frac{7}{5}\right)^2 - 2\left(\frac{6}{5}\right) - 2\left(\frac{7}{5}\right) = -\frac{8}{5}.$$

The minimum point of $z = f(x,y) = x^2 + y^2 - 2x - 2y$ subject to the constraint $x + 2y = 4$ is

$$\left(\frac{6}{5},\frac{7}{5},-\frac{8}{5}\right).$$

This seems to agree with our assumption that it would be "close" to the surface's minimum at $(1,1,-2)$, its component values each a little higher than those of the vertex.

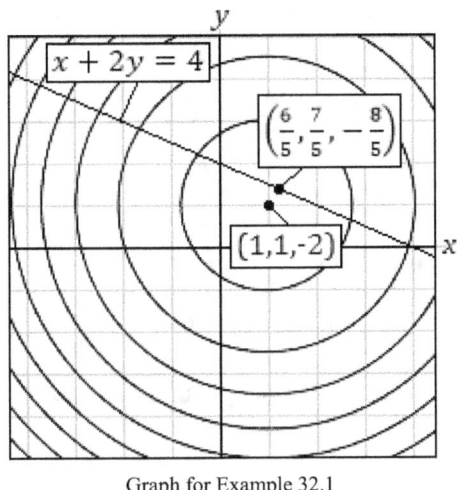

Graph for Example 32.1

◆ ◆ ◆ ◆ ◆ ◆ ◆ ◆ ◆

Example 32.2: Let $z = f(x, y) = x^2 + y^2 - 2x - 2y$. Find the minimum value of f subject to the constraint $x^2 + y^2 = 4$.

Solution: This is the same surface as in the previous example. However, the constraint path is a circle of radius 2 (as viewed on the xy-plane). When conformed to the surface f, the path will rise and fall along with the surface. Observing the path (in bold) in relation to the contours, we can estimate where the path's lowest point may be, and where its highest point may be:

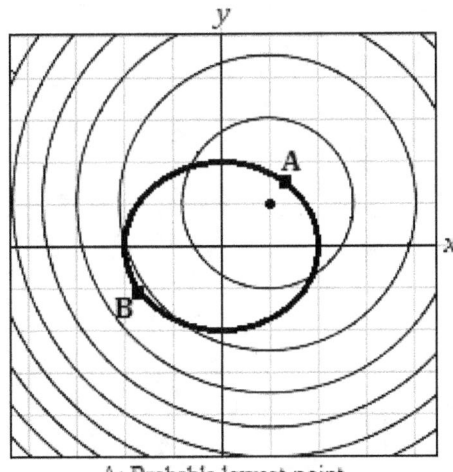

A: Probable lowest point
B: Probable highest point

Using the Method of Lagrange Multipliers, we start by building function L:

$$L(x, y, \lambda) = x^2 + y^2 - 2x - 2y - \lambda(x^2 + y^2 - 4)$$
$$L(x, y, \lambda) = x^2 + y^2 - 2x - 2y - \lambda x^2 - \lambda y^2 + 4\lambda. \quad \text{(Simplified)}$$

Now find the partial derivatives:

$$L_x = 2x - 2 - 2\lambda x$$
$$L_y = 2y - 2 - 2\lambda y$$
$$L_\lambda = -x^2 - y^2 + 4.$$

Set each partial derivative to 0. Again, note that $L_\lambda = 0$ (Equation **(3)**) is the constraint path:

$$2x - 2 - 2\lambda x = 0 \quad \textbf{(1)}$$
$$2y - 2 - 2\lambda y = 0 \quad \textbf{(2)}$$
$$x^2 + y^2 = 4. \quad \textbf{(3)}$$

Isolate λ in equations **(1)** and **(2)**, then equate. Note any restrictions on the variables:

$$\lambda = \frac{x-1}{x} \quad \text{and} \quad \lambda = \frac{y-1}{y}, \quad \text{so that} \quad \frac{x-1}{x} = \frac{y-1}{y} \quad (x, y \neq 0).$$

Clearing fractions, we have

$$y(x - 1) = x(y - 1)$$
$$xy - y = xy - x$$
$$y = x.$$

Substitute $y = x$ into **(3)**:

$$x^2 + x^2 = 4$$
$$2x^2 = 4$$
$$x^2 = 2$$
$$x = \pm\sqrt{2}.$$

Since $y = x$, we have $y = \sqrt{2}$ when $x = \sqrt{2}$, and $y = -\sqrt{2}$ when $x = -\sqrt{2}$. There are two critical points:

$$\left(\sqrt{2}, \sqrt{2}, f(\sqrt{2}, \sqrt{2})\right) \quad \text{and} \quad \left(-\sqrt{2}, -\sqrt{2}, f(-\sqrt{2}, -\sqrt{2})\right).$$

We now evaluate the function at each of these x and y values:

$$f(\sqrt{2},\sqrt{2}) = (\sqrt{2})^2 + (\sqrt{2})^2 - 2\sqrt{2} - 2\sqrt{2} = 4 - 4\sqrt{2} \approx -1.657,$$
$$f(-\sqrt{2}, -\sqrt{2}) = (-\sqrt{2})^2 + (-\sqrt{2})^2 + 2\sqrt{2} + 2\sqrt{2} = 4 + 4\sqrt{2} \approx 9.657.$$

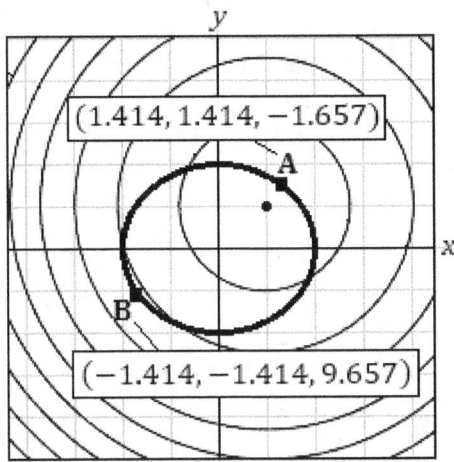

By observation,

$$\left(\sqrt{2}, \sqrt{2}, f(\sqrt{2}, \sqrt{2})\right) \approx (1.414, 1.414, -1.657)$$

is the minimum point (**A**) on the surface subject to the constraint, while

$$\left(-\sqrt{2}, -\sqrt{2}, f(-\sqrt{2}, -\sqrt{2})\right) \approx (-1.414, -1.414, 9.657)$$

is the maximum point (**B**) on the surface subject to the constraint. We also see that these points are where we surmised they would be: the minimum point on the path is closest to the minimum point of the entire surface, while the maximum point is farthest away.

The restrictions that $x \neq 0$ or $y \neq 0$ did not play a role in this example. In the next example, it does.

Example 32.3: Let $z = f(x, y) = x^2 + y^2 - 2x$. Find the minimum and maximum values of f subject to the constraint $x^2 + y^2 = 4$.

Solution: The surface as defined by f is a paraboloid with vertex at $(1, 0, -1)$. Since the paraboloid opens upward, the vertex is the absolute minimum point on the surface. We show the contour map and identify the path constraint (in bold), which is the circle of radius 2, centered at the origin. It is reasonable to infer that the minimum point on the surface subject to the constraint is probably the point closest to the vertex (denoted **A**), and the maximum point is farthest away from the vertex (denoted **B**) given that the surface rises the farther away one moves from the origin.

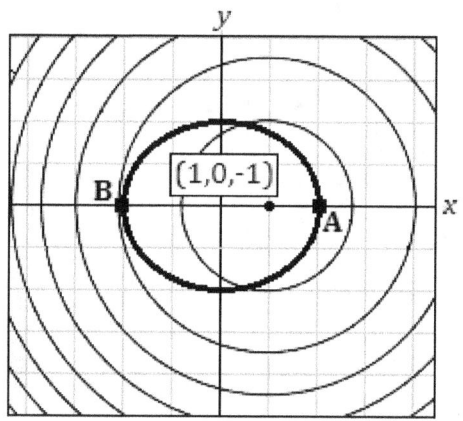

We build function L:

$$L(x, y, \lambda) = x^2 + y^2 - 2x - \lambda x^2 - \lambda y^2 + 4\lambda.$$

Now find the partial derivatives:

$$L_x = 2x - 2 - 2\lambda x$$
$$L_y = 2y - 2\lambda y$$
$$L_\lambda = -x^2 - y^2 + 4.$$

Set each partial derivative to 0:

$$2x - 2 - 2\lambda x = 0 \quad (1)$$
$$2y - 2\lambda y = 0 \quad (2)$$
$$x^2 + y^2 = 4. \quad (3)$$

Isolate λ in equations **(1)** and **(2)**, then equate. Note any restrictions on the variables:

$$\lambda = \frac{x-1}{x} \quad \text{and} \quad \lambda = \frac{y}{y} = 1, \quad \text{so that} \quad \frac{x-1}{x} = 1 \quad (x, y \neq 0)$$

Simplifying $\frac{x-1}{x} = 1$, we get $x - 1 = x$, or $0 = -1$, which is a false statement. It seems the process has stalled. However, it has not. The nature of the algebra in this step forces $x \neq 0$ and $y \neq 0$, but in truth, the surface and the constraint are defined when $x = 0$ or $y = 0$. In equation **(2)**, which is $2y - 2\lambda y = 0$, note that $y = 0$ is also a solution.

Substituting this into equation **(3)**, the original constraint, we can solve for x:

$$x^2 + 0^2 = 4$$
$$x^2 = 4$$
$$x = \pm 2.$$

Thus, we have two critical points, $(2, 0, f(2,0))$ and $(-2, 0, f(-2,0))$. The z values are

$$f(2,0) = 2^2 + 0^2 - 2(2) = 0 \quad \& \quad f(-2,0) = (-2)^2 + 0^2 - 2(-2) = 8.$$

The point $(2,0,0)$ is the minimum point (**A**) on the surface subject to the constraint, and the point $(-2,0,8)$ is the maximum point (**B**) on the surface subject to the constraint. This agrees with our original intuition.

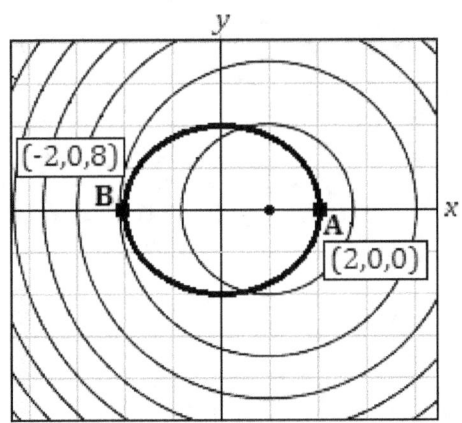

Graph for example 32.3

In the next example, the algebra itself poses a challenge.

Example 32.4: Let $z = f(x,y) = x^2 + y^2 + 4x - 2y$. Find the minimum and maximum values of f subject to the constraint $2x^2 + y^2 = 4$.

Solution: The surface is a parabolid opening upward. Its vertex, $(-2, 1, -5)$, is the absolute minimum point on this surface. The path is an ellipse centered at the origin with a major axis of 4 units in the y direction (± 2 units from the origin) and a minor axis of $2\sqrt{2}$ units in the x direction ($\pm\sqrt{2}$ units from the origin). We label what we think may be the location of the minimum point (**A**) of the surface subject to the constrain, and what we think may be the maximum point (**B**) of the surface, subject to the constraint.

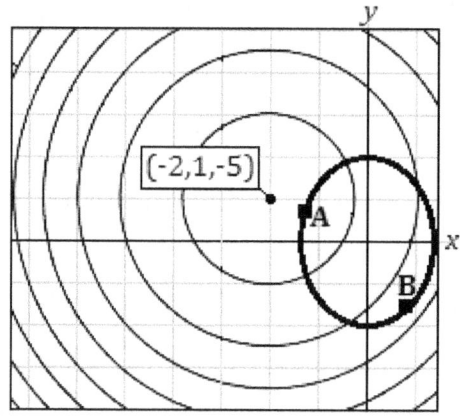

We follow the same steps as before. First, build the Lagrange function L:

$$L(x, y, \lambda) = x^2 + y^2 + 4x - 2y - 2\lambda x^2 - \lambda y^2 + 4\lambda.$$

The partial derivatives are

$$L_x = 2x + 4 - 4\lambda x$$
$$L_y = 2y - 2 - 2\lambda y$$
$$L_\lambda = -2x^2 - y^2 + 4.$$

Setting each to 0, we have a system:

$$2x + 4 - 4\lambda x = 0 \quad (1)$$
$$2y - 2 - 2\lambda y = 0 \quad (2)$$
$$2x^2 + y^2 = 4. \quad (3)$$

Isolate λ in equations **(1)** and **(2)**, then equate. Note any restrictions on the variables:

$$\lambda = \frac{x+2}{2x} \quad \text{and} \quad \lambda = \frac{y-1}{y}, \quad \text{so that} \quad \frac{x+2}{2x} = \frac{y-1}{y} \quad (x, y \neq 0)$$

Clearing fractions and noting restrictions, we have

$$y(x+2) = 2x(y-1)$$
$$xy + 2y = 2xy - 2x$$
$$2y - xy = -2x$$
$$y(2-x) = -2x$$
$$y = \frac{2x}{x-2} \quad (x \neq 2).$$

Substitute this into **(3)**:

$$2x^2 + \left(\frac{2x}{x-2}\right)^2 = 4.$$

Clear fractions:

$$(x-2)^2 2x^2 + (2x)^2 = 4(x-2)^2.$$

Expanding by multiplication and collecting terms, we have

$$x^4 - 4x^3 + 4x^2 + 8x - 8 = 0.$$

It is difficult to isolate x in a quartic polynomial. Instead, the roots are determined graphically. The roots are $x \approx -1.3$ and $x \approx 0.88$:

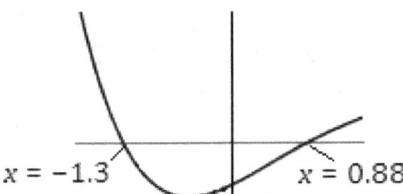

Now use the equation $y = \frac{2x}{x-2}$ to determine y at each x-value:

$$y = \frac{2(-1.3)}{(-1.3)-2} \approx 0.79 \quad \text{and} \quad y = \frac{2(0.88)}{(0.88)-2} \approx -1.57.$$

We then find the z-values:

$$z = f(-1.3, 0.79) = (-1.3)^2 + (0.79)^2 + 4(-1.3) - 2(0.79) \approx -4.47,$$
$$z = f(0.88, -1.57) = (0.88)^2 + (-1.57)^2 + 4(0.88) - 2(-1.57) \approx 9.89.$$

Thus, the point $(-1.3, 0.79, -4.47)$ is the minimum point (**A**) on the surface subject to the constraint, and the point $(0.88, -1.57, 9.89)$ is the maximum point (**B**) on the surface subject to the constraint.

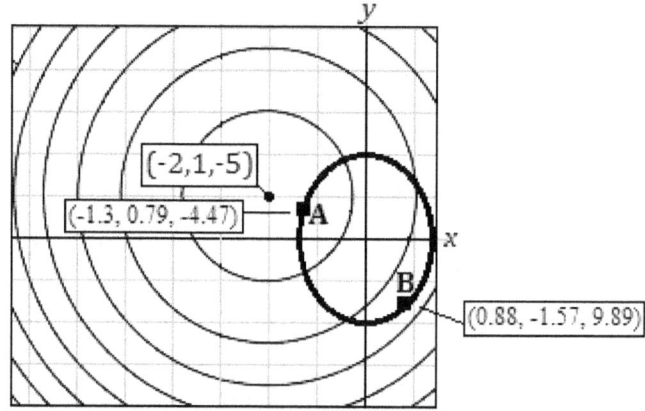

The restrictions imposed on x and y during the algebra steps did not play a role in finding the solutions.

Lagrange Multiplies can be extended into situations with three or more variables.

Example 32.5: Consider the portion of the plane $2x + 4y + 5z = 20$ in the first octant. Find the point on the plane closest to the origin. (This is the same as Example 30.4)

Solution: If (x, y, z) is a point on the plane, then its distance from the origin is $d(x, y, z) = \sqrt{x^2 + y^2 + z^2}$. Using the constraint $2x + 4y + 5z - 20 = 0$, we build the function L:

$$L(x, y, z, \lambda) = \sqrt{x^2 + y^2 + z^2} - \lambda(2x + 4y + 5z - 20).$$

Then we find partial derivatives:

$$L_x = \frac{x}{\sqrt{x^2+y^2+z^2}} - 2\lambda,$$

$$L_y = \frac{y}{\sqrt{x^2+y^2+z^2}} - 4\lambda,$$

$$L_z = \frac{z}{\sqrt{x^2+y^2+z^2}} - 5\lambda,$$

$$L_\lambda = -2x - 4y - 5z + 20.$$

These are then set equal to 0:

$$\frac{x}{\sqrt{x^2+y^2+z^2}} - 2\lambda = 0, \quad \text{so that} \quad \lambda = \frac{x}{2\sqrt{x^2+y^2+z^2}}, \quad (1)$$

$$\frac{y}{\sqrt{x^2+y^2+z^2}} - 4\lambda = 0, \quad \text{so that} \quad \lambda = \frac{y}{4\sqrt{x^2+y^2+z^2}}, \quad (2)$$

$$\frac{z}{\sqrt{x^2+y^2+z^2}} - 5\lambda = 0, \quad \text{so that} \quad \lambda = \frac{z}{5\sqrt{x^2+y^2+z^2}}, \quad (3)$$

$$-2x - 4y - 5z + 20 = 0, \quad \text{so that} \quad 2x + 4y + 5z = 20. \quad (4)$$

Equating (1) and (2), we have

$$\frac{x}{2\sqrt{x^2+y^2+z^2}} = \frac{y}{4\sqrt{x^2+y^2+z^2}}.$$

Clearing fractions, we have $y = 2x$. Then, equating (1) and (3) and clearing fractions, we have $z = \frac{5}{2}x$. These are substituted into equation (4):

$$2x + 4(2x) + 5\left(\frac{5}{2}x\right) = 20.$$

Simplifying, we have $\frac{45}{2}x = 20$, so that $x = \frac{40}{45} = \frac{8}{9}$. Since $y = 2x$, we have $y = 2\left(\frac{8}{9}\right) = \frac{16}{9}$, and since $z = \frac{5}{2}x$, we have $z = \frac{5}{2}\left(\frac{8}{9}\right) = \frac{40}{18} = \frac{20}{9}$.

The point on the plane $2x + 4y + 5z = 20$ closest to the origin is $\left(\frac{8}{9}, \frac{16}{9}, \frac{20}{9}\right)$.

Example 32.6: Consider the portion of the plane $2x + 4y + 5z = 20$ in the first octant. A rectangular box is situated with one corner at the origin and its opposite corner on the plane so that the box's edges lie along (or are parallel to) the x-axis, y-axis or z-axis. Find the largest possible volume of such a box, keeping the box to within the first octant. (This is the same as Example 30.5)

Solution: The volume of the box is given by $V(x, y, z) = xyz$, and along with the constraint $2x + 4y + 5z - 20 = 0$, we build function L:

$$L(x, y, z, \lambda) = xyz - \lambda(2x + 4y + 5z - 20).$$

Taking partial derivatives, we have

$$L_x = yz - 2\lambda,$$
$$L_y = xz - 4\lambda,$$
$$L_z = xy - 5\lambda,$$
$$L_\lambda = -2x - 4y - 5z + 20.$$

Setting each to 0, we have

$$yz - 2\lambda = 0, \quad \text{so that} \quad \lambda = \frac{yz}{2}, \quad (1)$$

$$xz - 4\lambda = 0, \quad \text{so that} \quad \lambda = \frac{xz}{4}, \quad (2)$$

$$xy - 5\lambda = 0, \quad \text{so that} \quad \lambda = \frac{xy}{5}, \quad (3)$$

$$-2x - 4y - 5z + 20 = 0, \quad \text{so that} \quad 2x + 4y + 5z = 20. \quad (4)$$

Equating **(1)** and **(2)**, we have

$$\frac{yz}{2} = \frac{xz}{4}, \quad \text{so that} \quad y = \frac{1}{2}x.$$

Equating **(1)** and **(3)**, we have

$$\frac{yz}{2} = \frac{xy}{5}, \quad \text{so that} \quad z = \frac{2}{5}x.$$

These are substituted into **(4)** and variable x is isolated:

$$2x + 4\left(\frac{1}{2}x\right) + 5\left(\frac{2}{5}x\right) = 20$$
$$2x + 2x + 2x = 20$$
$$6x = 20$$
$$x = \frac{10}{3}.$$

Since $y = \frac{1}{2}x$, we have $y = \frac{1}{2}\left(\frac{10}{3}\right) = \frac{5}{3}$, and since $z = \frac{2}{5}x$, we have $z = \frac{2}{5}\left(\frac{10}{3}\right) = \frac{4}{3}$. Thus, the dimensions of the largest box will be $\frac{10}{3} \times \frac{5}{3} \times \frac{4}{3}$, with the volume $\left(\frac{10}{3}\right)\left(\frac{5}{3}\right)\left(\frac{4}{3}\right) = \frac{200}{27}$.

◆ ◆ ◆ ◆ ◆ ◆ ◆ ◆ ◆ ◆

33. Riemann Summation over Rectangular Regions

A rectangular region R in the xy-plane can be defined using compound inequalities, where x and y are each bound by constants such that $a_1 \leq x \leq a_2$ and $b_1 \leq y \leq b_2$. Let $z = f(x, y)$ be a continuous function defined over a rectangular region R in the xy-plane. The notation

$$\iint_R f(x, y)\, dA$$

represents the **double integral** of $z = f(x, y)$ over R. The dA represents "area element", and is either $dy\, dx$ or $dx\, dy$. Thus, we can write

$$\iint_R f(x, y)\, dA = \int_{a_1}^{a_2} \int_{b_1}^{b_2} f(x, y)\, dy\, dx = \int_{b_1}^{b_2} \int_{a_1}^{a_2} f(x, y)\, dx\, dy.$$

Note that the bounds a_1 and a_2 correspond with the differential dx, and bounds b_1 and b_2 correspond with dy.

The value of a double integral can be approximated by **Riemann sums** adapted to the two-dimensional case. Interval $a_1 \leq x \leq a_2$ is subdivided into m subdivisions (not necessarily of equal size) and interval $b_1 \leq y \leq b_2$ is subdivided into n subdivisions (again, not necessarily of equal size). If we define indices $1 \leq i \leq m$ and $1 \leq j \leq n$, then we have a way to identify a particular subdivision within region R. For example, if $a_1 \leq x \leq a_2$ is subdivided into 4 subdivisions and $b_1 \leq y \leq b_2$ is subdivided into 5 subdivisions, then (x_2, y_3) is a representative point within the 2nd subdivision of the x-interval and the 3rd subdivision of the y-interval, and $f(x_2, y_3)$ is the function evaluated at (x_2, y_3).

Using this scheme, a double integral can be approximated by a double sum over i and j:

$$\iint_R f(x,y)\, dA \approx \sum_{i=1}^{m}\sum_{j=1}^{n} f(x_i, y_j)\, \Delta y\, \Delta x \text{ or } \sum_{j=1}^{n}\sum_{i=1}^{m} f(x_i, y_j)\, \Delta x\, \Delta y.$$

Example 33.1: Use Riemann Sums to approximate $\iint_R x^2 y\, dA$ where R is the rectangle $0 \leq x \leq 3$ and $1 \leq y \leq 5$ in the xy plane. Subdivide the region R into subregions each with length 1 to a side, and from each subregion, choose x and y to be the "upper right" corner.

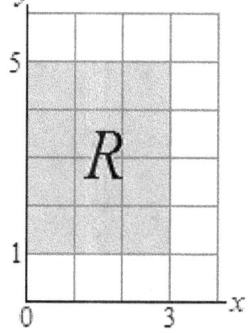

Solution: The rectangular region R is shown at right, subdivided into subregions, so that $\Delta A = \Delta x\, \Delta y = (1)(1) = 1$. There are 12 such subregions.

Then choose a representative point (x_i, y_j) within each subregion. In this example, we choose (x_i, y_j) to be the "upper right" point within each subregion (this is an arbitrary choice. We could choose the "lower left" or the "middle point", and so on). Here, $1 \leq i \leq 3$ and $2 \leq j \leq 5$, the bounds chosen for convenience.

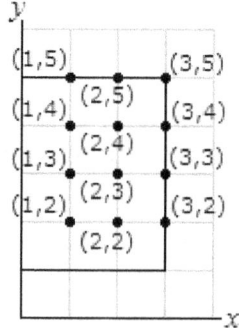

Next, evaluate the integrand $z = f(x, y) = x^2 y$ at the representative points (x_i, y_j):

$$\begin{array}{lll} f(1,5) = 5 & f(2,5) = 20 & f(3,5) = 45 \\ f(1,4) = 4 & f(2,4) = 16 & f(3,4) = 36 \\ f(1,3) = 3 & f(2,3) = 12 & f(3,3) = 27 \\ f(1,2) = 2 & f(2,2) = 8 & f(3,2) = 18 \end{array}$$

Visually, we have a surface $z = f(x, y) = x^2 y$ "above" the xy-plane. Each subregion in R is the base of a rectangular box whose height is the function value shown in the table above. Each box has a volume of $f(x_i, y_j) \, dA$. Since $dA = dx \, dy = (1)(1) = 1$ in each case, each box has volume $f(x_i, y_j) \times 1$, or simply $f(x_i, y_j)$. The value of $\iint_R x^2 y \, dA$ is approximated by the sum of the volumes of the rectangular boxes contained within it. Thus,

$$\iint_R x^2 y \, dA \approx \sum_{i=1}^{3} \sum_{j=2}^{5} f(x_i, y_j) \, \Delta y \, \Delta x$$
$$= 2 + 8 + 18 + 3 + 12 + 27 + 4 + 16 + 36 + 5 + 20 + 45$$
$$= 196.$$

Note that if we chose the representative point to be the lower-left corner of each subregion, we would find that the Riemann Sum is 50. The mean, $\frac{196+50}{2} = 123$, is a reasonable approximation of $\iint_R x^2 y \, dA$.

♦ • ♦ • ♦ • ♦ • ♦ • ♦

Example 33.2: Use Riemann Sums to approximate $\iint_R g(x, y) \, dA$, where g is shown by the contour map below. Let the region of integration R be given by $-4 \le x \le 4, -6 \le y \le 6$, and let $\Delta x = 2$ and $\Delta y = 2$. Use the middle point within each subregion.

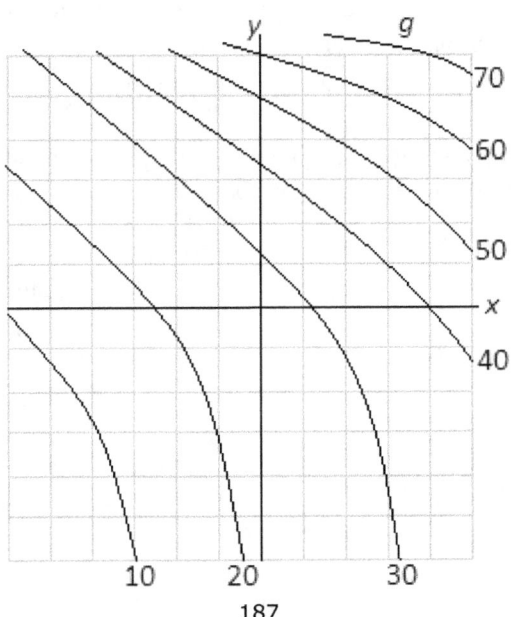

Solution: The region R is identified and then subdivided into 2×2 subregions (lower left, boldfaced). Then the middle point (x_i, y_j) from within each subregion is identified (lower right):

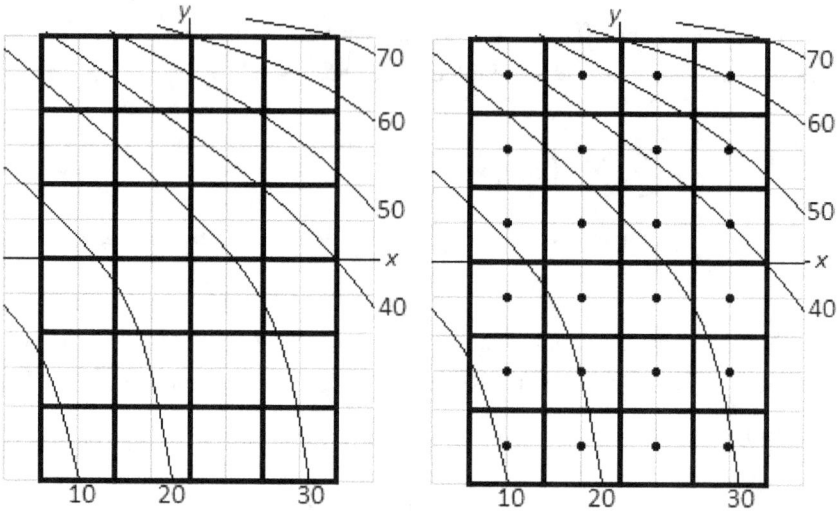

The values of $z = g(x, y)$ are estimated from the contour map. For example, in the top tier of subregions, reading left to right and using the middle points, the values of g are approximately $g(-3,5) = 37, g(-1,5) = 46, g(1,5) = 55$ and $g(3,5) = 60$.

Each of these subregions is the base of a rectangular box whose heights are given by the $z_i = g(x_i, y_j)$ values. Each box then has a volume of $g(x_i, y_j)\, dA$. Since $dA = (2)(2) = 4$, each box has a volume of $g(x_i, y_j) \times 4$.

The approximate values of $g(x_i, y_j)$ are shown below in an array that matches the orientation of the subregions in the above figure:

37	46	55	60
27	34	42	49
22	27	33	40
16	23	28	34
13	20	25	31
11	18	25	29

The approximate value of $\iint_R g(x,y)\, dA$ is the sum of the volumes of each rectangular box contained within it:

$$\iint_R g(x,y)\, dA \approx 37(4) + 46(4) + 55(4) + 60(4) + \cdots.$$

Note that the 4 can be factored to the front. Thus, the approximate value of $\iint_R g(x,y)\, dA$ is the sum of all the $g(x_i, y_j)$ values in the array above, multiplied by 4:

$$\iint_R g(x,y)\, dA$$
$$\approx 4 \begin{pmatrix} 37 + 46 + 55 + 60 + 27 + 34 + 42 + 49 + 22 + 27 + 33 + 40 \\ + 16 + 23 + 28 + 34 + 13 + 20 + 25 + 31 + 11 + 18 + 25 + 29 \end{pmatrix},$$

which is about 2,980 cubic units.

◆ ◆ ◆ ◆ ◆ ◆ ◆ ◆ ◆ ◆ ◆

34. Double Integration over Rectangular Regions

A double integral is evaluated "inside out"— the inside integral is evaluated first, then that result becomes the integrand of the outer integral, which is then evaluated.

Example 34.1: Evaluate $\iint_R x^2 y\, dA$ where R is the rectangle $0 \leq x \leq 3$ and $1 \leq y \leq 5$.

Solution: We can choose either the $dy\, dx$ ordering or the $dx\, dy$ ordering. Let's choose $dA = dx\, dy$. Thus, we have

$$\iint_R x^2 y\, dA = \int_1^5 \int_0^3 x^2 y\, dx\, dy.$$

Evaluate the inner integral with respect to x, treating y as a constant:

$$\int_0^3 x^2 y\, dx = \left[\frac{1}{3} x^3 y\right]_0^3 = \frac{1}{3} y[3^3 - 0^3] = 9y.$$

Now we integrate the result with respect to y:

$$\int_1^5 9y \, dy = \left[\tfrac{9}{2}y^2\right]_1^5 = \tfrac{9}{2}(5^2 - 1^2) = 108.$$

If we chose $dA = dy \, dx$, we have the following:

$$\int_0^3 \int_1^5 x^2 y \, dy \, dx.$$

The inner integral is evaluated first with respect to y, treating x as a constant:

$$\int_1^5 x^2 y \, dy = x^2 \left[\tfrac{1}{2}y^2\right]_1^5 = \tfrac{1}{2}x^2[(5)^2 - (1)^2] = \tfrac{1}{2}x^2(24) = 12x^2.$$

This result is now integrated with respect to x:

$$\int_0^3 12x^2 \, dx = [4x^3]_0^3 = 4[(3)^3 - (0)^3] = 4(27) = 108.$$

Both orderings of the differentials gives the same result, 108, as expected. This is the volume of the solid bounded below by the region of integration R and above by the surface $z = x^2 y$.

◆ • ◆ • • ◆ • • ◆ • ◆

> If the region is infinite in one direction, the integral is improper and may be evaluated using limits.

Example 34.2: Evaluate

$$\int_0^\infty \int_{-1}^2 \frac{x^3}{1+y^2} \, dx \, dy.$$

Solution: The inner integral is evaluated first, with $\frac{1}{1+y^2}$ moved outside the integral since it is a constant multiplier:

$$\int_{-1}^2 \frac{x^3}{1+y^2} \, dx = \frac{1}{1+y^2} \int_{-1}^2 x^3 \, dx = \frac{1}{1+y^2}\left[\tfrac{1}{4}x^4\right]_{-1}^2$$

$$= \frac{1}{4}\left(\frac{1}{1+y^2}\right)[(2)^4 - (-1)^4] = \frac{15}{4}\left(\frac{1}{1+y^2}\right).$$

This is then integrated with respect to y. The constant $\frac{15}{4}$ is moved outside the integral, and the upper bound, ∞, is replaced with b, where b approaches infinity as a limit:

$$\frac{15}{4}\int_0^\infty \frac{1}{1+y^2}\,dy = \lim_{b\to\infty}\left(\frac{15}{4}\int_0^b \frac{1}{1+y^2}\,dy\right)$$

$$= \lim_{b\to\infty}\frac{15}{4}[\arctan y]_0^b$$

$$= \frac{15}{4}\lim_{b\to\infty}[\arctan b - \arctan 0]$$

$$= \frac{15}{4}\left(\frac{\pi}{2}-0\right) = \frac{15\pi}{8}.$$

As an angle θ approaches $\frac{\pi}{2}$ radians from below, $\tan\theta$ approaches positive ∞. Thus, if $\theta = \arctan b$, then $\arctan b$ approaches $\frac{\pi}{2}$ as b approaches ∞.

♦ ♦ ♦ ♦ ♦ ♦ ♦ ♦ ♦ ♦ ♦

Example 34.3: Evaluate

$$\int_0^4 \int_{\pi/6}^{\pi/2} (x^2 + \cos 3y)\,dy\,dx.$$

Solution: The inner integral is determined first:

$$\int_{\pi/6}^{\pi/2} (x^2 + \cos 3y)\,dy = \left[x^2 y + \frac{1}{3}\sin 3y\right]_{\pi/6}^{\pi/2}$$

$$= \left[x^2\left(\frac{\pi}{2}\right) + \frac{1}{3}\sin\left(\frac{3\pi}{2}\right)\right] - \left[x^2\left(\frac{\pi}{6}\right) + \frac{1}{3}\sin\left(\frac{3\pi}{6}\right)\right].$$

Recall that $\sin\left(\frac{3\pi}{2}\right) = -1$ and that $\sin\left(\frac{3\pi}{6}\right) = \sin\left(\frac{\pi}{2}\right) = 1$. We have,

$$\left[x^2\left(\frac{\pi}{2}\right) + \frac{1}{3}(-1)\right] - \left[x^2\left(\frac{\pi}{6}\right) + \frac{1}{3}(1)\right] = x^2\left(\frac{\pi}{2}-\frac{\pi}{6}\right) - \frac{2}{3} = \frac{\pi}{3}x^2 - \frac{2}{3}.$$

This is then integrated:

$$\int_0^4 \left(\frac{\pi}{3}x^2 - \frac{2}{3}\right) dx = \left[\frac{\pi}{9}x^3 - \frac{2}{3}x\right]_0^4$$

$$= \left[\frac{\pi}{9}(4)^3 - \frac{2}{3}(4)\right] - \left[\frac{\pi}{9}(0)^3 - \frac{2}{3}(0)\right]$$

$$= \frac{64\pi}{9} - \frac{8}{3}.$$

◆ ● ◆ ● ● ◆ ● ● ◆ ● ● ◆

Example 34.4: Evaluate

$$\int_0^2 \int_1^3 xye^{x+y^2} \, dx \, dy.$$

Solution: We can simplify the integrand using algebra first: $xye^{x+y^2} = xye^x e^{y^2} = xe^x y e^{y^2}$. Note that since this is a single term, we may group the factors as desired. The factor xe^x will be integrated using integration by parts, while the factor ye^{y^2} can be integrated using u-du substitution. It does not make a difference in which order we integrate, but it may be simpler to integrate with respect to y first. Thus, we rewrite the iterated integral as

$$\int_1^3 \int_0^2 xe^x y e^{y^2} \, dy \, dx.$$

Integrating the inside integral with respect to y, treating xe^x as a constant:

$$\int_0^2 xe^x y e^{y^2} \, dy = xe^x \left[\frac{1}{2}e^{y^2}\right]_0^2$$

$$= \frac{1}{2}xe^x \left[e^{(2)^2} - e^{(0)^2}\right]$$

$$= \frac{1}{2}xe^x [e^4 - 1].$$

This is now integrated with respect to x. Note that $\frac{1}{2}(e^4 - 1)$ is a constant and can be moved outside the integral:

$$\frac{e^4 - 1}{2} \int_1^3 xe^x \, dx.$$

To antidifferentiate xe^x, use integration by parts. Let $u = x$ and $dv = e^x dx$. Thus, $du = dx$ and $v = e^x$. Since $\int u\, dv = uv - \int v\, du$, we have

$$\frac{e^4 - 1}{2} \int_1^3 xe^x\, dx = \frac{e^4 - 1}{2}\left[xe^x - \int e^x dx\right]$$

$$= \frac{e^4 - 1}{2}[xe^x - e^x]_1^3$$

$$= \frac{e^4 - 1}{2}[(3e^3 - e^3) - (e^1 - e^1)]$$

$$= \frac{2e^3(e^4 - 1)}{2}$$

$$= e^7 - e^3.$$

◆ ◆ ◆ ◆ • ◆ • ◆ • ◆ • ◆

Example 34.5: The density of a city's population is $P(x, y) = 0.2x^2 + 0.1y^3$, where x and y are in miles, and P is in thousands of people per square mile. Assume that the city is a rectangle measuring 6 miles east to west (x), and 4 miles north to south (y), and that $x = 0$ and $y = 0$ is the southwestern corner of the city's boundaries. Find the city's population.

Solution: The city's population is given by the double integral:

$$\int_0^4 \int_0^6 (0.2x^2 + 0.1y^3)\, dx\, dy.$$

Evaluating the inside integral with respect to x first, we have

$$\int_0^6 (0.2x^2 + 0.1y^3)\, dx = \left[\frac{0.2}{3}x^3 + 0.1xy^3\right]_0^6$$

$$= \left(\frac{0.2}{3}(6)^3 + 0.1(6)y^3\right) - \left(\frac{0.2}{3}(0)^3 + 0.1(0)y^3\right)$$

$$= 14.4 + 0.6y^3.$$

This is then integrated with respect to y:

$$\int_0^4 (14.4 + 0.6y^3)\, dy = \left[14.4y + \frac{0.6}{4}y^4\right]_0^4$$

$$= \left(14.4(4) + \frac{0.6}{4}(4)^4\right) - \left(14.4(0) + \frac{0.6}{4}(0)^4\right)$$

$$= 96.$$

Multiply the result by 1000. Thus, the city has about 96,000 people within its boundaries.

♦ • ♦ • ♦ • ♦ • ♦ • ♦

> The **average value** of a multivariable function $z = f(x,y)$ over a region R is given by $f_{av} = \frac{1}{A(R)} \iint_R f(x,y)\, dA$, where $A(R)$ is the area of region R.

Example 34.6: Find the average value of the result in the previous example and explain its meaning in context.

Solution: The region R has an area of $(6)(4) = 24$ square miles. Thus, the average value of $P(x,y) = 0.2x^2 + 0.1y^3$ over R is $P_{av} = \frac{1}{24}(96) = 4$. The city has an average density of 4,000 people per square mile.

♦ • ♦ • ♦ • ♦ • ♦ • ♦

See an error? Have a suggestion?
Please see www.surgent.net/vcbook

35. Double Integration over Non-Rectangular Regions of Type I

Consider the region R shown below.

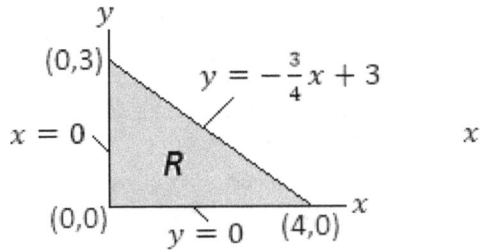
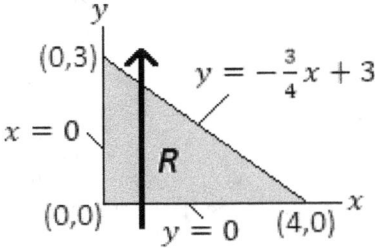

The region is bounded by the lines $y = 0$ (the x-axis), $x = 0$ (the y-axis), and $y = -\frac{3}{4}x + 3$. If we set up a double integral is the $dy\,dx$ ordering of integration, we draw an arrow in the positive y direction (see image, above right). It enters the region at $y_1 = 0$ and exits through $y_2 = -\frac{3}{4}x + 3$, where the subscripts help us remember the order in which the boundaries are crossed. The double integral is

$$\int_0^4 \int_{y_1=0}^{y_2=-(3/4)x+3} f(x,y)\,dy\,dx = \int_0^4 \int_0^{-(3/4)x+3} f(x,y)\,dy\,dx.$$

As a $dx\,dy$ integral, draw an arrow drawn in the positive x direction (see image at right). It enters the region at $x_1 = 0$ and exits through $x_2 = -\frac{4}{3}y + 4$ (which is the equation $y = -\frac{3}{4}x + 3$ that has been solved for x). The resulting y bounds are 0 to 3, and the double integral is

$$\int_0^3 \int_0^{-(4/3)y+4} f(x,y)\,dx\,dy.$$

There is *no* ambiguity where an arrow enters or exits the region. Such a region is called a **Type I** region. If there is ambiguity, then the region is called a Type II region.

195

Below, at left, are regions of Type I in the *dy dx* order of integration. At right are regions of Type II also in the *dy dx* order. Some may be Type I in the *dx dy* ordering.

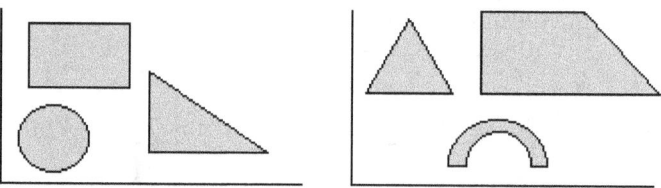

Integrals over a region of Type I usually require one iterated integral. For regions of Type II, more than one iterated integral is required.

Example 35.1: Find the volume below $z = f(x, y) = xy + x^2$ over the region R, which is a triangle with vertices (0,0), (5,0) and (0,10).

Solution. The sketch below shows this to be a region of Type I. Identify all vertex points and the equation of all boundaries.

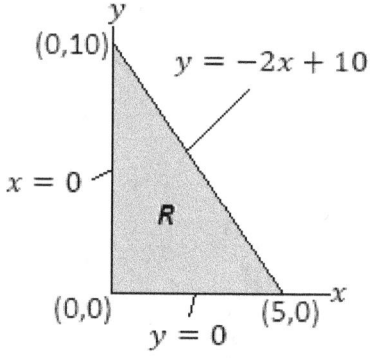 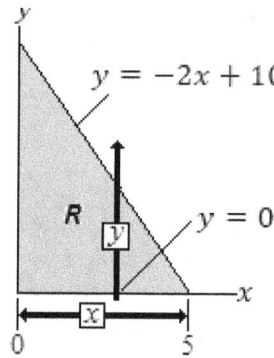

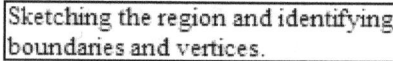

| Sketching the region and identifying boundaries and vertices. | The *dy dx* order of integration: Find the bounds for *y* first, then *x*. |

If we choose to integrate in the *dy dx* ordering, visualize an arrow drawn in the positive *y* direction. It enters the region at the *x*-axis, which is $y_1 = 0$, and exits through $y_2 = -2x + 10$. The *x* bounds are 0 to 5, and the iterated integral is

$$\int_0^5 \int_0^{-2x+10} (xy + x^2) \, dy \, dx.$$

Integrating with respect to y, we have

$$\int_0^{-2x+10} (xy + x^2)\, dy = \left[\frac{1}{2}xy^2 + x^2 y\right]_0^{-2x+10}$$

$$= \left(\frac{1}{2}x(-2x+10)^2 + x^2(-2x+10)\right) - \left(\frac{1}{2}x(0)^2 + x^2(0)\right).$$

The expression above simplifies to $-10x^2 + 50x$. This is the integrand to be integrated with respect to x now:

$$\int_0^5 (-10x^2 + 50x)\, dx = \left[-\frac{10}{3}x^3 + 25x^2\right]_0^5$$

$$= \left(-\frac{10}{3}(5)^3 + 25(5)^2\right) - 0$$

$$= \frac{625}{3}.$$

♦ ♦ ♦ ♦ ♦ ♦ ♦ ♦ ♦ ♦

Example 35.2: Evaluate

$$\iint_R 2xy^2\, dA,$$

where R is in the first quadrant bounded by the x-axis, the y-axis and the parabola $y = 25 - x^2$.

Solution: Sketch the region and decide on an ordering of integration. If we choose a $dy\, dx$ ordering, visualize an arrow drawn in the positive y direction. It enters the region at the x-axis, which is $y_1 = 0$, and exits through the parabola $y_2 = 25 - x^2$. The bounds for x are 0 to 5.

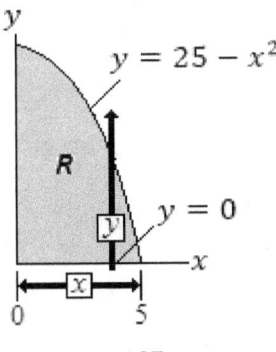

The double integral is

$$\int_0^5 \int_0^{25-x^2} 2xy^2 \, dy \, dx.$$

The inside integral is determined:

$$\int_0^{25-x^2} 2xy^2 \, dy = \left[\frac{2}{3}xy^3\right]_0^{25-x^2} = \frac{2}{3}x(25-x^2)^3.$$

This is integrated with respect to x using a u-du substitution, with $u = 25 - x^2$:

$$\int_0^5 \frac{2}{3}x(25-x^2)^3 \, dx = \left[-\frac{1}{12}(25-x^2)^4\right]_0^5$$

$$= \left(-\frac{1}{12}(25-(5)^2)^4\right) - \left(-\frac{1}{12}(25-(0)^2)^4\right)$$

$$= 0 - \left(-\frac{1}{12}(25)^4\right) = \frac{390{,}625}{12} \approx 32{,}552.083.$$

If we use a $dx \, dy$ ordering, the double integral is written

$$\int_0^{25} \int_0^{\sqrt{25-y}} 2xy^2 \, dx \, dy,$$

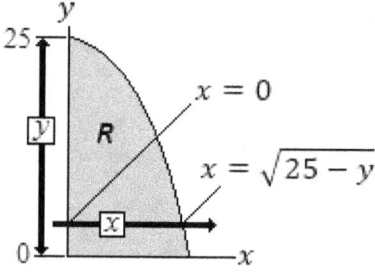

Where the bounds are written such that x has been isolated. This double integral also evaluates to $\frac{390{,}625}{12}$, or 32,522.083.

◆ ◆ ◆ ◆ ◆ ◆ ◆ ◆ ◆ ◆ ◆

Example 35.3: Given

$$\int_0^5 \int_{e^x}^{e^5} g(x,y) \, dy \, dx,$$

Reverse the order of integration (rewrite this double integral as a $dx \, dy$ integral).

Solution: The ordering of integration tells us that if we visualize an arrow in the positive y direction, it will enter the region at $y_1 = e^x$ and exit at the line $y = e^5$, with the x bounds being 0 to 5. The region is shown below, with all vertices and boundaries identified:

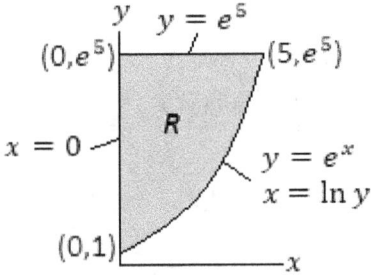

To reverse the ordering, now visualize an arrow in the positive x direction. It enters at $x_1 = 0$ (the y-axis) and exits at $x_2 = \ln y$. The bounds for y are 1 to e^5. We have

$$\int_1^{e^5} \int_0^{\ln y} g(x,y)\, dx\, dy.$$

♦ • ♦ ♦ • • • ♦ • • ♦

Example 35.4: Reverse the order of integration of

$$\int_0^9 \int_{y/3}^{\sqrt{y}} h(x,y)\, dx\, dy.$$

Solution: Visualize an arrow in the positive x direction. It enters the region R at $x_1 = \frac{1}{3}y$ and exits at $x_2 = \sqrt{y}$. The two graphs meet at $(0,0)$ and $(3,9)$, and the region is shown below:

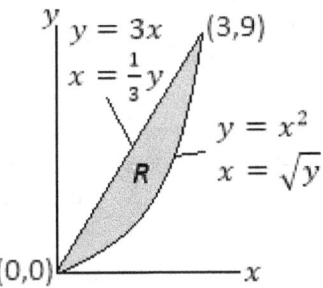

Observe that we redefined the bounds in y as a function in terms of x.

Viewing the region, now visualize an arrow in the positive y direction. It will enter R at $y_1 = x^2$ and exit at $y_2 = 3x$. These become the bounds for the dy integral. The bounds for x are 0 to 3, and the equivalent double integral in the $dy\,dx$ ordering is

$$\int_0^3 \int_{x^2}^{3x} h(x,y)\, dy\, dx.$$

◆ ◆ ◆ ◆ ◆ ◆ ◆ ◆ ◆ ◆ ◆

Example 35.5: Evaluate

$$\int_0^2 \int_y^2 \sqrt{1+x^2}\, dx\, dy.$$

Solution: If we attempt to evaluate the integrals as written (inside first with respect to x, then outside with respect to y), we discover that finding the antiderivative of $\sqrt{1+x^2}$ with respect to x is challenging (it would require a trigonometric substitution). Instead, we reverse the order of integration.

The double integral, as written, suggests that the region R is bounded by the line $x = y$ and the line $x = 2$, with the bounds for y being 0 to 2. This region is sketched below, and all vertices and boundaries are identified:

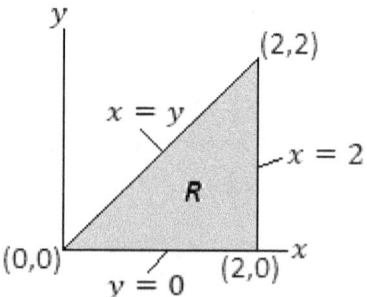

Reversing the order of integration, we visualize an arrow in the positive y-direction. It enters R at $y_1 = 0$ and exits at $y_2 = x$. The bounds for x will be 0 to 2, and the double integral in the $dy\,dx$ ordering is

$$\int_0^2 \int_0^x \sqrt{1+x^2}\, dy\, dx.$$

Now, the inside integral is determined. Note that the antiderivative of $\sqrt{1+x^2}$ with respect to y is $y\sqrt{1+x^2}$. Thus, we have

$$\int_0^x \sqrt{1+x^2}\, dy = \left[y\sqrt{1+x^2}\right]_0^x = x\sqrt{1+x^2}.$$

Now we integrate $x\sqrt{1+x^2}$ with respect to x. The antiderivative of $x\sqrt{1+x^2}$ is found by a *u-du* substitution. We have

$$\int_0^2 x\sqrt{1+x^2}\, dx = \left[\frac{1}{3}(1+x^2)^{3/2}\right]_0^2 = \frac{1}{3}\left(5^{3/2}-1\right).$$

◆ ◆ ◆ ◆ ◆ ◆ ◆ ◆ ◆ ◆

36. Double Integration over Non-Rectangular Regions of Type II

When establishing the bounds of a double integral, visualize an arrow initially in the positive x direction or the positive y direction. A region of **Type II** is one in which there may be ambiguity as to where this arrow enters or exits the region. In such cases, more than one double integral will be required. It is possible that a region may be Type II in one ordering of integration, but Type I in another ordering.

Example 36.1: Rewrite the following integral expression

$$\int_0^2 \int_0^{x^2} f(x,y)\, dy\, dx + \int_2^6 \int_0^{6-x} f(x,y)\, dy\, dx$$

in the *dx dy* order of integration.

Solution: Let's look at one double integral at a time. We start with the first,

$$\int_0^2 \int_0^{x^2} f(x,y)\, dy\, dx.$$

It may be helpful to write in equations for each bound:

$$\int_{x=0}^{x=2} \int_{y=0}^{y=x^2} f(x,y)\, dy\, dx.$$

The bounds of the inner integral suggests a region with a lower bound of $y = 0$ (the x-axis), and an upper bound of $y = x^2$. Then, the bounds of the outer integral suggest that x must be contained in the interval $0 \leq x \leq 2$. We obtain the following region:

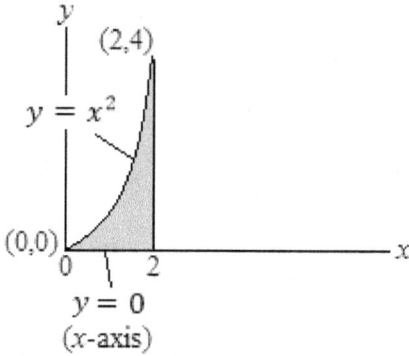

Now, we examine the second double integral, where the bounds have been written as equations:

$$\int_{x=2}^{x=6} \int_{y=0}^{y=6-x} f(x, y)\, dy\, dx$$

This suggests a region bounded below by $y = 0$ (the x-axis) and above by the line $y = 6 - x$, then the bounds of the outer integral suggest that $2 \leq x \leq 6$. This region is sketched alongside the one already sketched. All vertex or extreme points are also noted:

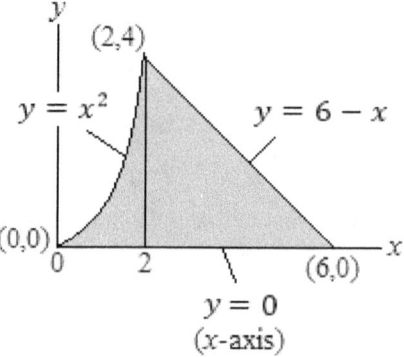

In the $dy\, dx$ ordering of integration, this is a Type II region. An arrow drawn in the positive y-direction enters the region at the x-axis ($y = 0$), but may exit through the parabola $y = x^2$ or the line $y = 6 - x$. This ambiguity is why this region is considered Type II in the $dy\,dx$ ordering, and why two double integrals are necessary to describe the region properly.

If the order of integration is reversed to $dx\,dy$, then there is no ambiguity as to where an arrow drawn in the positive x-direction would enter and exit the region. The same region is drawn below, but now the boundaries are stated with variable x isolated:

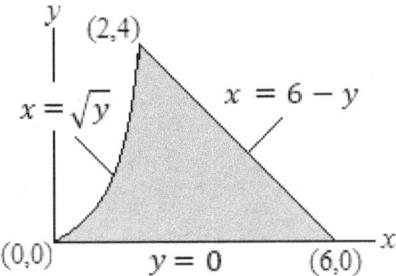

An arrow drawn in the positive x direction enters the region at $x = \sqrt{y}$ and exits at $x = 6 - y$. The bounds for y are 0 to 4. In this ordering, the region is Type I, and one double integral is sufficient to describe this region:

$$\int_0^4 \int_{\sqrt{y}}^{6-y} f(x, y)\, dx\, dy.$$

♦ • ♦ • ♦ • ♦ • ♦ • ♦

Example 36.2: Set up a double integral over region R that is outside a circle of radius 2 centered at the origin, inside a circle of radius 5 centered at the origin, such that y is non-negative. Use the $dy\,dx$ ordering and use $f(x,y)$ as the integrand.

Solution: A sketch shows that this region has the following appearance and is Type II. Vertical lines are placed where R is split into smaller Type I regions, labeled A, B and C, reading left to right:

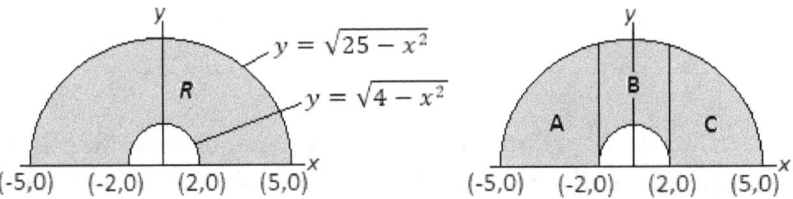

203

For region A, we have $\int_{-5}^{-2}\int_{0}^{\sqrt{25-x^2}} f(x,y)\, dy\, dx$.

For region B, we have $\int_{-2}^{2}\int_{\sqrt{4-x^2}}^{\sqrt{25-x^2}} f(x,y)\, dy\, dx$.

For region C, we have $\int_{2}^{5}\int_{0}^{\sqrt{25-x^2}} f(x,y)\, dy\, dx$.

Summing, we have

$$\int_{-5}^{-2}\int_{0}^{\sqrt{25-x^2}} f(x,y)\, dy\, dx + \int_{-2}^{2}\int_{\sqrt{4-x^2}}^{\sqrt{25-x^2}} f(x,y)\, dy\, dx + \int_{2}^{5}\int_{0}^{\sqrt{25-x^2}} f(x,y)\, dy\, dx.$$

Regions that are formed by circles are better solved using polar and cylindrical coordinates, which are discussed later.

♦ ♦ ♦ ♦ ♦ ♦ ♦ ♦ ♦ ♦ ♦

Example 36.3: Reverse the order of integration:

$$\int_{-1}^{4}\int_{x^2}^{3x+4} g(x,y)\, dy\, dx.$$

Solution: The region is shown below. In the $dy\, dx$ ordering of integration, it is Type I (at left).

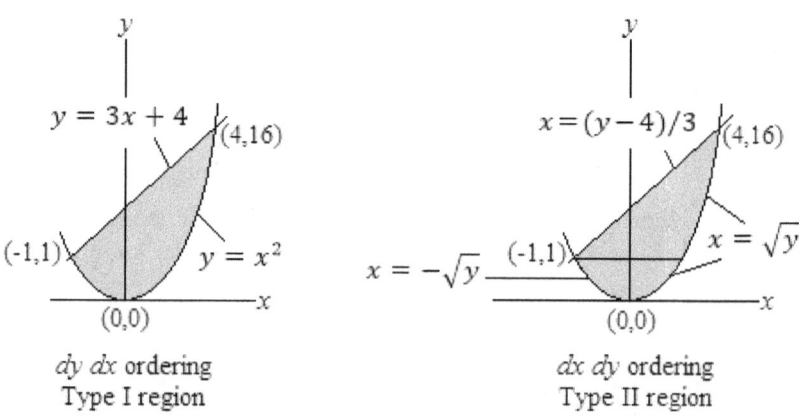

dy dx ordering
Type I region

dx dy ordering
Type II region

However, as a $dx\, dy$ ordering of integration, an arrow drawn in the positive x direction may enter the region through the line $x = \frac{y-4}{3}$ (assuming $1 < y \le 16$)

or through the curve $x = -\sqrt{y}$ (assuming $0 \leq y < 1$). Because of the ambiguity as to where such an arrow could enter the region, this is a Type II region.

We split the region at $y = 1$, forming two smaller Type I regions, as shown above right, with the bounding curves now written with x isolated. Thus, in the $dx\,dy$ ordering of integration, we have

$$\int_0^1 \int_{-\sqrt{y}}^{\sqrt{y}} g(x,y)\,dx\,dy + \int_1^{16} \int_{(y-4)/3}^{\sqrt{y}} g(x,y)\,dx\,dy.$$

◆ ◆ ◆ ◆ ◆ ◆ ◆ ◆ ◆ ◆ ◆

37. Double Integration in Polar Coordinates

Regions that are formed by circles are better described using polar coordinates. If (r, θ) represents a point in the plane, then r is the distance from the point to the origin, and θ represents the angle that a ray from the origin to the point makes with the positive x-axis. The usual conversion formulas between rectangular (x, y) coordinates to polar (r, θ) coordinates are:

(x,y) to (r,θ): $\begin{cases} r^2 = x^2 + y^2 \\ \theta = \arctan\left(\frac{y}{x}\right) \end{cases}$ (r,θ) to (x,y): $\begin{cases} x = r\cos\theta \\ y = r\sin\theta \end{cases}$

Circular regions (or portions thereof) in the xy-plane can be described using polar coordinates where $a \leq r \leq b$ and $c \leq \theta \leq d$, and a, b, c and d are constants. Such regions are called **polar rectangles**.

To establish bounds for r, visualize an arrow that starts at the origin and extends outward. The value of r at which the arrow enters the region is the lower bound a, and the value of r at which this arrow exits the region is the upper bound b. If the region includes the origin, then the lower bound for r is 0. Since r is a radius, it is never negative.

To establish bounds for θ, visualize a ray attached at the origin that "sweeps" through the region in a counterclockwise manner. The angle at which θ enters the region is the lower bound c, and the value at which θ exits the region is the upper bound d. The "sweep" must always be done in a counterclockwise manner, so it may be necessary to allow θ to be negative in order to preserve the ordering of the values c and d.

Example 37.1: Describe the following regions using polar coordinates.

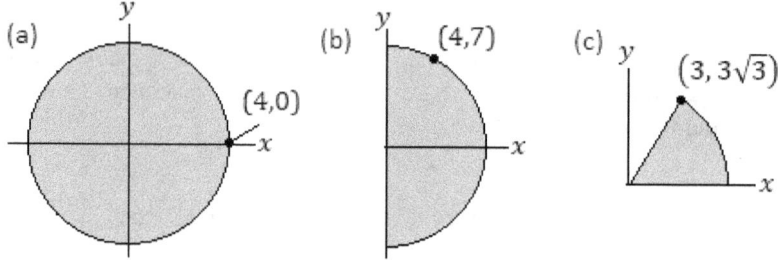

Solution:

(a) The point (4,0) in rectangular coordinates suggests that this circle has a radius of 4. Since the region includes the origin, we have a lower bound of $r = 0$, and since the circle has radius 4, we have an upper bound of $r = 4$. Thus, the interval for r is $0 \leq r \leq 4$. Since this is an entire circle, we have the interval for θ as $0 \leq \theta \leq 2\pi$.

(b) The point (4,7) in rectangular coordinates allows us to determine the radius, which is $r = \sqrt{4^2 + 7^2} = \sqrt{65}$. The region includes the origin. Thus, $0 \leq r \leq \sqrt{65}$. Meanwhile, this semicircle is swept by a ray that would start at the negative y-axis and sweep counterclockwise to the positive y-axis. Thus, the interval for θ is $-\frac{\pi}{2} \leq \theta \leq \frac{\pi}{2}$.

(c) The point $(3, 3\sqrt{3})$ in rectangular coordinates allows us to find the radius, which is $r = \sqrt{3^2 + (3\sqrt{3})^2} = \sqrt{9 + 27} = \sqrt{36} = 6$. Since the region also contains the origin, we have $0 \leq r \leq 6$. The region is swept by a ray that starts at the positive x-axis, so $\theta = 0$. To find the upper bound for θ, observe that the point $(3, 3\sqrt{3})$ lies on this ray. Since $x = 3$ and $y = 3\sqrt{3}$, we have $\theta = \arctan\left(\frac{3\sqrt{3}}{3}\right) = \arctan\sqrt{3} = \frac{\pi}{3}$ radians. Thus, $0 \leq \theta \leq \frac{\pi}{3}$.

◆ ● ◆ ● ◆ ● ◆ ● ◆ ● ◆

Integration in Polar Coordinates: The standard form for integration in polar coordinates is

$$\int_c^d \int_a^b f(r \cos\theta, r \sin\theta) \, r \, dr \, d\theta,$$

where $a \leq r \leq b$ and $c \leq \theta \leq d$. The area element is $r \, dr \, d\theta$, the r being the **Jacobian** of integration (Jacobians are discussed in Section 42).

Example 37.2: Evaluate

$$\int_0^3 \int_0^{\sqrt{9-x^2}} xy \, dy \, dx.$$

Solution: The region of integration is a quarter circle in the first quadrant, center at the origin, radius 3.

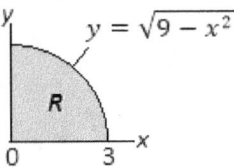

This integral is solved using polar coordinates. The bounds of integration are $0 \le r \le 3$ and $0 \le \theta \le \frac{\pi}{2}$. Furthermore, we substitute $x = r\cos\theta$ and $y = r\sin\theta$, and exchange $dy\, dx$ with $r\, dr\, d\theta$:

$$\int_0^3 \int_0^{\sqrt{9-x^2}} xy \, dy \, dx = \int_0^{\pi/2} \int_0^3 (r\cos\theta)(r\sin\theta) \, r \, dr \, d\theta$$

$$= \int_0^{\pi/2} \int_0^3 r^3 \cos\theta \sin\theta \, dr \, d\theta.$$

The inside integral is evaluated first:

$$\int_0^3 r^3 \cos\theta \sin\theta \, dr = \cos\theta \sin\theta \left[\frac{1}{4} r^4\right]_0^3$$

$$= \frac{81}{4} \cos\theta \sin\theta.$$

This is then integrated with respect to θ, using u-du substitution, with $u = \sin\theta$ and $du = \cos\theta$:

$$\frac{81}{4} \int_0^{\pi/2} \cos\theta \sin\theta \, d\theta = \left[\frac{81}{8} \sin^2\theta\right]_0^{\pi/2}$$

$$= \frac{81}{8} [1^2 - 0]$$

$$= \frac{81}{8}.$$

Example 37.3: Evaluate

$$\int_{-5}^{-2}\int_{0}^{\sqrt{25-x^2}} x^2\,dy\,dx + \int_{-2}^{2}\int_{\sqrt{4-x^2}}^{\sqrt{25-x^2}} x^2\,dy\,dx + \int_{2}^{5}\int_{0}^{\sqrt{25-x^2}} x^2\,dy\,dx.$$

Solution: The region of integration as suggested by the bounds in the three integrals is shown below (See Example 36.2):

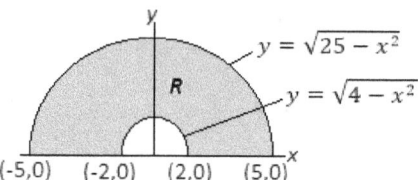

This region is better described using polar coordinates, where $2 \leq r \leq 5$ and $0 \leq \theta \leq \pi$. Replace the integrand x^2 with $(r\cos\theta)^2 = r^2\cos^2\theta$, and the area element $dy\,dx$ with $r\,dr\,d\theta$. In doing so, the three double integrals above, in rectangular coordinates, are equivalent to one double integral, in polar coordinates, with constant bounds:

$$\int_{0}^{\pi}\int_{2}^{5} r^2 \cos^2\theta\, r\, dr\, d\theta, \quad \text{which simplifies to} \quad \int_{0}^{\pi}\int_{2}^{5} r^3 \cos^2\theta\, dr\, d\theta.$$

The inside integral with respect to r is evaluated first:

$$\int_{2}^{5} r^3 \cos^2\theta\, dr = \cos^2\theta \left[\frac{1}{4}r^4\right]_{2}^{5}$$

$$= \frac{1}{4}\cos^2\theta\,[(5)^4 - (2)^4]$$

$$= \frac{609}{4}\cos^2\theta.$$

This expression is next integrated with respect to θ. To antidifferentiate $\cos^2\theta$, use the identity $\cos^2\theta = \frac{1}{2}(1+\cos 2\theta)$:

$$\int_{0}^{\pi}\frac{609}{4}\cos^2\theta\,d\theta = \frac{609}{4}\int_{0}^{\pi}\left(\frac{1}{2}(1+\cos 2\theta)\right)d\theta$$

$$= \frac{609}{8}\int_{0}^{\pi} 1+\cos 2\theta\, d\theta$$

$$= \frac{609}{8}\left[\theta + \frac{1}{2}\sin 2\theta\right]_0^\pi$$

$$= \frac{609}{8}\left[\left(\pi + \frac{1}{2}\sin 2(\pi)\right) - \left(0 + \frac{1}{2}\sin 2(0)\right)\right]$$

$$= \frac{609}{8}\pi.$$

In this example, it is faster and preferable to integrate using polar coordinates rather than perform three double integrals using rectangular coordinates.

Example 37.4: Evaluate

$$\iint_R \frac{1}{(1+x^2+y^2)^2}\, dA,$$

where R is the region in the xy-plane outside the circle of radius 1 that is centered at the origin.

Solution: Below is a sketch of the region of integration R.

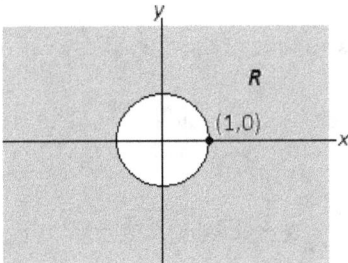

An arrow drawn from the origin outward enters region R when $r = 1$. Since the region extends forever, use ∞ as the upper bound for r. Thus, $1 \le r < \infty$. The bounds for θ are $0 \le \theta \le 2\pi$.

The double integral, using polar coordinates, is

$$\int_0^{2\pi}\int_1^\infty \frac{1}{(1+r^2)^2}\, r\, dr\, d\theta.$$

The inside integral is evaluated using u-du substitution, where $u = 1 + r^2$ and $du = 2r\, dr$. We have

$$\int_1^\infty \frac{1}{(1+r^2)^2} r\, dr = \lim_{d\to\infty} \int_1^d \frac{r}{(1+r^2)^2}\, dr$$

$$= \lim_{d\to\infty} \left[\frac{1}{2}\int_1^d \frac{1}{(1+r^2)^2} 2r\, dr\right]$$

$$= \frac{1}{2}\lim_{d\to\infty}\left[-\frac{1}{1+r^2}\right]_1^d$$

$$= \frac{1}{2}\lim_{d\to\infty}\left[\left(-\frac{1}{1+(d)^2}\right) - \left(-\frac{1}{1+(1)^2}\right)\right]$$

$$= \frac{1}{2}\left[0 - \left(-\frac{1}{2}\right)\right] = \frac{1}{4}.$$

The integral with respect to θ is now evaluated:

$$\frac{1}{4}\int_0^{2\pi} d\theta = \frac{1}{4}[\theta]_0^{2\pi} = \frac{2\pi}{4} = \frac{\pi}{2}.$$

◆ ◆ ◆ ◆ ◆ ◆ ◆ ◆ ◆ ◆ ◆

Example 37.5: Find the volume of the ellipsoid $x^2 + \frac{1}{9}y^2 + z^2 = 1$.

Solution: We must decide which variable should be isolated. Note that solving for y yields

$$y = \pm 3\sqrt{1 - x^2 - z^2}.$$

Furthermore, when $y = 0$, then the ellipsoid forms a circle, $x^2 + z^2 = 1$, on the xz-plane. We will integrate $y = f(x, z)$ over a region of integration R that is the disk $x^2 + z^2 \leq 1$. In rectangular coordinates, we have the double integral

$$\int_{-1}^{1}\int_{-\sqrt{1-x^2}}^{\sqrt{1-x^2}} \left(3\sqrt{1-x^2-z^2} - \left(-3\sqrt{1-x^2-z^2}\right)\right) dz\, dx.$$

This looks challenging, so instead, we use polar coordinates in place of variables x and z. Region R is now defined by $0 \leq r \leq 1$ and $0 \leq \theta \leq 2\pi$, and the integrand is now written $y = \pm 3\sqrt{1-r^2}$.

The double integral in polar coordinates is now

$$\int_0^{2\pi} \int_0^1 \left(3\sqrt{1-r^2} - \left(-3\sqrt{1-r^2}\right)\right) r\, dr\, d\theta.$$

This simplifies to

$$6 \int_0^{2\pi} \int_0^1 \sqrt{1-r^2}\, r\, dr\, d\theta.$$

Note: It is also possible to use symmetry to set up the integral, noting that the volume between the xz-plane and $y = 3\sqrt{1-r^2}$ is the same as the volume between the xz-plane and $y = -3\sqrt{1-r^2}$. Thus, we could use $3\sqrt{1-r^2}$ as the integrand, and double the result. Either way, the $\sqrt{1-r^2}$ remains in the integrand, the 3 moves to the front and is doubled, so that we arrive at the same integral as above.

The inside integral is evaluated using u-du substitution, with $u = 1 - r^2$ and $du = -2r\, dr$:

$$\int_0^1 \sqrt{1-r^2}\, r\, dr = \left[-\frac{1}{3}(1-r^2)^{3/2}\right]_0^1$$

$$= 0 - \left(-\frac{1}{3}\right)$$

$$= \frac{1}{3}.$$

This is then integrated with respect to θ:

$$6 \int_0^{2\pi} \left(\frac{1}{3}\right) d\theta = 2 \int_0^{2\pi} d\theta = 4\pi.$$

Example 37.6: Find the volume of the solid bounded by $z = 2x^2 + 2y^2$ and $z = 9 - 2x^2 - 2y^2$.

Solution: Set the two functions equal and simplify:

$$2x^2 + 2y^2 = 9 - 2x^2 - 2y^2$$
$$4x^2 + 4y^2 = 9$$
$$x^2 + y^2 = \frac{9}{4}.$$

Thus, the region of integration is the disk $x^2 + y^2 \leq \frac{9}{4}$, which can be described in polar coordinates as $0 \leq r \leq \frac{3}{2}$ and $0 \leq \theta \leq 2\pi$. Below is a sketch of the solid along with the region of integration:

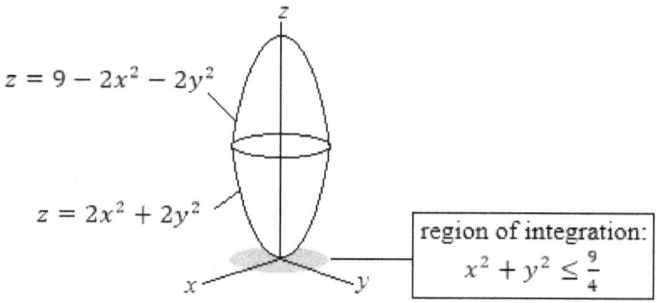

The integrand is the "top" boundary $(z = 9 - 2x^2 - 2y^2)$ subtracted by the "bottom" boundary $(z = 2x^2 + 2y^2)$. This is

$$9 - 2x^2 - 2y^2 - (2x^2 + 2y^2) = 9 - 4x^2 - 4y^2$$
$$= 9 - 4(x^2 + y^2)$$
$$= 9 - 4r^2.$$

The volume is found by evaluating

$$\int_0^{2\pi} \int_0^{3/2} (9 - 4r^2) \, r \, dr \, d\theta,$$

or simplified as: $\int_0^{2\pi} \int_0^{3/2} (9r - 4r^3) \, dr \, d\theta.$

For the inner integral, we have

$$\int_0^{3/2} (9r - 4r^3)\, dr = \left[\frac{9}{2}r^2 - r^4\right]_0^{3/2}$$

$$= \frac{9}{2}\left(\frac{3}{2}\right)^2 - \left(\frac{3}{2}\right)^4 - 0$$

$$= \frac{243}{16}.$$

Finally, the volume is

$$\int_0^{2\pi} \frac{243}{16}\, d\theta = \frac{243}{16}\int_0^{2\pi} d\theta$$

$$= \frac{243}{16}(2\pi)$$

$$= \frac{243\pi}{8}\ \text{units}^3.$$

◆ ◆ ◆ ◆ ◆ ◆ ◆ ◆ ◆ ◆

Example 37.7: A sphere of radius 10 is intersected by a circular cylinder of radius 6 such that the cylinder and the sphere share a common axis of symmetry (that is, the cylinder's axis of symmetry intersects the sphere through one of the sphere's diameters). Find the volume of the outer ring-shaped solid, which consists of the material inside the sphere but outside of the cylinder.

Solution: Centering everything at the origin and using the z-axis as the line of symmetry, we can define the sphere as $x^2 + y^2 + z^2 = 100$, or $z = \sqrt{100 - x^2 - y^2}$. The sphere intersects the xy-plane and creates a disk $x^2 + y^2 \leq 100$, but the cylinder then removes the inner portion, everything inside a circle of radius 6. Thus, using polar coordinates, the bounds are $6 \leq r \leq 10$ and $0 \leq \theta \leq 2\pi$. The integrand is $z = \sqrt{100 - x^2 - y^2} = \sqrt{100 - r^2}$. The volume of this solid is given by

$$2\int_0^{2\pi}\int_6^{10} \sqrt{100 - r^2}\ r\, dr\, d\theta.$$

The leading 2 represents the fact that the integral, as shown, will determine the volume between the sphere's surface and the xy-plane. We need to double the result to find the entire solid's volume.

The inside integral is evaluated with respect to r:

$$\int_6^{10} \sqrt{100-r^2}\, r\, dr = \left[-\frac{1}{3}(100-r^2)^{3/2}\right]_6^{10}$$

$$= \left(-\frac{1}{3}(100-(10)^2)^{3/2}\right) - \left(-\frac{1}{3}(100-(6)^2)^{3/2}\right)$$

$$= \left(-\frac{1}{3}(100-100)^{3/2}\right) - \left(-\frac{1}{3}(100-36)^{3/2}\right)$$

$$= 0 - \left(-\frac{1}{3}(64)^{3/2}\right)$$

$$= \frac{512}{3}.$$

Then, the outer integral is evaluated.

$$2\int_0^{2\pi} \frac{512}{3}\, d\theta = \frac{1024}{3}(2\pi)$$

$$= \frac{2048\pi}{3}\text{ units}^3.$$

◆ ◆ ◆ ◆ ◆ ◆ ◆ ◆ ◆ ◆

Example 37.8: Find the area inside a circle or radius 1 centered at the origin, to the right of the vertical line $x = \frac{1}{2}$.

Solution: The region is shown below:

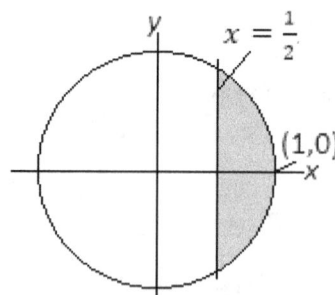

The circle's equation is $x^2 + y^2 = 1$, or $y = \pm\sqrt{1-x^2}$. Using a single integral, the shaded area is given by

$$2\int_{1/2}^{1} \sqrt{1-x^2}\, dx.$$

However, this integral requires a trigonometric substitution. Instead, we try a different approach, using a double integral in polar coordinates. The boundaries are redefined: the circle is $r = 1$, and for the line $x = \frac{1}{2}$, and using the substitution $x = r\cos\theta$, we have $r\cos\theta = \frac{1}{2}$, or $r = \frac{1}{2\cos\theta}$. Setting the two equations equal, we solve to determine the bounds for θ:

$$1 = \frac{1}{2\cos\theta} \quad \text{so that} \quad \cos\theta = \frac{1}{2}. \quad \text{Thus,} \quad \theta = \pm\arccos\left(\frac{1}{2}\right).$$

Since the region is symmetric with the positive x-axis, we use 0 as a lower bound for θ and add a leading factor of 2 to the integral to double the result. Thus, the bounds are $0 \leq \theta \leq \arccos\left(\frac{1}{2}\right)$.

An arrow drawn outward from the origin enters the region at the line $r = \frac{1}{2\cos\theta}$, and exits at the circle $r = 1$, so the bounds for r are $\frac{1}{2\cos\theta} \leq r \leq 1$. Since this is an area integral, we use a 1 for the integrand. However, recall that in polar coordinates, the Jacobian r will also be present in the integrand.

$$2\int_{0}^{\arccos(1/2)} \int_{1/(2\cos\theta)}^{1} 1\, r\, dr\, d\theta.$$

The inside integral is evaluated:

$$\int_{1/(2\cos\theta)}^{1} r\, dr = \left[\frac{1}{2}r^2\right]_{1/(2\cos\theta)}^{1}$$

$$= \frac{1}{2}\left(1^2 - \left(\frac{1}{2\cos\theta}\right)^2\right)$$

$$= \frac{1}{2}\left(1 - \frac{1}{4}\sec^2\theta\right).$$

Recall that $\int \sec^2 u \, du = \tan u + C$. Thus, we now integrate this expression with respect to θ:

$$2 \int_0^{\arccos(1/2)} \frac{1}{2}\left(1 - \frac{1}{4}\sec^2 \theta\right) d\theta = \int_0^{\arccos(1/2)} \left(1 - \frac{1}{4}\sec^2 \theta\right) d\theta$$

$$= \left[\theta - \frac{1}{4}\tan \theta\right]_0^{\arccos(1/2)}$$

$$= \left(\arccos\left(\frac{1}{2}\right) - \frac{1}{4}\tan\left(\arccos\left(\frac{1}{2}\right)\right)\right) - 0.$$

The expression $\tan\left(\arccos\left(\frac{1}{2}\right)\right)$ can be simplified. If $\theta = \arccos\left(\frac{1}{2}\right)$, then this suggests a right triangle with hypotenuse of length 2, adjacent leg of length 1, and by Pythagoras' Theorem, an opposite leg of length $\sqrt{3}$.

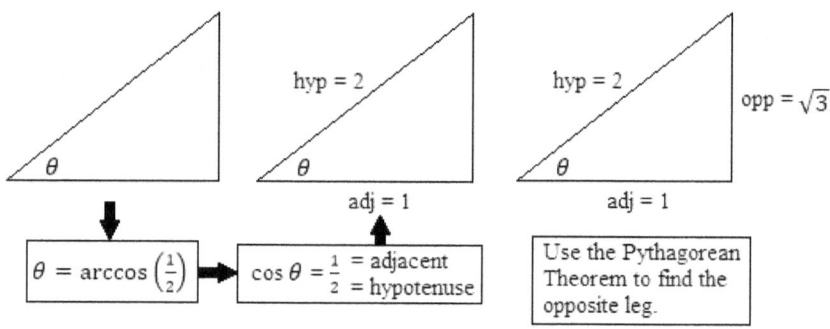

Thus, we have $\tan \theta = \frac{\text{opposite leg}}{\text{adjacent leg}} = \frac{\sqrt{3}}{1} = \sqrt{3}$, so therefore, $\tan\left(\arccos\left(\frac{1}{2}\right)\right) = \sqrt{3}$. The area of the region is $\arccos\left(\frac{1}{2}\right) - \frac{\sqrt{3}}{4}$, or about 0.614 units².

38. Triple Integration over Rectangular Regions

A rectangular solid region S in R^3 is defined by three compound inequalities,

$$a_1 \leq x \leq a_2, \quad b_1 \leq y \leq b_2, \quad c_1 \leq z \leq c_2,$$

where a_1, a_2, b_1, b_2, c_1 and c_2 are constants. A function of three variables $w = f(x, y, z)$ that is continuous over S can be integrated as a **triple integral**:

$$\iiint_S f(x,y,z) \, dV = \int_{a_1}^{a_2} \int_{b_1}^{b_2} \int_{c_1}^{c_2} f(x,y,z) \, dz \, dy \, dx.$$

Observe that the integrals are nested: the inside integral, labeled dz, is associated with the bounds $c_1 \leq z \leq c_2$, and similarly as one works outward.

The volume element is labeled dV and there are six possible orderings of the differentials dx, dy and dz, whose product is equivalent to dV:

$$dz \, dy \, dx, \quad dz \, dx \, dy, \quad dy \, dz \, dx,$$
$$dy \, dx \, dz, \quad dx \, dz \, dy, \quad dx \, dy \, dz.$$

When all bounds are constant, no ordering is more advantageous than any other.

♦ ♦ ♦ ♦ ♦ ♦ ♦ ♦

Example 38.1: Evaluate

$$\int_{-1}^{2} \int_{1}^{3} \int_{-2}^{4} (x + 2yz^2) \, dz \, dy \, dx.$$

Solution: The inner-most integral is evaluated. Since the integrand is being antidifferentiated with respect to z, the variables x and y are treated as constants or coefficients for the moment:

$$\int_{-2}^{4} (x + 2yz^2) \, dz = \left[xz + \frac{2}{3} y z^3 \right]_{-2}^{4}$$
$$= \left(x(4) + \frac{2}{3} y(4)^3 \right) - \left(x(-2) + \frac{2}{3} y(-2)^3 \right)$$
$$= \left(4x + \frac{128}{3} y \right) - \left(-2x - \frac{16}{3} y \right)$$
$$= 6x + 48y.$$

This is now integrated with respect to y (the "middle" integral). The x is still treated as a constant or coefficient in this step:

$$\int_1^3 (6x + 48y) \, dy = [6xy + 24y^2]_1^3$$
$$= (6x(3) + 24(3)^2) - (6x(1) + 24(1)^2)$$
$$= 12x + 192.$$

Lastly, this is integrated with respect to x, the "outer" integral:

$$\int_{-1}^2 (12x + 192) \, dx = [6x^2 + 192x]_{-1}^2$$
$$= (6(2)^2 + 192(2)) - (6(-1)^2 + 192(-1))$$
$$= 594.$$

◆ · ◆ · ◆ · ◆ · ◆

How do we interpret answers obtained from a triple integral? Analogous to a single-variable integral (the definite integral is the *area* between a curve and over an interval on the input axis) and a two-variable double integral (the definite double integral is the *volume* between a surface and a region in the input plane), a three-variable continuous function $w = f(x, y, z)$ evaluated over a triple integral gives a "hyper-volume" between the graph of f and the region S in R^3 over which it is being integrated. However, the graph of $w = f(x, y, z)$ is embedded within R^4, so it is not easy to visualize this four-dimensional analog to area or volume. Nevertheless, it is a reasonable interpretation.

One corollary is to allow the integrand to be 1. In such a case, we get a volume integral, where $\iiint_S 1 \, dV$ is the volume of S.

Example 38.2: Evaluate

$$\int_{-3}^5 \int_2^4 \int_{-1}^8 1 \, dz \, dy \, dx.$$

Solution: Working inside out, we have $\int_{-1}^8 1 \, dz = [z]_{-1}^8 = 8 - (-1) = 9$. Then, we have $9 \int_2^4 dy = 9[y]_2^4 = 9(4-2) = 18$. Lastly, we have $18 \int_{-3}^5 dx = 18[x]_{-3}^5 = 18(5 - (-3)) = 144.$

This is the volume of the rectangular solid region in R^3 in which length x is 8 units, length y is 2 units, and length z is 9 units. Not surprisingly, $(8)(2)(9) = 144$ cubic units.

Example 38.3: Evaluate

$$\int_2^5 \int_0^4 \int_{-1}^3 x^2 yz^3 \, dx \, dy \, dz.$$

Solution: Note the order of integration. The inside integral is integrated with respect to x. The yz^3 factors are treated as a constant and moved outside the integral:

$$\int_{-1}^3 x^2 yz^3 \, dx = yz^3 \int_{-1}^3 x^2 \, dx$$

$$= yz^3 \left[\frac{1}{3}x^3\right]_{-1}^3$$

$$= yz^3 \left(\left(\frac{1}{3}(3)^3\right) - \left(\frac{1}{3}(-1)^3\right)\right) = \frac{28}{3}yz^3.$$

This expression is now integrated with respect to y, the middle integral. We can move the $\frac{28}{3}z^3$ to the front of the integral:

$$\int_0^4 \left(\frac{28}{3}yz^3\right) dy = \frac{28}{3}z^3 \int_0^4 y \, dy$$

$$= \frac{28}{3}z^3 \left[\frac{1}{2}y^2\right]_0^4$$

$$= \frac{28}{3}z^3 (8)$$

$$= \frac{224}{3}z^3.$$

Lastly, this expression is integrated with respect to z:

$$\int_2^5 \left(\frac{224}{3}z^3\right) dz = \frac{224}{3} \int_2^5 z^3 \, dz$$

$$= \frac{224}{3} \left[\frac{1}{4}z^4\right]_2^5$$

$$= \frac{224}{3} \left(\left(\frac{1}{4}(5)^4\right) - \left(\frac{1}{4}(2)^4\right)\right)$$

$$= \frac{224}{3}\left(\frac{609}{4}\right)$$

$$= \frac{136{,}416}{12} = 11{,}368.$$

If the integrand is held by multiplication so that it can be written as $f(x, y, z) = g(x)h(y)k(z)$, and the bounds are constants, then

$$\int_{a_1}^{a_2} \int_{b_1}^{b_2} \int_{c_1}^{c_2} f(x, y, z) \, dz \, dy \, dx = \int_{a_1}^{a_2} \int_{b_1}^{b_2} \int_{c_1}^{c_2} g(x)h(y)k(z) \, dz \, dy \, dx$$

$$= \left(\int_{a_1}^{a_2} g(x) \, dx \right) \left(\int_{b_1}^{b_2} h(y) \, dy \right) \left(\int_{c_1}^{c_2} k(z) \, dz \right).$$

◆ • ◆ • ◆ • ◆ • ◆

Example 38.4: Use the shortcut shown above to evaluate

$$\int_2^5 \int_0^4 \int_{-1}^3 x^2 y z^3 \, dx \, dy \, dz.$$

Solution: Since the bounds are constants and the integrand is held by multiplication, the above triple integral can be rewritten as a product of three single-variable integrals, and evaluated individually:

$$\left(\int_{-1}^3 x^2 \, dx \right) \left(\int_0^4 y \, dy \right) \left(\int_2^5 z^3 \, dz \right) = \left(\left[\tfrac{1}{3} x^3 \right]_{-1}^3 \right) \left(\left[\tfrac{1}{2} y^2 \right]_0^4 \right) \left(\left[\tfrac{1}{4} z^4 \right]_2^5 \right)$$

$$= \left(\tfrac{1}{3}(3^3 - (-1)^3) \right) \left(\tfrac{1}{2}(4^2 - 0^2) \right) \left(\tfrac{1}{4}(5^4 - 2^4) \right)$$

$$= \left(\tfrac{28}{3} \right) (8) \left(\tfrac{609}{4} \right) = 11{,}368.$$

Note that in Example 38.1, the integral $\int_{-1}^2 \int_1^3 \int_{-2}^4 (x + 2yz^2) \, dz \, dy \, dx$ would not be solvable using the above shortcut.

◆ • ◆ • ◆ • ◆ • ◆

See an error? Have a suggestion?
Please see www.surgent.net/vcbook

39. Triple Integration over Non-Rectangular Regions of Type I

A solid region S in R^3 is **Type I** if there is no ambiguity as to any of its bounds of integration in such a way that one triple integral is sufficient to describe S. Because there are six possible orderings of the variables of integration, it is possible that one ordering may result in a non-Type I (called **Type-II**) region, while another ordering may result in a Type I region. Whenever possible, choose a Type I ordering of integration.

For example, all rectangular solid regions in the previous examples are Type I, in any ordering of the differentials.

◆ ◆ ◆ ◆ ◆ ◆ ◆ ◆ ◆

Example 39.1: Set up the integral for $\iiint_S f(x,y,z)\,dV$, where S is a solid hemisphere, centered at the origin, of radius 2 such that $z \geq 0$.

Solution: Sketch the solid. The restriction $z \geq 0$ means all points are on or above the xy-plane:

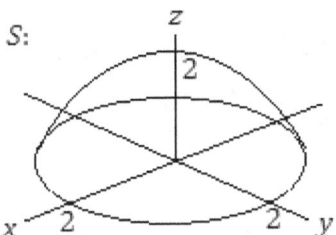

Now, select an ordering of integration. Let's try $dz\,dy\,dx$, so that the first integral is evaluated with respect to z. Sketch an arrow in the positive z direction so that it enters the solid through one surface, and exits through another. It is important to observe that in this case, there is *no* ambiguity as to where such an arrow would enter or exit the solid: it *must* enter through the surface $z_1 = 0$ (the xy-plane) and *must* exit through $z_2 = \sqrt{4 - x^2 - y^2}$, the hemisphere. These will be the bounds for the dz integral.

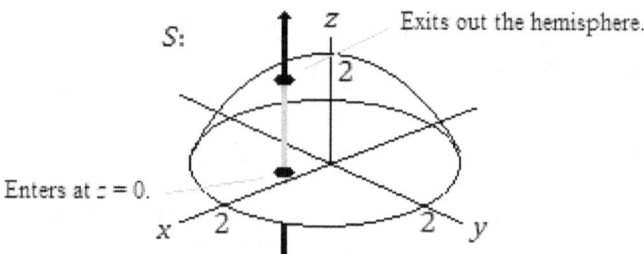

Now, we concentrate on the region defined by the x and y variables. This is the "footprint" of the solid on the xy-plane, and is a disk of radius 2, centered at the origin:

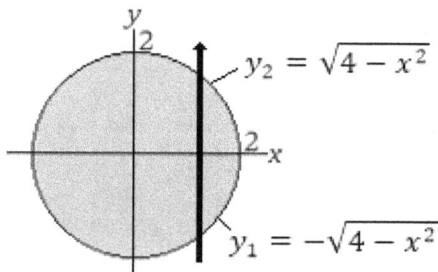

If we next choose to integrate with respect to y, we draw an arrow in the positive y direction. It will enter the region through the lower half of the circle, $y_1 = -\sqrt{4-x^2}$, and exit through the upper half, $y_2 = \sqrt{4-x^2}$. There is no ambiguity as to where this arrow enters or exits the region. It is of Type I as well.

Lastly, the bounds for x are constants: $-2 \leq x \leq 2$. The triple integral is

$$\iiint_S f(x,y,z)\, dV = \int_{-2}^{2} \int_{-\sqrt{4-x^2}}^{\sqrt{4-x^2}} \int_{0}^{\sqrt{4-x^2-y^2}} f(x,y,z)\, dz\, dy\, dx.$$

If $f(x,y,z) = 1$, then $\iiint_S dV$ represents the volume of the hemisphere of radius 2 (the region of integration S). Using geometry, the volume of a hemisphere is $\frac{1}{2}\left(\frac{4}{3}\pi r^3\right)$. When $r = 2$, we have $\frac{1}{2}\left(\frac{4}{3}\pi r^3\right) = \frac{2}{3}\pi(2)^3 = \frac{16}{3}\pi$:

$$\int_{-2}^{2} \int_{-\sqrt{4-x^2}}^{\sqrt{4-x^2}} \int_{0}^{\sqrt{4-x^2-y^2}} dz\, dy\, dx = \frac{16}{3}\pi.$$

Triple integrals defined over spheres (or portions thereof) such as the one above are better set up and evaluated using spherical coordinates, in Section 40.

The Legal Form of a Triple Integral

Triple integrals follow the form shown below:

$$\int_a^b \int_{y_1(x)}^{y_2(x)} \int_{z_1(x,y)}^{z_2(x,y)} f(x,y,z) \, dz \, dy \, dx.$$

Note the ordering of integration: z first, then y, then x. If this ordering is chosen, then the innermost integral will have bounds that may contain x and y, possibly both:

$$z_1(x,y) \leq z \leq z_2(x,y).$$

The next integral, with respect to y, may have bounds that contain x, but not z:

$$y_1(x) \leq y \leq y_2(x).$$

The last (outermost) integral with respect to x, has bounds that are constants:

$$a \leq x \leq b.$$

The ordering of integration "drives" the bounds, so to speak. The following is a legal triple integral but in a different ordering of integration:

$$\int_{-1}^{4} \int_{-x}^{3x} \int_{0}^{x+z} (x^2 + z) \, dy \, dz \, dx.$$

Note that the innermost integral with respect to y has bounds that may contain x or z (or both), while the middle integral, with respect to z, has bounds that may contain x, but not y. The outer integral's bounds must be constant.

This is an "illegal" triple integral:

$$\int_0^2 \int_{x+z}^{2y} \int_{-y^2}^{x} (\sin(xyz) + x) \, dz \, dy \, dx.$$

The innermost integral is legal: the bounds with respect to z may contain x or y (or both). However, the middle integral, with respect to y, cannot contain itself as a variable, nor z, since z is "done" by the time we evaluate this middle integral.

Example 39.2: Solid S is shown below. Let $f(x, y, z)$ be a generic integrand.

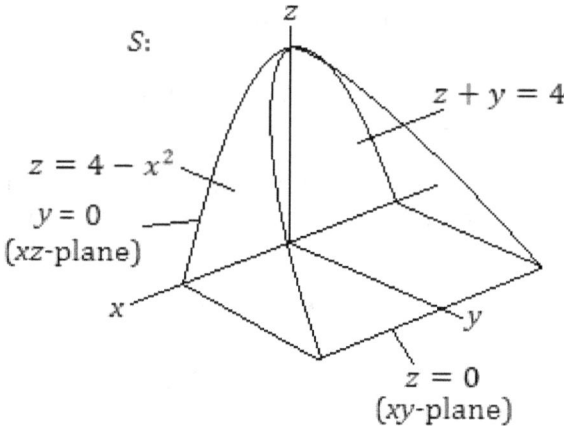

a) Set up a triple integral over S in the $dy\, dz\, dx$ ordering.
b) Set up a triple integral over S in the $dx\, dy\, dz$ ordering.
c) Explain why any ordering starting with dz is not of Type I.

Solution:

a) Sketch an arrow in the positive y direction:

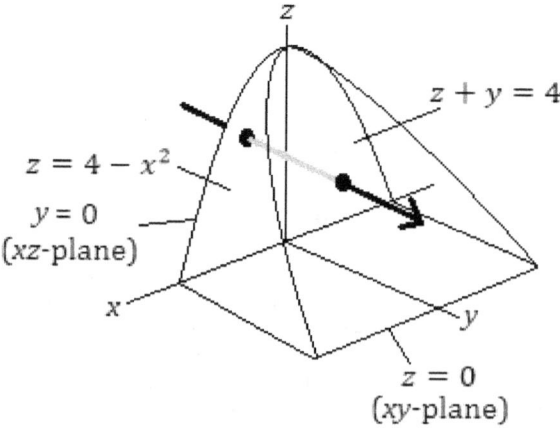

This arrow enters the solid at the xz-plane ($y_1 = 0$), passes through the interior (gray), and exits out the plane $z + y = 4$, or $y_2 = 4 - z$. These are the bounds for y.

Next, we look at the footprint of the solid as projected onto the xz-plane. Variable y is no longer needed.

224

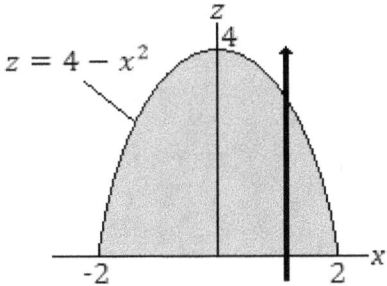

This region is Type I. The z-bounds, as shown by the arrow above, are $0 \leq z \leq 4 - x^2$, and the x bounds are constants, $-2 \leq x \leq 2$. Thus, the triple integral is

$$\int_{-2}^{2} \int_{0}^{4-x^2} \int_{0}^{4-z} f(x,y,z) \, dy \, dz \, dx.$$

Note that this integral is "legal". Do you agree?

b) For the $dx \, dy \, dz$ ordering, draw an arrow in the positive x direction. It enters the region through the parabolic sheet $x_1 = -\sqrt{4-z}$ and exits through $x_2 = \sqrt{4-z}$.

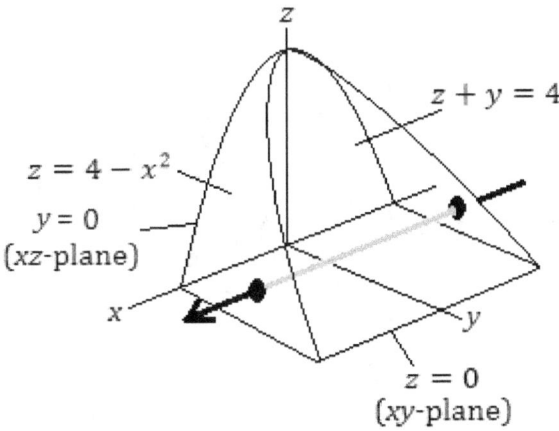

Variable x is "done". We now look at the footprint of the solid projected onto the yz plane, and since the middle integral will be with respect to y, we sketch an arrow in the positive y direction.

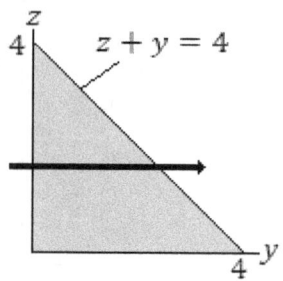

This region is also Type I. An arrow drawn in the positive y direction enters it at $y_1 = 0$ (the z axis) and exits through the line $y_2 = 4 - z$. Finally, the bounds on z are $0 \leq z \leq 4$. The triple integral is

$$\int_0^4 \int_0^{4-z} \int_{-\sqrt{4-z}}^{\sqrt{4-z}} f(x, y, z) \, dx \, dy \, dz.$$

Study this integral to convince yourself it is legal.

c) Any ordering starting with dz is not of Type I because an arrow drawn in the positive z direction may exit through the plane $z = 4 - y$, or the parabolic sheet $z = 4 - x^2$. Because there is ambiguity as to z's bounds, this solid is not of Type I if starting the integration with respect to z. In such a case, it's wise to find a different ordering.

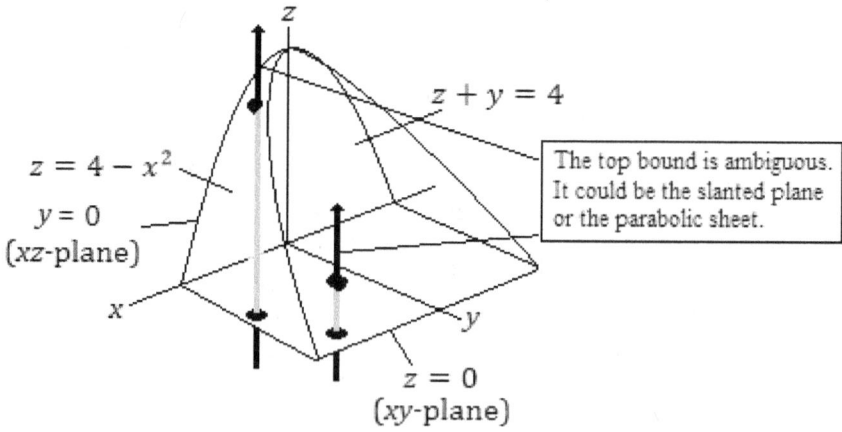

Example 39.3: Solid S is bounded by the surface $z = 4 - x^2 - y^2$, the plane $y = x$, the xy-plane and the xz-plane in the first octant. Find this solid's volume.

Solution: It is important to visualize the solid. The surface $z = 4 - x^2 - y^2$ is a paraboloid with vertex (0,0,4) that opens downward (left image below). The plane $y = x$ can be seen as the line $y = x$ in R^2, then extended into the z-direction (middle image, below).

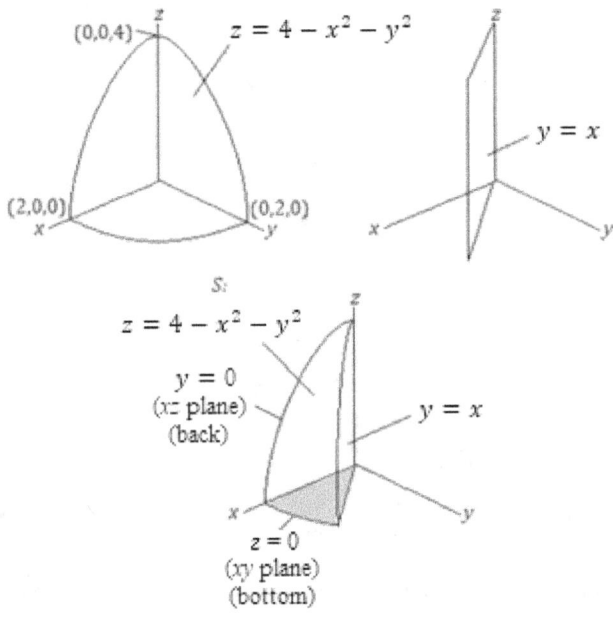

If we choose to integrate with respect to z first, there will be no ambiguity in the bounds. The bounds for z will be $0 \leq z \leq 4 - x^2 - y^2$. The footprint of this region on the xy-plane is a circular wedge:

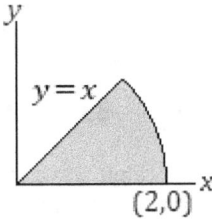

We use polar coordinates to describe this region. This bounds are $0 \leq r \leq 2$ and $0 \leq \theta \leq \frac{\pi}{4}$. However, since we have replaced variables x and y with r and θ, the top bound for z, which is $4 - x^2 - y^2$, is rewritten as $4 - (x^2 + y^2) = 4 - r^2$.

Thus, the volume is given by the triple integral below, with 1 as the integrand. Note the Jacobian r is also present in the integral.

$$\int_0^{\pi/4} \int_0^2 \int_0^{4-r^2} 1 \, dz \, r \, dr \, d\theta.$$

The inside integral is evaluated first:

$$\int_0^{4-r^2} 1 \, dz = 4 - r^2.$$

This is then integrated with respect to r:

$$\int_0^2 (4 - r^2) r \, dr = \int_0^2 (4r - r^3) \, dr = \left[2r^2 - \frac{1}{4} r^4 \right]_0^2 = 8 - 4 = 4.$$

Lastly, the outside integral is evaluated:

$$\int_0^{\pi/4} 4 \, d\theta = 4 \left(\frac{\pi}{4} \right) = \pi.$$

The solid has a volume of π cubic units.

The previous example, in which the variables x and y were replaced with r and θ, is an example of integrating in **cylindrical coordinates**. Note that the variable z was left unchanged, but its bounds, which included variables x and y, had to be adjusted to include the new variables r and θ. In general, such a triple integral in cylindrical coordinates is given by

$$\int_{\theta_1}^{\theta_2} \int_{r_1}^{r_2} \int_{z_1(r,\theta)}^{z_2(r,\theta)} f(r,\theta,z) \, dz \, r \, dr \, d\theta.$$

Typically, the inside integral, with respect to z, is integrated first.

This does not exclude situations where two of the other variables may be exchanged for r and θ. For example, if variables y and z are defined over a region that is better described using polar coordinates, then x is left alone, but the bounds for x are adjusted to include r and θ, and a triple integral in cylindrical coordinates would be given by

$$\int_{\theta_1}^{\theta_2} \int_{r_1}^{r_2} \int_{x_1(r,\theta)}^{x_2(r,\theta)} f(x,r,\theta) \, dx \, r \, dr \, d\theta.$$

Furthermore, the transformation is arbitrary: we can declare that $y = r\cos\theta$ and $z = r\sin\theta$, or that $y = r\sin\theta$ and $z = r\cos\theta$. As long as the bounds are handled correctly, either transformation is acceptable.

◆ • ◆ • ◆ • ◆ • ◆

Example 39.4: A cylinder, $x^2 + z^2 = 1$, is intersected by the planes $y + z = 1$ and $y - z = -1$. Find the volume of this intersecting region.

Solution: Below is a sketch of the region. Note that the cylinder $x^2 + z^2 = 1$ can be viewed as a circle of radius 1, centered at the origin, on the xz-plane, then extended into the positive and negative y directions. The planes $y + z = 1$ and $y - z = -1$ can be viewed as lines on the yz-plane, then extended into the positive and negative x directions.

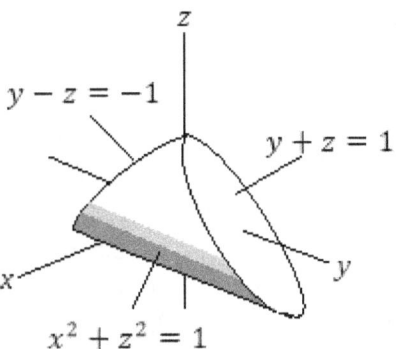

Visualizing an arrow in the positive y direction, it enters the solid through the plane $y - z = -1$, or $y_1 = z - 1$, then exits the solid through the plane $y + z = 1$, or $y_2 = 1 - z$. Note that variables x and z form a circular region on the xz-plane, and this suggests we may want to exchange them for r and θ, and integrate with respect to y first. The bounds for r are $0 \le r \le 1$ and the bounds for θ are $0 \le \theta \le 2\pi$. An initial triple integral in cylindrical coordinates is given by

$$\int_0^{2\pi} \int_0^1 \int_{z-1}^{1-z} (1)\, dy\, r\, dr\, d\theta.$$

However, this is not quite correct. The bounds for y need to be written in terms of r and θ. If we define $x = r\cos\theta$ and $z = r\sin\theta$, the triple integral is now properly written as

$$\int_0^{2\pi}\int_0^1\int_{r\sin\theta-1}^{1-r\sin\theta} (1)\, dy\, r\, dr\, d\theta.$$

The inside integral is evaluated first:

$$\int_{r\sin\theta-1}^{1-r\sin\theta} (1)\, dy = [y]_{r\sin\theta-1}^{1-r\sin\theta} = (1 - r\sin\theta) - (r\sin\theta - 1)$$
$$= 2 - 2r\sin\theta.$$

This is integrated with respect to r:

$$\int_0^1 (2 - 2r\sin\theta)\, r\, dr = \int_0^1 (2r - 2r^2\sin\theta)\, dr$$
$$= \left[r^2 - \frac{2}{3}r^3\sin\theta\right]_0^1$$
$$= 1 - \frac{2}{3}\sin\theta.$$

Finally, this is integrated with respect to θ:

$$\int_0^{2\pi} \left(1 - \frac{2}{3}\sin\theta\right) d\theta = \left[\theta + \frac{2}{3}\cos\theta\right]_0^{2\pi}$$
$$= \left((2\pi) + \frac{2}{3}\cos(2\pi)\right)$$
$$- \left((0) + \frac{2}{3}\cos(0)\right) \quad \begin{cases}\text{Recall that } \cos(2\pi) = 1 \\ \text{and } \cos(0) = 1.\end{cases}$$
$$= 2\pi + \frac{2}{3} - \frac{2}{3}$$
$$= 2\pi.$$

Finding Volumes using Double Integrals and Triple Integrals. What's the Difference?

Any subtraction expression can be written as a definite integral:

$$5 - 2 = \int_2^5 dt.$$

The variable of integration is chosen arbitrarily. Similarly, a single integral in one variable can be rewritten as a double integral in two variables:

$$\int_1^4 x^2 - 2x - 1 \, dx = \int_1^4 \int_0^{x^2-2x-1} dy \, dx = \int_1^4 \int_{2x+1}^{x^2} dy \, dx, \text{ etc.}$$

Suppose we want to determine the volume contained between the surface (graph) of $z = f(x, y)$ and the plane $z = 0$ (the xy-plane), where the region of integration in the xy-plane is defined by $y_1(x) \leq y \leq y_2(x)$ and $a \leq x \leq b$. Using a double integral, we would write

$$\int_a^b \int_{y_1(x)}^{y_2(x)} f(x,y) \, dy \, dx.$$

Using a triple integral, we would write

$$\int_a^b \int_{y_1(x)}^{y_2(x)} \int_0^{f(x,y)} dz \, dy \, dx.$$

Observe that the innermost integral is $\int_0^{f(x,y)} dz = [z]_0^{f(x,y)} = f(x,y)$.

This is a common tactic, in which the integrand can be rewritten as the bound(s) of an entirely new integral. For example, if we wanted to find the volume between the paraboloids $z = 8 - x^2 - y^2$ and $z = x^2 + y^2$, we could represent this volume by a double integral:

$$\int_{-2}^2 \int_{-\sqrt{4-x^2}}^{\sqrt{4-x^2}} ((8 - x^2 - y^2) - (x^2 + y^2)) \, dy \, dx,$$

where the region of integration in the xy-plane is a circle of radius 2, and the integrand is written as "top surface" minus "bottom surface". As a triple integral, these expressions become bounds, and we have

$$\int_{-2}^2 \int_{-\sqrt{4-x^2}}^{\sqrt{4-x^2}} \int_{x^2+y^2}^{8-x^2-y^2} dz \, dy \, dx.$$

Example 39.5: Consider the triple integral

$$\int_{-4}^{4} \int_{0}^{16-y^2} \int_{0}^{\frac{1}{2}z} f(x,y,z) \, dx \, dz \, dy.$$

Rewrite this integral in the $dy \, dz \, dx$ ordering.

Solution: From the bounds, we can develop the solid S over which the integral is defined. Working inside out, we see that the bounds for x are $0 \leq x \leq \frac{1}{2}z$. This suggests that one bounding surface is the yz-plane, since $x = 0$. The other bounding surface is the plane, $x = \frac{1}{2}z$. It is important to remember that the bounding surfaces exist in R^3. Note that $x = \frac{1}{2}z$ is the same as $z = 2x$.

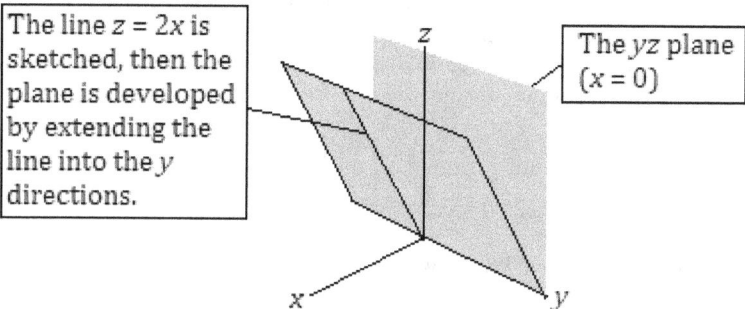

The line $z = 2x$ is sketched, then the plane is developed by extending the line into the y directions.

The yz plane $(x = 0)$

Now, the middle integral suggests that the bounds for z are the xy-plane ($z = 0$) and the parabolic sheet, $z = 16 - y^2$:

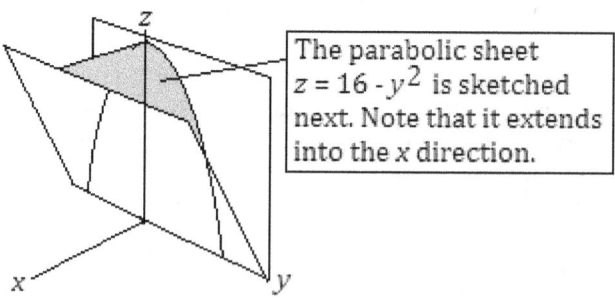

The parabolic sheet $z = 16 - y^2$ is sketched next. Note that it extends into the x direction.

From this, the shape of the solid can be inferred. Strategic points are identified.

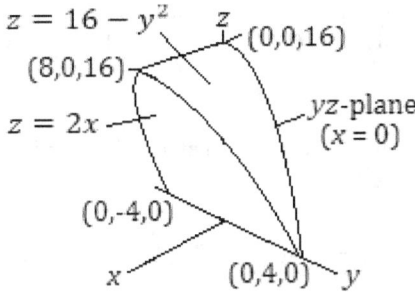

To rewrite the integral in the *dy dz dx* ordering, visualize an arrow in the positive *y* direction. There is no ambiguity where it enters or exits the solid. It enters through one half of the parabolic sheet $y_1 = -\sqrt{16-z}$ and exits through the other half, $y_2 = \sqrt{16-z}$. These are the bounds for *y*.

Now, we view the footprint of the solid as it appears projected onto the *xz*-plane. It will appear as a triangle, as shown below:

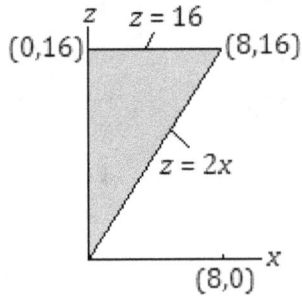

Integrating next with respect to *z*, the lower bound is $z_1 = 2x$ and the upper bound is $z_2 = 16$. Lastly, the bounds for *x* are $0 \le x \le 8$. Thus, the triple integral

$$\int_{-4}^{4} \int_{0}^{16-y^2} \int_{0}^{\frac{1}{2}z} f(x,y,z)\, dx\, dz\, dy$$

is equivalent to

$$\int_{0}^{8} \int_{2x}^{16} \int_{-\sqrt{16-z}}^{\sqrt{16-z}} f(x,y,z)\, dy\, dz\, dx.$$

Example 39.6: Let solid S be a tetrahedron in the first octant with vertices $(0,0,0)$, $(2,0,0)$, $(0,4,0)$ and $(0,0,8)$. Set up all six possible triple integrals $\iiint_S f(x,y,z) \, dV$.

Solution: The equation of the plane that passes through the points $(a,0,0)$, $(0,b,0)$ and $(0,0,c)$ is given by

$$\frac{x}{a} + \frac{y}{b} + \frac{z}{c} = 1. \quad \text{(See Example 13.6)}$$

Thus, the equation of the plane passing through $(2,0,0)$, $(0,4,0)$ and $(0,0,8)$ is

$$\frac{x}{2} + \frac{y}{4} + \frac{z}{8} = 1.$$

If the inside integral is chosen to be evaluated with respect to z, then solve for z, getting $z = 8 - 4x - 2y$. The bounds are $0 \leq z \leq 8 - 4x - 2y$. This leaves a triangular region in the xy-plane with vertices $(0,0)$, $(2,0)$ and $(0,4)$, shown below.

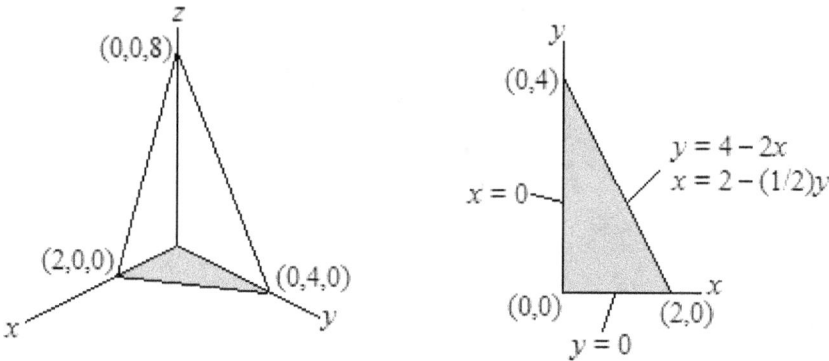

Integrating next with respect to y, the bounds are $0 \leq y \leq 4 - 2x$, and lastly, the bounds on x are $0 \leq x \leq 2$. The triple integral is

$$\int_0^2 \int_0^{4-2x} \int_0^{8-4x-2y} f(x,y,z) \, dz \, dy \, dx.$$

Suppose we chose to integrate next with respect to x instead of y. The bounds on x would be $0 \leq x \leq 2 - \frac{1}{2}y$, with $0 \leq y \leq 4$, and the triple integral is:

$$\int_0^4 \int_0^{2-(1/2)y} \int_0^{8-4x-2y} f(x,y,z) \, dz \, dx \, dy.$$

Repeating this process from the start, suppose that the inside integral is chosen to be evaluated with respect to y. Solve for y, getting $y = 4 - 2x - \frac{1}{2}z$. The bounds of this integral are $0 \leq y \leq 4 - 2x - \frac{1}{2}z$. This leaves a triangular region in the xz-plane with vertices $(0,0)$, $(2,0)$ and $(0,8)$, shown below.

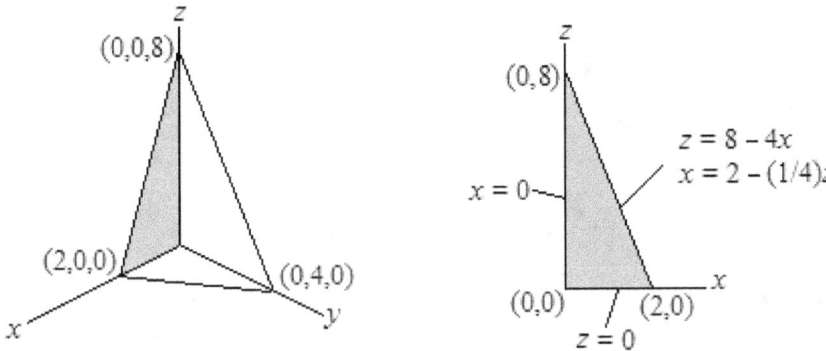

Integrating next with respect to z, the bounds are $0 \leq z \leq 8 - 4x$, where the bounds on x are $0 \leq x \leq 2$. The triple integral is

$$\int_0^2 \int_0^{8-4x} \int_0^{4-2x-(1/2)z} f(x,y,z)\, dy\, dz\, dx.$$

But if we chose to integrate next with respect to x, the bounds on x would be $0 \leq x \leq 2 - \frac{1}{4}z$, where $0 \leq z \leq 8$. The triple integral would be

$$\int_0^8 \int_0^{2-(1/4)z} \int_0^{4-2x-(1/2)z} f(x,y,z)\, dy\, dx\, dz.$$

You should verify that the triple integral in the $dx\, dy\, dz$ ordering is

$$\int_0^8 \int_0^{4-(1/2)z} \int_0^{2-(1/2)y-(1/4)z} f(x,y,z)\, dx\, dy\, dz,$$

and that the triple integral in the $dx\, dz\, dy$ ordering is

$$\int_0^4 \int_0^{8-2y} \int_0^{2-(1/2)y-(1/4)z} f(x,y,z)\, dx\, dz\, dy.$$

These six possible triple integrals all describe the same object and all would evaluate to the same numerical value.

40. Spherical Coordinate System

A point $P = (x, y, z)$ described by rectangular coordinates in R^3 can also be described by three independent variables, ρ (rho), θ and ϕ (phi), whose meanings are given below:

ρ: the distance from the origin to P.

θ: the angle from the positive x-axis to the line connecting the origin to the point $(x, y, 0)$.

ϕ: the angle from the positive z-axis to the line connecting the origin to P.

Descriptively, ρ (rho) is the spherical radius, θ is the "sweep" or "azimuth" angle of the point's projection onto the xy-plane relative to the positive x axis, and ϕ (phi) is the "lean" angle of the point relative to the positive z-axis.

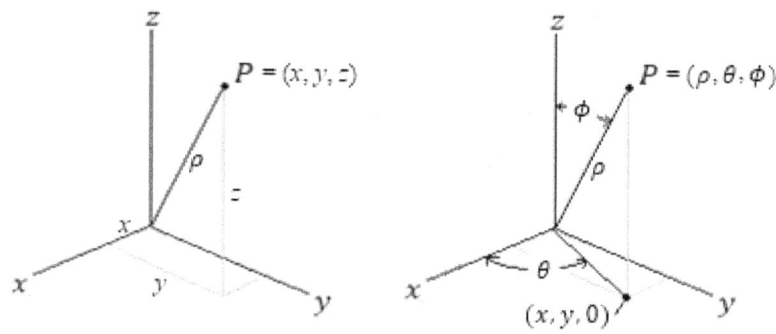

A point P is shown in R^3. A line connects the origin to P (above left). This line has length ρ. This same line also "leans" at an angle of ϕ radians, relative to the positive z-axis (above right). Furthermore, a line is drawn from the origin to the point's projection onto the xy-plane, $(x, y, 0)$. This line is at an angle of θ radians relative to the positive x-axis in the counterclockwise manner. The use of θ here is identical to its usage in the polar and cylindrical coordinate systems and is confined to the xy-plane.

These three variables comprise the **spherical coordinate system** and are best used to describe regions in R^3 that are spheres, or parts of a sphere. For such regions, the bounds of ρ, θ and ϕ will be constants. The common restrictions on ρ, θ and ϕ are:

$$\rho \geq 0, \quad 0 \leq \theta \leq 2\pi, \quad 0 \leq \phi \leq \pi.$$

Any point in R^3 can be described by spherical coordinates (ρ, θ, ϕ) that meet the restrictions stated above.

The variable ϕ can be thought of as the "lean" of the line connecting the origin to P relative to the positive z-axis. If $\phi = 0$, then P lies on the positive z-axis. If $\phi = \frac{\pi}{2}$, then P lies on the xy-plane, which is at right angles to the positive z-axis, and if $\phi = \pi$, then P lies on the negative z-axis.

The conversion formulas between rectangular coordinates (x, y, z) and spherical coordinates (ρ, θ, ϕ) are:

$$\rho = \sqrt{x^2 + y^2 + z^2}, \quad \theta = \arctan\left(\frac{y}{x}\right), \quad \phi = \arctan\left(\frac{\sqrt{x^2 + y^2}}{z}\right).$$

$$x = \rho \sin\phi \cos\theta, \quad y = \rho \sin\phi \sin\theta, \quad z = \rho \cos\phi.$$

◆ ◆ ◆ ◆ ◆ ◆ ◆ ◆ ◆

A Review of the Arctangent Operator, or Getting your θ and ϕ angles correct!

Assume x and y are rectangular coordinates in the xy-plane, and that θ is the angle from the positive x-axis to the line connecting the origin to the point (x,y), given by $\theta = \arctan(y/x)$. When calculating the angle θ, we must be careful in handling the result as given by a calculator. Recall that the arctangent operator only returns values in the first quadrant (if y/x is positive) or fourth quadrants (if y/x is negative).

If both x and y are positive (in Quadrant 1), then y/x is positive, so that $\theta = \arctan\left(\frac{y}{x}\right)$ is in the interval $0 \le \theta \le \frac{\pi}{2}$ (the first quadrant).

Example: $x = 3, y = 2$, so that $\theta = \tan^{-1}\left(\frac{2}{3}\right) \approx 0.59$ radians.

What needs to be done: Nothing. The arctan operator returned a value in the Quadrant I, as expected. The correct answer is $\theta = 0.59$ radians.

If x is negative and y is positive (in Quadrant II), then y/x is negative, but $\theta = \arctan(y/x)$ is in the interval $-\frac{\pi}{2} \leq \theta \leq 0$ (the fourth quadrant). We must add π to this result to place the angle in the interval $\frac{\pi}{2} \leq \theta \leq \pi$, the second quadrant.

Example: $x = -3, y = 2$, so that $\theta = \tan^{-1}\left(\frac{2}{-3}\right) \approx -0.59$ radians. This result is in Quadrant-IV. We need to get it in Quadrant II.

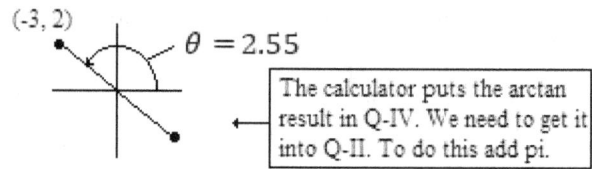

What needs to be done: The value –0.59 is in Quadrant IV, so we add π to place the angle into Quadrant II. The correct answer is $\theta = -0.59 + 3.14 = 2.55$ radians.

If both x and y are negative (in Quadrant III), then y/x is positive, but $\theta = \arctan(y/x)$ is in the interval $0 \leq \theta \leq \frac{\pi}{2}$ (the first quadrant). Thus, we add π to this result to place the angle in the interval $\pi \leq \theta \leq \frac{3\pi}{2}$, the third quadrant.

Example: $x = -3, y = -2$, so that $\theta = \tan^{-1}\left(\frac{-2}{-3}\right) = \tan^{-1}\left(\frac{2}{3}\right) \approx 0.59$. This result is in Quadrant-I. We need to get it in Quadrant III.

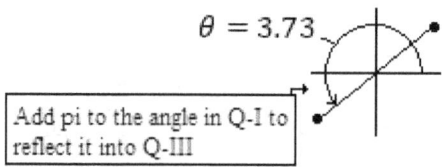

What needs to be done: The value 0.59 is in Quadrant-I, so add π to place the angle into Quadrant III. The correct answer is $\theta = 0.59 + 3.14 = 3.73$ radians.

If x is positive and y is negative (in Quadrant IV), then y/x is negative, so that $\theta = \arctan(y/x)$ is in the interval $-\frac{\pi}{2} \leq \theta \leq 0$ (the fourth quadrant). This is usually acceptable. However, if we desire that the angle be positive, then we add 2π to the result.

Example: $x = 3, y = -2$, so that $\theta = \tan^{-1}\left(\frac{-2}{3}\right) \approx -0.59$. This result is inn Quadrant-IV.

$$\theta = 5.69$$
$$\theta = -0.59$$

What needs to be done: Nothing. The value is in Quadrant IV, as expected. However, if the angle is to be stated as a positive number, add 2π. Thus, $\theta = -0.59 + 6.28 = 5.69$ radians.

For ϕ, the process is simpler. If the point lies above the xy-plane (that is, z is positive), then the result given by $\phi = \arctan(\sqrt{x^2 + y^2}/z)$ will be in the interval $0 \leq \phi \leq \frac{\pi}{2}$ and no adjustments need to be made. If z is negative, then ϕ must be in the interval $\frac{\pi}{2} \leq \phi \leq \pi$. However, the expression $\arctan(\sqrt{x^2 + y^2}/z)$ will give a value in $-\frac{\pi}{2} \leq \phi \leq 0$, in which case, add π to place the angle ϕ in the desired interval of $\frac{\pi}{2} \leq \phi \leq \pi$.

◆ ◆ ◆ ◆ ◆ ◆ ◆ ◆ ◆ ◆

Example 40.1: Convert the rectangular coordinate (2,5,3) into spherical coordinates.

Solution: This point lies above the first quadrant of the xy-plane. Thus, we expect that both θ and ϕ will be in the intervals $0 < \theta < \frac{\pi}{2}$ and $0 < \phi < \frac{\pi}{2}$.

$$\rho = \sqrt{2^2 + 5^2 + 3^2} = \sqrt{38},$$
$$\theta = \arctan\left(\frac{5}{2}\right) \approx 1.1903 \text{ radians,}$$
$$\phi = \arctan\left(\frac{\sqrt{2^2 + 5^2}}{3}\right) \approx 1.0625 \text{ radians.}$$

Since $\frac{\pi}{2} \approx 1.571$, the values for θ and ϕ are plausible.

Example 40.2: Convert the rectangular coordinate $(-3, -4, -2)$ into spherical coordinates.

Solution: This point lies below the third quadrant of the xy-plane. We expect that θ will be in the interval $\pi < \theta < \frac{3\pi}{2}$ and that ϕ will be in the interval $\frac{\pi}{2} < \phi < \pi$. We have

$$\rho = \sqrt{(-3)^2 + (-4)^2 + (-2)^2} = \sqrt{29},$$
$$\theta = \arctan\left(\frac{-4}{-3}\right) = \arctan\left(\frac{4}{3}\right) \approx 0.9273 \text{ radians},$$
$$\phi = \arctan\left(\frac{\sqrt{(-3)^2 + (-4)^2}}{-2}\right) = \arctan\left(-\frac{5}{2}\right) \approx -1.1903 \text{ radians}.$$

The current value for θ is incorrect. The value of 0.9273 radians places θ in the first quadrant. Thus, add π, so that the correct value for θ is $0.9273 + 3.1416 \approx 4.0689$ radians, which is in the in the interval $\pi < \theta < \frac{3\pi}{2}$, as desired.

Furthermore, we can rewrite ϕ so that it is in the interval $\frac{\pi}{2} < \phi < \pi$. We add π to $\phi \approx -1.1903$, getting $-1.1903 + 3.1416 \approx 1.9513$ radians, which is an angle in the desired interval.

To summarize, the point $(-3, -4, -2)$ in rectangular coordinates is equivalent to the point $(\rho, \theta, \phi) = (\sqrt{29}, 4.0689, 1.9513)$ in spherical coordinates.

◆ ◆ ◆ ◆ ◆ ◆ ◆ ◆ ◆ ◆

Example 40.3: Describe the solid sphere of radius 2 centered at the origin using spherical coordinates.

Solution: The solid sphere of radius 2 is described by

$$0 \leq \rho \leq 2, \quad 0 \leq \theta \leq 2\pi, \quad 0 \leq \phi \leq \pi.$$

Example 40.4: Describe the solid hemisphere of radius 4, bounded by the xy-plane, extending into the negative z direction.

Solution: We have

$$0 \le \rho \le 4, \quad 0 \le \theta \le 2\pi, \quad \frac{\pi}{2} \le \phi \le \pi.$$

Note that the bounds $\frac{\pi}{2} \le \phi \le \pi$ indicate that points in this region lie below the xy-plane.

♦ ● ♦ ● ♦ ● ♦ ● ♦

Example 40.5: Describe $\rho = 3$, with $0 \le \theta \le 2\pi$ and $0 \le \phi \le \pi$.

Solution: This is a sphere of radius 3, centered at the origin. Had we set $0 \le \rho \le 3$, this would describe the solid sphere of radius 3.

Converting back to rectangular coordinates, this same spherical surface is given by

$$x = 3 \sin \phi \cos \theta$$
$$y = 3 \sin \phi \sin \theta$$
$$z = 3 \cos \phi,$$

with $0 \le \theta \le 2\pi$ and $0 \le \phi \le \pi$.

♦ ● ♦ ● ♦ ● ♦ ● ♦

Example 40.6: Describe the solid given by $9 \le x^2 + y^2 + z^2 \le 25$, where both $x \ge 0$ and $y \ge 0$, using spherical coordinates.

Solution: Note that x and y are restricted to the first quadrant in the xy-plane, so that θ cannot be greater than $\frac{\pi}{2}$. The object is two nested spheres, one of radius 3 and the other of radius 5, lying above and below the first quadrant of the xy-plane. Note that z still may be positive or negative. We have $3 \le \rho \le 5$, $0 \le \theta \le \frac{\pi}{2}$, $0 \le \phi \le \pi$.

Example 40.7: Given the point P defined by spherical coordinates $(\rho, \theta, \phi) = \left(3, \frac{\pi}{6}, \frac{\pi}{5}\right)$, find the reflection of P (a) cross the xy-plane, (b) across the yz-plane, and (c) across the xz-plane.

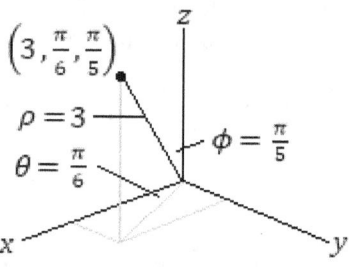

Solution: Unlike the rectangular coordinate axis system, where we can negate certain values within the ordered triple to achieve a reflection, we must remember that the point in spherical coordinates is partially described by angle measures.

a) When P is reflected across the xy-plane, the ρ and θ values do not change. However, the new ϕ value is now the supplement of the original value. Thus, the reflection of P across the xy-plane is given by $\left(3, \frac{\pi}{6}, \pi - \frac{\pi}{5}\right) = \left(3, \frac{\pi}{6}, \frac{4\pi}{5}\right)$.

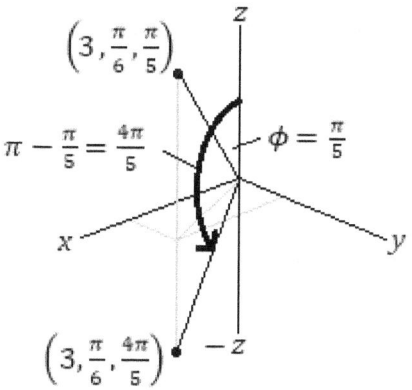

b) When P is reflected across the yz-plane, the ρ and ϕ values do not change. However, the new θ value is now the supplement of the original value. Thus, the reflection of P across the yz-plane is given by $\left(3, \pi - \frac{\pi}{6}, \frac{\pi}{5}\right) = \left(3, \frac{5\pi}{6}, \frac{\pi}{5}\right)$.

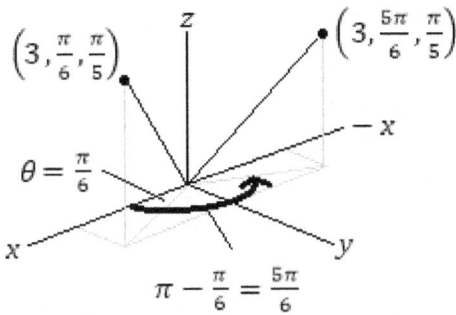

c) When P is reflected across the xz-plane, the ρ and φ values do not change. However, the new θ value is now the negation of the original value. Thus, the reflection of P across the xz-plane is given by $\left(3, -\frac{\pi}{6}, \frac{\pi}{5}\right)$. If we require that θ be positive, then this point is also described by $\left(3, 2\pi - \frac{\pi}{6}, \frac{\pi}{5}\right) = \left(3, \frac{11\pi}{6}, \frac{\pi}{5}\right)$.

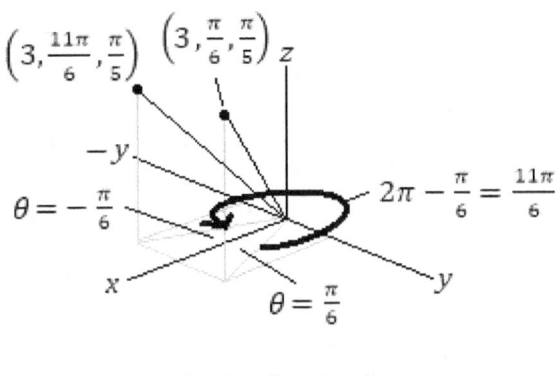

♦ ♦ ♦ ♦ ♦ ♦ ♦ ♦ ♦

Example 40.8: Rewrite the point P given by spherical coordinates $(\rho, \theta, \phi) = \left(-3, \frac{\pi}{6}, \frac{\pi}{5}\right)$ so that all values are positive.

Solution: Note that the angles $\theta = \frac{\pi}{6}$ and $\phi = \frac{\pi}{5}$ describe a ray extending from the origin into the positive octant, where x, y and z are all positive. Any point on this ray would have a positive value for ρ, being the point's distance from the origin.

If the ray is extended in the opposite direction, it extends into the octant where x, y and z are all negative. Thus, a "distance" of $\rho = -3$ is interpreted as a point that lies on this ray such that its x, y and z coordinates would all be negative

Adjust the angle values so that $\pi \leq \theta \leq \frac{3\pi}{2}$, making both the x and y coordinates negative, and $\frac{\pi}{2} \leq \phi \leq \pi$, making the z coordinate negative. For θ, add π to the original angle measure, and for ϕ, we use the supplement of the original angle measure. Thus, $\left(-3, \frac{\pi}{6}, \frac{\pi}{5}\right)$ is equivalent to $\left(-3, \pi + \frac{\pi}{6}, \pi - \frac{\pi}{5}\right) = \left(3, \frac{7\pi}{6}, \frac{4\pi}{5}\right)$.

◆ ◆ ◆ ◆ ◆ ◆ ◆ ◆ ◆

41. Integration with Spherical Coordinates

A function $f(x, y, z)$ integrated over a region R can be integrated in spherical coordinates, where $\rho^2 \sin \phi$ is the Jacobian, and present in all integrals defined in spherical coordinates.

$$\iiint_R f(x, y, z) \, dV$$
$$= \int_{\theta_1}^{\theta_2} \int_{\phi_1}^{\phi_2} \int_{\rho_1}^{\rho_2} f(x(\rho, \theta, \phi), y(\rho, \theta, \phi), z(\rho, \theta, \phi)) \, \rho^2 \sin \phi \, d\rho \, d\phi \, d\theta.$$

Here, $f(x(\rho, \theta, \phi), y(\rho, \theta, \phi), z(\rho, \theta, \phi))$ represents the integrand after the variables x, y and z have been converted into spherical coordinates.

◆ ◆ ◆ ◆ ◆ ◆ ◆ ◆ ◆

Example 41.1: Use spherical coordinates to find the volume of a sphere of radius 1.

Solution. The sphere is described in spherical coordinates by

$$0 \leq \rho \leq 1, \quad 0 \leq \theta \leq 2\pi, \quad 0 \leq \phi \leq \pi.$$

The integrand is $f(x, y, z) = 1$, since this is a volume integral in R^3. The volume element is

$$dV = \rho^2 \sin \phi \, d\rho \, d\phi \, d\theta.$$

Thus, we have

$$\iiint_R 1 \, dV = \int_0^{2\pi} \int_0^{\pi} \int_0^1 1 \rho^2 \sin \phi \, d\rho \, d\phi \, d\theta.$$

The inner-most integral is evaluated first with respect to ρ:

$$\int_0^1 \rho^2 \sin\phi \, d\rho = \sin\phi \int_0^1 \rho^2 \, d\rho$$

$$= \sin\phi \left[\frac{\rho^3}{3}\right]_0^1$$

$$= \frac{1}{3}\sin\phi.$$

This is then integrated with respect to ϕ:

$$\int_0^\pi \left(\frac{1}{3}\sin\phi\right) d\phi = \frac{1}{3}\int_0^\pi \sin\phi \, d\phi$$

$$= \frac{1}{3}[-\cos\phi]_0^\pi$$

$$= \frac{1}{3}[(-\cos\pi) - (-\cos 0)] \quad \begin{cases} -\cos\pi = -(-1) = 1 \\ -\cos 0 = -1 \end{cases}$$

$$= \frac{2}{3}.$$

Finally, we integrate with respect to θ:

$$\int_0^{2\pi} \left(\frac{2}{3}\right) d\theta = \frac{2}{3}(2\pi) = \frac{4\pi}{3}.$$

From geometry, we know that the volume of a sphere of radius 1 is $\frac{4}{3}\pi(1)^3 = \frac{4\pi}{3}$. This is a check our work.

◆ • ◆ • ◆ • ◆ • ◆ • ◆

Example 41.2: Evaluate $\iiint_R \sqrt{x^2 + y^2 + z^2} \, dV$, where R is a hemisphere of radius 5, centered at the origin and above the xy-plane.

Solution: In rectangular coordinates, the triple integral is

$$\int_{-5}^{5} \int_{-\sqrt{25-x^2}}^{\sqrt{25-x^2}} \int_0^{\sqrt{25-x^2-y^2}} \sqrt{x^2 + y^2 + z^2} \, dz \, dy \, dx.$$

In spherical coordinates, the integrand is rewritten as $\sqrt{x^2 + y^2 + z^2} = \sqrt{\rho^2} = \rho$, then multiplied by the Jacobian $\rho^2 \sin \phi$. This same integral in spherical coordinates is

$$\int_0^{2\pi} \int_0^{\pi/2} \int_0^5 (\rho) \rho^2 \sin \phi \, d\rho \, d\phi \, d\theta = \int_0^{2\pi} \int_0^{\pi/2} \int_0^5 \rho^3 \sin \phi \, d\rho \, d\phi \, d\theta.$$

This integral has constant bounds and is more easily solved using spherical coordinates than in rectangular coordinates.

The inner-most integral is evaluated first with respect to ρ:

$$\int_0^5 \rho^3 \sin \phi \, d\rho = \sin \phi \int_0^5 \rho^3 \, d\rho$$

$$= \sin \phi \left[\frac{\rho^4}{4}\right]_0^5$$

$$= \frac{625}{4} \sin \phi.$$

Then, this is integrated with respect to ϕ:

$$\int_0^{\pi/2} \left(\frac{625}{4} \sin \phi\right) d\phi = \frac{625}{4} [-\cos \phi]_0^{\pi/2}$$

$$= \frac{625}{4} (0 - (-1))$$

$$= \frac{625}{4}.$$

Lastly, integrate with respect to θ:

$$\int_0^{2\pi} \left(\frac{625}{4}\right) d\theta = \frac{625}{4} (2\pi) = \frac{625\pi}{2}.$$

Comment: When using spherical coordinates to find the volume of a solid in R^3 (or any situation where the variables in the integrand can be isolated as separate factors) and assuming all bounds are constants, then the triple integral can be written as the product of three single integrals:

$$\text{Volume} = \int_{\theta_1}^{\theta_2} \int_{\phi_1}^{\phi_2} \int_{\rho_1}^{\rho_2} \rho^2 \sin\phi \, d\rho \, d\phi \, d\theta$$

$$= \left(\int_{\theta_1}^{\theta_2} d\theta \right) \left(\int_{\phi_1}^{\phi_2} \sin\phi \, d\phi \right) \left(\int_{\rho_1}^{\rho_2} \rho^2 \, d\rho \right).$$

◆ ◆ ◆ ◆ ◆ ◆ ◆ ◆ ◆

Example 41.3: Let Q be a sphere centered at the origin, and R be a cone whose vertex is at the origin and opens in the positive z direction. The solid S bounded inside the cone and the sphere is called a *spherical sector*. Suppose the point (4,5,7) in rectangular coordinates lies on the "lip", where the sphere and the cone intersect. Find the volume of S.

Solution: We determine bounds for ρ, θ and ϕ by sketching the solid and the point on its rim:

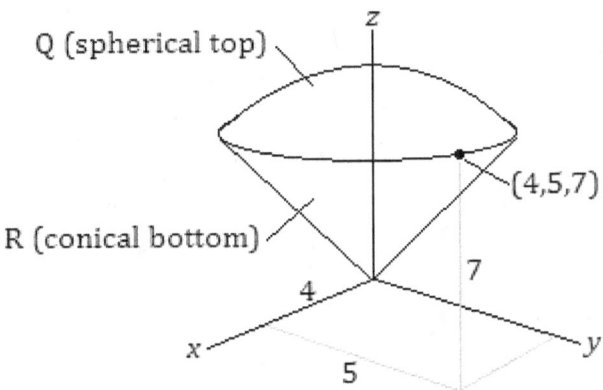

The distance from (0,0,0) to (4,5,7) is $\sqrt{4^2 + 5^2 + 7^2} = \sqrt{90} = 3\sqrt{10}$. Since the solid includes the origin, the bounds of ρ are $0 \leq \rho \leq 3\sqrt{10}$.

The solid includes the positive z-axis, so the lower bound for ϕ is 0. The upper bound is found by observing a right triangle with the adjacent leg on the z-axis, and the hypotenuse corresponding to a line from the origin to the point (4,5,7). From this, we see that for an upper bound, we have $\phi = \arccos\left(\frac{7}{3\sqrt{10}}\right)$. Lastly, the solid encircles the z-axis. Thus, the bounds of θ are $0 \leq \theta \leq 2\pi$.

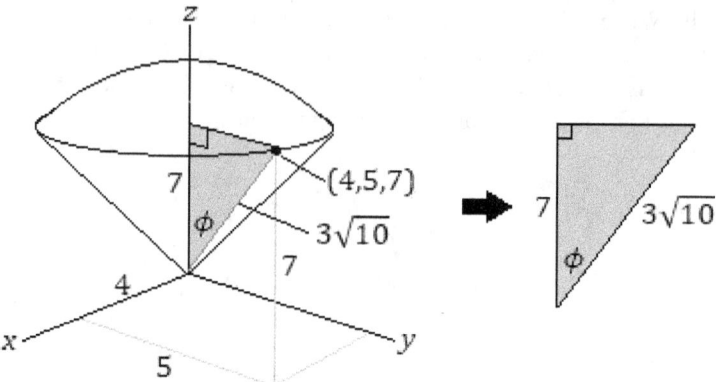

The volume integral in spherical coordinates is

$$\int_0^{2\pi} \int_0^{\arccos\left(\frac{7}{3\sqrt{10}}\right)} \int_0^{3\sqrt{10}} \rho^2 \sin\phi \, d\rho \, d\phi \, d\theta.$$

The inner-most integral is evaluated with respect to ρ:

$$\int_0^{3\sqrt{10}} \rho^2 \sin\phi \, d\rho = \sin\phi \int_0^{3\sqrt{10}} \rho^2 \, d\rho$$

$$= \sin\phi \left[\frac{\rho^3}{3}\right]_0^{3\sqrt{10}}$$

$$= \frac{(3\sqrt{10})^3}{3} \sin\phi$$

$$= 9(10)^{3/2} \sin\phi.$$

This is integrated with respect to ϕ:

$$\int_0^{\arccos\left(\frac{7}{3\sqrt{10}}\right)} (9(10)^{3/2} \sin\phi) \, d\phi = 9(10)^{3/2} \int_0^{\arccos\left(\frac{7}{3\sqrt{10}}\right)} \sin\phi \, d\phi$$

$$= 9(10)^{3/2} [-\cos\phi]_0^{\arccos\left(\frac{7}{3\sqrt{10}}\right)}$$

$$= 9(10)^{3/2} \left[-\cos\left(\arccos\left(\frac{7}{3\sqrt{10}}\right)\right) - (-\cos 0)\right]$$

$$= 9(10)^{3/2} \left(1 - \frac{7}{3\sqrt{10}}\right).$$

Finally, evaluate the outer-most integral with respect to θ:

$$\int_0^{2\pi} 9(10)^{3/2}\left(1 - \frac{7}{3\sqrt{10}}\right) d\theta = 9(10)^{3/2}\left(1 - \frac{7}{3\sqrt{10}}\right) 2\pi$$

$$= 18\pi(10)^{3/2}\left(1 - \frac{7}{3\sqrt{10}}\right)$$

$$\approx 468.76 \text{ cubic units.}$$

Using the short form

$$\int_{\theta_1}^{\theta_2}\int_{\phi_1}^{\phi_2}\int_{\rho_1}^{\rho_2} \rho^2 \sin\phi \, d\rho \, d\phi \, d\theta = \left(\int_{\theta_1}^{\theta_2} d\theta\right)\left(\int_{\phi_1}^{\phi_2} \sin\phi \, d\phi\right)\left(\int_{\rho_1}^{\rho_2} \rho^2 \, d\rho\right),$$

we have

$$\text{Volume} = \left(\int_0^{2\pi} d\theta\right)\left(\int_0^{\arccos\left(\frac{7}{3\sqrt{10}}\right)} \sin\phi \, d\phi\right)\left(\int_0^{3\sqrt{10}} \rho^2 \, d\rho\right)$$

$$= \left([\theta]_0^{2\pi}\right)\left([-\cos\theta]_0^{\arccos\left(\frac{7}{3\sqrt{10}}\right)}\right)\left(\left[\frac{\rho^3}{3}\right]_0^{3\sqrt{10}}\right)$$

$$= (2\pi)\left(-\frac{7}{3\sqrt{10}} - (-1)\right)\left(\frac{(3\sqrt{10})^3}{3}\right)$$

$$= 9(10)^{3/2}\left(1 - \frac{7}{3\sqrt{10}}\right) 2\pi$$

$$= 18\pi(10)^{3/2}\left(1 - \frac{7}{3\sqrt{10}}\right) \approx 468.76 \text{ cubic units.}$$

◆ ◆ ◆ ◆ ◆ ◆ ◆ ◆ ◆

Example 41.4: Use spherical coordinates to find the volume contained within the cone $z = \sqrt{x^2 + y^2}$ and below the plane $z = 6$.

Solution: First, observe that the solid is not a spherical sector as in the previous example. The value of ρ will vary as a function of ϕ.

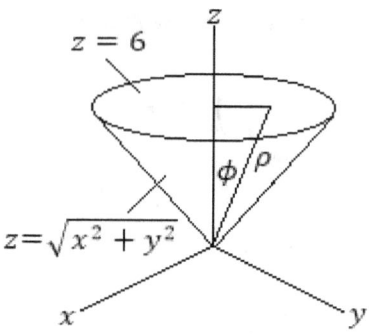

The top is flat, not spherical. Thus, ρ will vary as a function of ϕ.

The bounds for θ and ϕ are easy to determine. The "sweep" angle θ encompasses a full counter-clockwise rotation around the *xy*-plane from the positive *x*-axis back to the positive *x*-axis, so that $0 \leq \theta \leq 2\pi$. The "lean" angle ϕ varies from 0 (the positive *z*-axis) to $\frac{\pi}{4}$ (the side of the cone, which is 45 degrees from both the positive *x*-axis and the positive *y*-axis).

For the plane $z = 6$, substitute $z = \rho \cos \phi$, getting $\rho \cos \phi = 6$. Then solving for ρ, we have $\rho = 6/\cos \phi = 6 \sec \phi$. Since the object is a solid and includes the origin, the lower bound for ρ is 0, while the upper bound is the plane, so that the bounds for ρ are $0 \leq \rho \leq 6 \sec \phi$. Thus, the volume integral is

$$\int_0^{2\pi} \int_0^{\pi/4} \int_0^{6 \sec \phi} \rho^2 \sin \phi \, d\rho \, d\phi \, d\theta.$$

The inner-most integral is integrated with respect to ρ:

$$\int_0^{6 \sec \phi} \rho^2 \sin \phi \, d\rho = \sin \phi \left[\frac{\rho^3}{3} \right]_0^{6 \sec \phi}$$

$$= \sin \phi \left(\frac{216 \sec^3 \phi}{3} \right)$$

$$= 72 \tan \phi \sec^2 \phi.$$

Note that $\sin \phi \sec^3 \phi = \frac{\sin \phi}{\cos \phi} \sec^2 \phi = \tan \phi \sec^2 \phi$.

This is now integrated with respect to ϕ. Note that $72 \tan \phi \sec^2 \phi$ can be antidifferentiated by a *u-du* substitution, where $u = \tan \phi$ so that $du = \sec^2 \phi \, d\phi$. This results in a power-rule form, $\int 72u \, du = 36u^2$:

$$\int_0^{\pi/4} 72 \tan\phi \sec^2\phi \, d\phi = [36 \tan^2 \phi]_0^{\pi/4}$$
$$= 36(1)^2 - (36(0)^2)$$
$$= 36.$$

Lastly, we integrate with respect to θ:

$$\int_0^{2\pi} 36 \, d\theta = 36(2\pi) = 72\pi \text{ cubic units.}$$

◆ ◆ ◆ ◆ ◆ ◆ ◆ ◆ ◆

42. Change of Variables: The Jacobian

It is common to change the variable(s) of integration, the goal being to rewrite a complicated integrand into a simpler equivalent form. However, in doing so, the underlying geometry of the problem may be altered. This is seen often in single-variable integrals:

Example 42.1: Evaluate

$$\int_2^4 (3x+1)^3 \, dx.$$

Solution: Let $u = 3x + 1$, so that $du = 3 \, dx$. Then substitutions are made. Note that the expression $du = 3 \, dx$ is rewritten as $dx = \frac{1}{3} du$; and that the bounds $2 \le x \le 4$ are recalculated using $u = 3x + 1$, so that when $x = 2$, then $u = 3(2) + 1 = 7$, and when $x = 4$, then $u = 3(4) + 1 = 13$. The bounds with respect to u are now $7 \le u \le 13$.

$$\int_{x=2}^{x=4} (3x+1)^3 \, dx = \int_{u=3(2)+1=7}^{u=3(4)+1=13} u^3 \left(\frac{1}{3} du\right).$$

The integral in terms of u is simplified:

$$\int_{u=3(2)+1=7}^{u=3(4)+1=13} u^3 \left(\frac{1}{3} du\right) = \frac{1}{3} \int_7^{13} u^3 \, du.$$

It is important to note that the two integrals, $\int_2^4 (3x+1)^3 \, dx$ and $\frac{1}{3}\int_7^{13} u^3 \, du$, represent the identical problem. The $\frac{1}{3}$ that remains after simplification is called the **Jacobian**.

The integral in variable x is over an interval of length 2 units, while the integral in u is over an interval of length 6 units. Roughly speaking, variable u covers its interval of integration (length 6) three times "as fast" as that of x (length 2), and since u and x are linearly related, the leading $\frac{1}{3}$ adjusts for the change in the underlying geometry of the intervals.

For double integrals in R^2, we assume that a region of integration defined in terms of variables x and y are substituted for new variables u and v through two functions:

$$u = f_1(x, y)$$
$$v = f_2(x, y).$$

Note that the pair of equations are written so that u and v are written in terms of x and y. This is called a **transformation**. Such a "variable change" should be reversible. That is, we should be able to, through algebraic means, isolate x and y in terms of u and v:

$$x = g_1(u, v)$$
$$y = g_2(u, v).$$

The **Jacobian** is then defined as a determinant of a 2 by 2 matrix:

$$J(u,v) = \begin{vmatrix} \partial g_1/\partial u & \partial g_1/\partial v \\ \partial g_2/\partial u & \partial g_2/\partial v \end{vmatrix}, \quad \text{or} \quad J(u,v) = \begin{vmatrix} \partial x/\partial u & \partial x/\partial v \\ \partial y/\partial u & \partial y/\partial v \end{vmatrix}.$$

This can be extended into higher dimensions as well.

◆ • ◆ • ◆ • ◆ • ◆

Example 42.2: Find the Jacobian for the transformation

$$x = r \cos \theta$$
$$y = r \sin \theta.$$

Solution: We have

$$J(r, \theta) = \begin{vmatrix} \partial x/\partial r & \partial x/\partial \theta \\ \partial y/\partial r & \partial y/\partial \theta \end{vmatrix}$$

$$= \begin{vmatrix} \cos \theta & -r \sin \theta \\ \sin \theta & r \cos \theta \end{vmatrix}$$

$$= r\cos^2\theta + r\sin^2\theta$$
$$= r(\cos^2\theta + \sin^2\theta)$$
$$= r.$$

This is the common Jacobian when rectangular coordinates x and y are rewritten in polar coordinates r and θ.

◆ ◆ ◆ ◆ • • ◆ • • ◆

Example 42.3: Evaluate $\iint_R (x - 2y)\, dA$, where R is the parallelogram in the xy-plane with vertices $(0,0)$, $(4,1)$, $(6,4)$ and $(2,3)$.

Solution: A sketch of this region shows that it is a Type II region and would require three separate double integrals in either the $dy\, dx$ or the $dx\, dy$ ordering of integration. Instead, we observe that the region consists of two pairs of parallel sides, so we find equations for each of these sides:

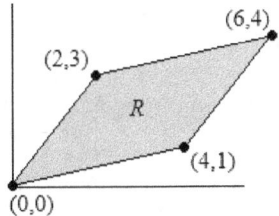

- From $(0,0)$ to $(4,1)$, we have $y = \frac{1}{4}x$, or $-x + 4y = 0$.
- From $(2,3)$ to $(6,4)$, we have $y = \frac{1}{4}x + \frac{5}{2}$, or $-x + 4y = 10$.
- From $(0,0)$ to $(2,3)$, we have $y = \frac{3}{2}x$, or $-3x + 2y = 0$.
- From $(4,1)$ to $(6,4)$, we have $y = \frac{3}{2}x - 5$, or $-3x + 2y = -10$.

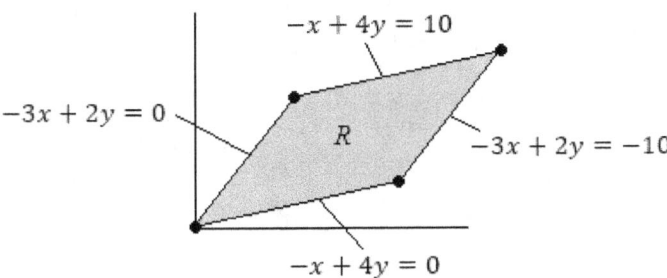

Thus, we can define a transformation from x and y into new variables u and v as

$$\begin{array}{ll} u = -x + 4y & \text{so that} \quad 0 \le u \le 10, \\ v = -3x + 2y & \text{so that} \quad -10 \le v \le 0. \end{array}$$

The region of integration R transformed into the uv-plane is a rectangle, so that u and v have constant bounds.

We now rewrite this transformation so that x and y are isolated. We start with the system

$$u = -x + 4y$$
$$v = -3x + 2y.$$

Multiply the bottom row by -2:

$$u = -x + 4y$$
$$-2v = -2(-3x + 2y).$$

Simplifying, we have

$$u = -x + 4y$$
$$-2v = 6x - 4y.$$

Summing, the y terms sum to 0 and we have

$$u - 2v = 5x, \quad \text{so that} \quad x = \frac{1}{5}u - \frac{2}{5}v.$$

Substituting this back into the system, we have

$$y = \frac{3}{10}u - \frac{1}{10}v.$$

We can now find the Jacobian:

$$J(u,v) = \begin{vmatrix} \partial x/\partial u & \partial x/\partial v \\ \partial y/\partial u & \partial y/\partial v \end{vmatrix}$$

$$= \begin{vmatrix} 1/5 & -2/5 \\ 3/10 & -1/10 \end{vmatrix}$$

$$= \frac{1}{5}\left(-\frac{1}{10}\right) - \frac{3}{10}\left(-\frac{2}{5}\right)$$

$$= -\frac{1}{50} + \frac{6}{50}$$

$$= \frac{5}{50}, \quad \text{or} \quad \frac{1}{10}.$$

We revisit the original double integral, and make substitutions:

$$\iint_R (x - 2y) \, dA = \int_{-10}^{0} \int_{0}^{10} \left(\left(\frac{1}{5}u - \frac{2}{5}v \right) - 2 \left(\frac{3}{10}u - \frac{1}{10}v \right) \right) \left(\frac{1}{10} \right) du \, dv.$$

$$= \frac{1}{10} \int_{-10}^{0} \int_{0}^{10} \left(-\frac{2}{5}u - \frac{1}{5}v \right) du \, dv.$$

The rest of the integration is routine calculation. The inside integral is evaluated first:

$$\int_{0}^{10} \left(-\frac{2}{5}u - \frac{1}{5}v \right) du = \left[-\frac{1}{5}u^2 - \frac{1}{5}uv \right]_{0}^{10} = -20 - 2v.$$

This is then integrated with respect to v. The Jacobian, $\frac{1}{10}$, has been moved to the front.

$$\frac{1}{10} \int_{-10}^{0} (-20 - 2v) \, dv = \frac{1}{10} [-20v - v^2]_{-10}^{0} = \frac{1}{10}(200 - 100) = 10.$$

This was an example of a *linear* transformation, in which the equations transforming x and y into u and v were linear, as were the equations reversing the transformation. In such a case, the Jacobian will be a constant.

We can also see how the geometry changed: The original region in the xy-plane has an area of 10 square units, while the region in the uv-plane, a rectangle where $0 \leq u \leq 10$ and $-10 \leq v \leq 0$, has an area of 100 square units. Thus, the region in the uv-plane is 10 times as large as the region in the xy-plane, so the Jacobian, $\frac{1}{10}$, "balances" this change in underlying area.

The Jacobian is usually taken to be a positive quantity. This is because the naming (and ordering) of the functions transforming x and y into u and v, then in reverse, is arbitrary. Since the Jacobian is a determinant, it is possible that two rows may be swapped depending on the original naming of the functions, which may introduce a factor of -1 into the result, which can be ignored.

Geometrical proofs for the Jacobians of the polar and spherical coordinates are given in the book's appendix.

43. Vector Fields

A vector field is a function **F** that assigns to each ordered pair (x, y) in R^2 a vector of the form $\langle M(x,y), N(x,y) \rangle$. We write

$$\mathbf{F}(x, y) = \langle M(x,y), N(x,y) \rangle.$$

This can be extended into higher dimensions. For example. In R^3, we would write

$$\mathbf{F}(x, y, z) = \langle M(x,y,z), N(x,y,z), P(x,y,z) \rangle.$$

◆ • ◆ • ◆ • ◆ • ◆ • ◆

Example 43.1: Sketch $\mathbf{F}(x,y) = \langle x, y \rangle$.

Solution: Using an input-output table, we can show some of the vectors in the vector field **F**:

Ordered pair (x,y)	Vector $\langle x,y \rangle$	Ordered pair (x,y)	Vector $\langle x,y \rangle$
(0,0)	$\langle 0,0 \rangle$	(−1,0)	$\langle -1,0 \rangle$
(1,0)	$\langle 1,0 \rangle$	(−1,1)	$\langle -1,1 \rangle$
(1,1)	$\langle 1,1 \rangle$	(1,−1)	$\langle 1,-1 \rangle$
(0,1)	$\langle 0,1 \rangle$	(2,1)	$\langle 2,1 \rangle$
(1,2)	$\langle 1,2 \rangle$	(2,2)	$\langle 2,2 \rangle$

The vector $\langle x, y \rangle$ is drawn so that its foot is at the point described by the ordered pair (x, y). Below (left) are a sample of vectors of **F**, and at right, a slightly-more complete rendering of the vector field. In this example, the vectors point radially (along straight lines) away from the origin.

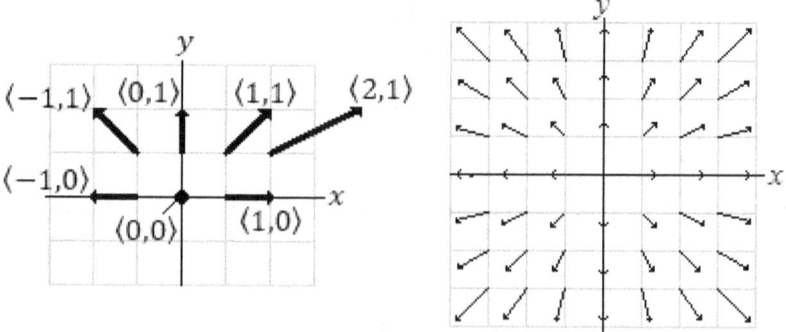

It is time consuming to sketch a vector field. Also, the vectors themselves "cover up" other vectors, resulting in a cluttered, unreadable image. Artistic liberties are allowed. For example, the vectors may be scaled down in size to show relative magnitudes rather than true magnitudes. Often, it is more important to see the "flow" created by the vectors, rather than the actual magnitudes.

◆ • ◆ • ◆ • ◆ • ◆

Example 43.2: Sketch $F(x, y) = \langle y, -x \rangle$.

Solution: An input-output table shows some of the vectors, followed by an image of the vector field.

Ordered pair (x, y)	Vector $\langle x, y \rangle$	Ordered pair (x, y)	Vector $\langle x, y \rangle$
(0,0)	$\langle 0,0 \rangle$	(−1,0)	$\langle 0,1 \rangle$
(1,0)	$\langle 0,-1 \rangle$	(−1,1)	$\langle 1,1 \rangle$
(1,1)	$\langle 1,-1 \rangle$	(1,−1)	$\langle -1,-1 \rangle$
(0,1)	$\langle 1,0 \rangle$	(2,1)	$\langle 1,-2 \rangle$
(1,2)	$\langle 2,-1 \rangle$	(2,2)	$\langle 2,-2 \rangle$

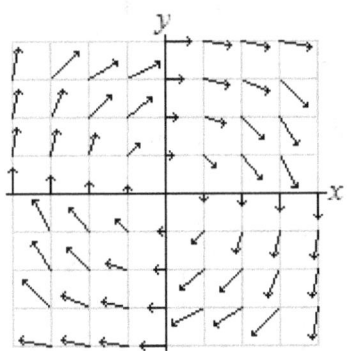

The vectors suggest a clockwise rotation around the origin.

Example 43.3: Sketch $\mathbf{F}(x,y) = \langle 1,2 \rangle$.

Solution: This is a constant vector field. All vectors are identical in magnitude and orientation. In the image below, each vector is shown at half-scale so as not to clutter the image too severely.

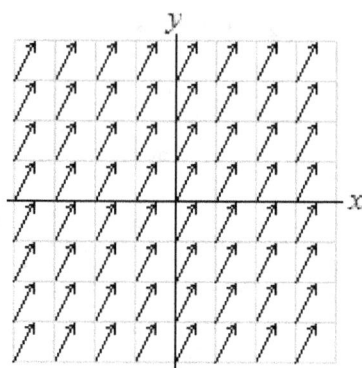

This vector field is not radial nor does it suggest any rotation.

◆ • ◆ • ◆ • ◆ • ◆

Example 43.4: Sketch $\mathbf{F}(x,y) = \langle x+y, y-x \rangle$.

Solution: The vector field is shown below:

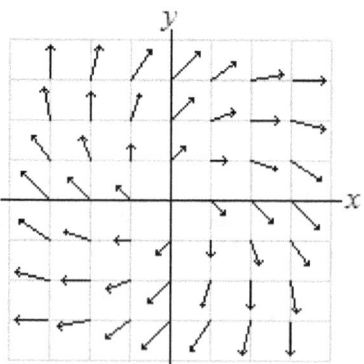

This vector field appears to have both radial and rotational aspects in its appearance.

The radial aspect of a vector field is known as **divergence**, and the rotational aspect known as **curl**. Both are discussed in Section 52.

Gradient Vector Fields

Given a function $z = f(x, y)$, its gradient is $\nabla f = \langle f_x(x,y), f_y(x,y) \rangle$. This is called a **gradient vector field** (or just **gradient field**). It is also called a **conservative vector field** and is discussed in depth in Section 47. In such a case, the vector field is written as $\mathbf{F}(x,y) = \nabla f = \langle f_x, f_y \rangle$.

Gradient vector fields have an interesting visual property: the vectors in the vector field lie orthogonal to the contours of f.

• • ◆ • • ◆ • • ◆

Example 43.5: Given $f(x,y) = \frac{1}{2}x^2 + \frac{1}{2}y^2$, find $\mathbf{F}(x,y) = \nabla f$ and sketch it along with the contour map of f.

Solution: The vector field is $\mathbf{F}(x,y) = \nabla f = \langle f_x, f_y \rangle = \langle x, y \rangle$. The contours of f are concentric circles $\frac{1}{2}x^2 + \frac{1}{2}y^2 = k$ centered at the origin, the surface being a paraboloid with its vertex at $(0,0,0)$ and opening upward. Note that the vectors in $\mathbf{F}$ are orthogonal to the contours of f. This is the same vector field as seen in Example 43.1. The vectors point in the direction of increasing z.

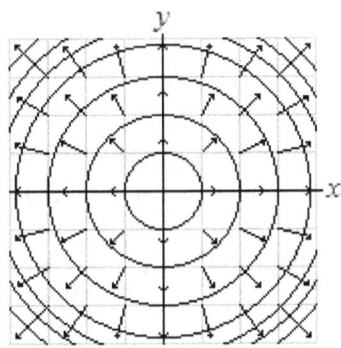

• • ◆ • • ◆ • • ◆

Example 43.6: Given $f(x,y) = x + 2y$, find $\mathbf{F}(x,y) = \nabla f$ and sketch it along with the contour map of f.

Solution: The vector field is $\mathbf{F}(x,y) = \nabla f = \langle f_x, f_y \rangle = \langle 1, 2 \rangle$. The surface of f is a plane tilting "upward" as x and y both increase in value. Note that the contours of f are all lines of the form $x + 2y = k$, or $y = -\frac{1}{2}x + \frac{k}{2}$, and that the vectors in $\mathbf{F}$ are orthogonal to the contours of f, pointing in the direction of increasing z. This is the same vector field as in Example 43.3.

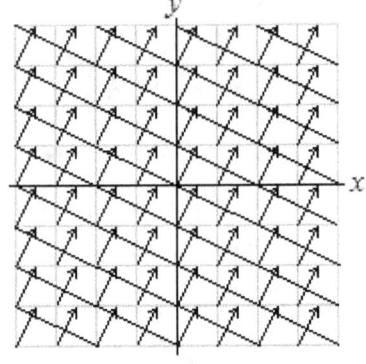

Not all vector fields are gradient fields. Those in Examples 43.2, and 43.4 are not gradient fields. There do not exist functions $z = f(x,y)$ such that $\mathbf{F}(x,y) = \nabla f$ in these two examples.

> If $\mathbf{F}$ is a gradient field, then there exists a function f such that $\mathbf{F}(x,y) = \nabla f$. This function f is called a **potential function**, discussed later in Section 47.

All constant vector fields $\mathbf{F}(x,y) = \langle a, b \rangle$ are gradient fields, where $f(x,y) = ax + by$ is a potential function. In R^3, we would have $\mathbf{F}(x,y,z) = \langle a, b, c \rangle$, with potential function $f(x,y,z) = ax + by + cz$.

All vector fields of the form $\mathbf{F}(x,y) = \langle M(x), N(y) \rangle$ are gradient fields, where a potential function is $f(x,y) = \int M(x)\,dx + \int N(y)\,dy$.

◆ • ◆ • ◆ • ◆ • ◆

Example 43.7: Find the potential functions for $\mathbf{F}(x,y,z) = \langle -1, 4, 2 \rangle$ and for $\mathbf{G}(x,y) = \langle 2x, y^4 \rangle$.

Solution: For $\mathbf{F}$, a potential function is $f(x,y,z) = -x + 4y + 2z$, and for $\mathbf{G}$, a potential function is $g(x,y) = \int 2x\,dx + \int y^4\,dy = x^2 + \frac{1}{5}y^5$.

Constants of integration are not necessary. If $\mathbf{F}$ is a gradient field, then it has infinitely-many potential functions, all equivalent up to its constant of integration. Note that for $\mathbf{F}$ above, $f(x,y,z) = -x + 4y + 2z + 7$ is also a valid potential function. Usually, we let any such constant be 0.

◆ • ◆ • ◆ • ◆ • ◆

Paths that are orthogonal to the contours for each point in the path are called **streams**, or **streamlines**. In Example 43.5, streamlines would be of the form $y = kx$, and in Example 43.6, of the form $y = 2x + k$.

On a surface, a stream of water would flow orthogonally to the contours of the surface, always in the direction of steepest descent.

◆ • ◆ • ◆ • ◆ • ◆

Example 43.8: Given $f(x,y) = x^3 + y^3 - 3x - 3y$, find $\mathbf{F}(x,y) = \nabla f$ and sketch it along with the contour map of f.

Solution: The vector field is

$$\mathbf{F}(x,y) = \nabla f = \langle f_x, f_y \rangle = \langle 3x^2 - 3, 3y^2 - 3 \rangle.$$

The vector field **F** is shown below with the contours of f:

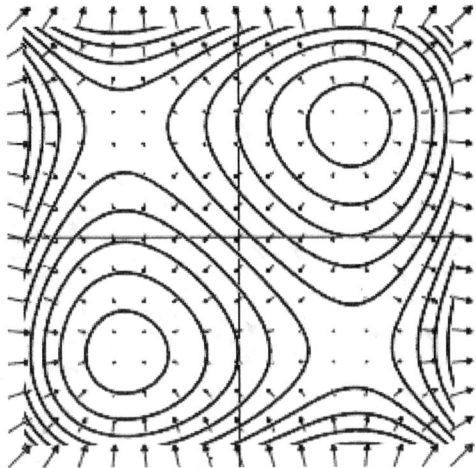

Using techniques of unconstrained optimization (Section 29), there are four critical points. They are: $(1,1,-4)$, a minimum; $(-1,-1,4)$, a maximum; and $(1,-1,0)$ and $(-1,1,0)$, both saddle points. Observe a few things:

- The vectors in **F** always point in the direction of increasing z, or "up".
- Note that the vectors point "up" *toward* the maximum at $(-1,-1,4)$ and "up" *away* from the minimum at $(1,1,-4)$.
- At each critical point, $\nabla f = \langle 0,0 \rangle$. In the image above, the vectors near these points have very small magnitudes, while the vectors at these critical points have no magnitude.

Example 43.9: Sketch $\mathbf{F}(x, y, z) = \langle x, y, z \rangle$.

Solution: The vector field is sketched below, for $x > 0$, $y > 0$ and $z > 0$:

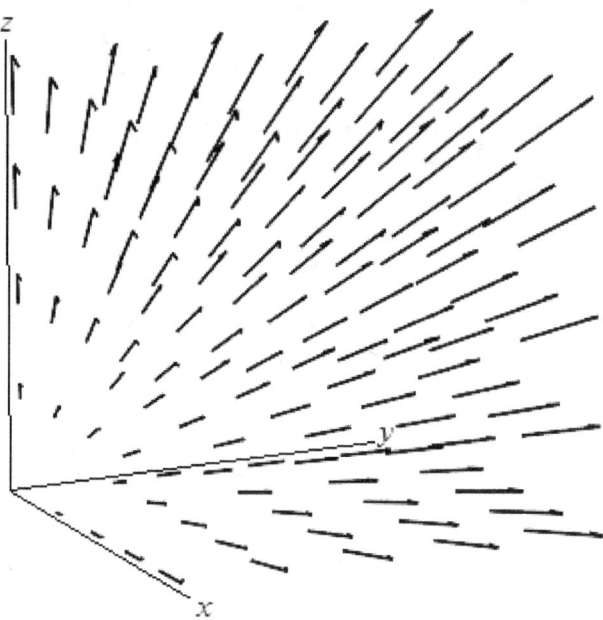

The vectors in **F** all point radially away from the origin, increasing in magnitude the farther away from the origin.

Sketching a vector field in R^3 is nearly impossible to do manually. A computer program is an essential tool to render such fields.

◆ • ◆ • ◆ • ◆ • ◆

Given a function $w = f(x, y, z)$, then a gradient field in R^3 can be defined by the gradient of f:

$$\mathbf{F}(x, y, z) = \nabla f = \langle f_x, f_y, f_z \rangle.$$

A function such as $w = f(x, y, z)$ exists in R^4 (three independent variables, one dependent variable). Its contours will be surfaces in R^3, and the vectors in the gradient field given by ∇f will be orthogonal to the contours, which in this case are surfaces.

Example 43.10: Show that $w = f(x, y, z) = \frac{1}{2}x^2 + \frac{1}{2}y^2 + \frac{1}{2}z^2$ is a potential function of $\mathbf{F}(x, y, z) = \langle x, y, z \rangle$. Discuss how the vectors in $\mathbf{F}$ compare to the contours of f.

Solution: The gradient of f is shown below:

$$\nabla f = \langle f_x, f_y, f_z \rangle = \langle x, y, z \rangle.$$

The contours of f are found by setting w equal to various constants. For example, when $w = 1$, then we have

$$1 = \frac{1}{2}x^2 + \frac{1}{2}y^2 + \frac{1}{2}z^2, \quad \text{so that} \quad x^2 + y^2 + z^2 = 2.$$

This is a sphere centered at the origin with radius $\sqrt{2}$. In fact, all contours of f are spheres. As w increases in value, its contour at $w = k$ is given by the sphere $x^2 + y^2 + z^2 = 2k$.

For example, the vector whose foot lies at $(1,2,3)$ is given by $\langle 1,2,3 \rangle$. The point $(1,2,3)$ itself lies on a sphere with radius $\sqrt{14}$. The vector $\langle 1,2,3 \rangle$ has its foot on this sphere, oriented orthogonally to this sphere, pointing directly away from the origin.

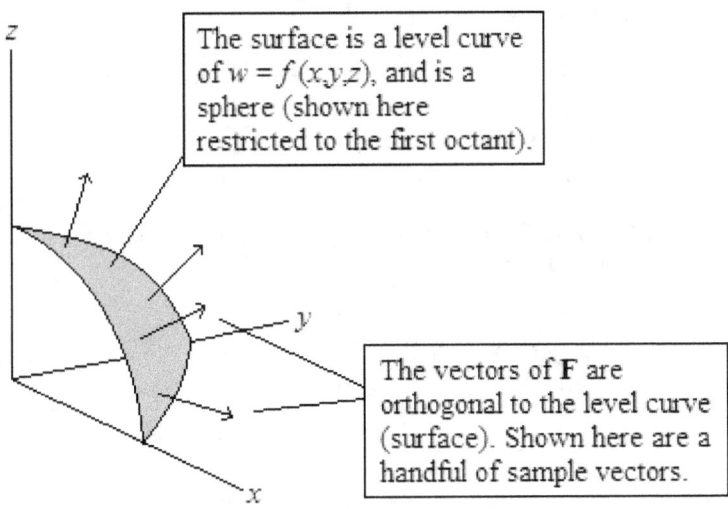

The surface is a level curve of $w = f(x,y,z)$, and is a sphere (shown here restricted to the first octant).

The vectors of $\mathbf{F}$ are orthogonal to the level curve (surface). Shown here are a handful of sample vectors.

44. Scalar Line Integrals

Let $z = f(x,y)$ be a continuous function (surface) in R^3 and C a path on the xy-plane. If C is parametrized by $\mathbf{r}(t) = \langle x(t), y(t) \rangle$ for $a \le t \le b$, then the **scalar line integral** of f along C is given by

$$\int_C f(x,y)\, ds,$$

where $ds = |\mathbf{r}'(t)|\, dt$. Thus, the integral is in variable t and can be written

$$\int_a^b f(x(t), y(t))\, |\mathbf{r}'(t)|\, dt.$$

The value of a scalar line integral is the area of a "sheet" above the path C to the surface f.

♦ • ♦ • ♦ • ♦ • ♦

Example 44.1: Find $\int_C x^2 y\, ds$, where C is the straight line from (2,1) to (6,4).

Solution: Parametrize the path C first, noting that $\langle 6 - 2, 4 - 1 \rangle = \langle 4, 3 \rangle$ is the direction vector of the line segment:

$$\mathbf{r}(t) = \langle 2, 1 \rangle + t \langle 4, 3 \rangle = \langle 2 + 4t, 1 + 3t \rangle, \quad \text{for} \quad 0 \le t \le 1.$$

Thus, we have $\mathbf{r}'(t) = \langle 4, 3 \rangle$ and $|\mathbf{r}'(t)| = \sqrt{4^2 + 3^2} = \sqrt{25} = 5$, so that $ds = |\mathbf{r}'(t)|\, dt = 5\, dt$.

From $\mathbf{r}(t)$, we obtain $x(t) = 2 + 4t$ and $y(t) = 1 + 3t$. These are substituted into the integrand, and simplified:

$$\int_C x^2 y\, ds = \int_C (2 + 4t)^2 (1 + 3t)\, ds$$
$$= \int_C 4(12t^3 + 16t^2 + 7t + 1)\, 5\, dt$$
$$= 20 \int_0^1 (12t^3 + 16t^2 + 7t + 1)\, dt$$
$$= 20 \left[3t^4 + \frac{16}{3}t^3 + \frac{7}{2}t^2 + t \right]_0^1$$
$$= \frac{770}{3}.$$

Example 44.2: Find $\int_C x \, ds$, where C is the arc of the parabola $y = x^2$ from $(-1,1)$ to $(3,9)$.

Solution: Path C is parametrized:

$$\mathbf{r}(t) = \langle t, t^2 \rangle, \quad \text{for} \quad -1 \leq t \leq 3.$$

We have $\mathbf{r}'(t) = \langle 1, 2t \rangle$ and $|\mathbf{r}'(t)| = \sqrt{1^2 + (2t)^2} = \sqrt{1 + 4t^2}$, so that $ds = \sqrt{1 + 4t^2} \, dt$. The integrand is now written in terms of t and evaluated using u-du substitution:

$$\int_C x \, ds = \int_{-1}^{3} t\sqrt{1 + 4t^2} \, dt$$

$$= \left[\frac{1}{12}(1 + 4t^2)^{3/2} \right]_{-1}^{3}$$

$$= \left(\frac{1}{12}(1 + 4(3)^2)^{3/2} \right) - \left(\frac{1}{12}(1 + 4(-1)^2)^{3/2} \right)$$

$$= \frac{1}{12}(37^{3/2} - 5^{3/2}) \approx 17.82.$$

◆ • ◆ • ◆ • ◆ • ◆

In some cases, a numerical method needs to be used to evaluate the integral.

Example 44.3: Find $\int_C x^3 y^2 \, ds$, where C is the curve $y = x^3$ from $(1,1)$ to $(2,8)$.

Solution: Path C is parametrized as:

$$\mathbf{r}(t) = \langle t, t^3 \rangle, \text{ for } 1 \leq t \leq 2.$$

We have $\mathbf{r}'(t) = \langle 1, 3t^2 \rangle$ and $|\mathbf{r}'(t)| = \sqrt{1^2 + (3t^2)^2} = \sqrt{1 + 9t^4}$. The integrand is now written in terms of t:

$$x^3 y^2 \, ds = (t)^3 (t^3)^2 \sqrt{1 + 9t^4} \, dt = t^9 \sqrt{1 + 9t^4} \, dt.$$

The integral is

$$\int_C x^3 y^2 \, ds = \int_1^2 t^9 \sqrt{1 + 9t^4} \, dt.$$

Using numerical methods, this integral evaluates to approximately 1029.1 units.

Example 44.4: The roof of a building is a paraboloid $z = 10 - \frac{1}{8}x^2 - \frac{1}{12}y^2$, with the apex (highest point) above the origin. Assume x and y represent distances from the origin along the floor (the xy-plane) orthogonal to one another, and that all measurements are in meters. A wall is to be built extending from the origin along the line $y = 3x$. The wall will rise from the floor up to the roof. What is the area of this wall?

Solution: The path C is the line $y = 3x$, which is parametrized as $\mathbf{r}(t) = \langle t, 3t \rangle$. The lower bound for t is 0. The upper bound of t is where the roof meets the floor along this line. Since $z = 0$ at this point, and since $y = 3x$, we make substitutions and solve for x:

$$0 = 10 - \frac{1}{8}x^2 - \frac{1}{12}(3x)^2$$

$$0 = 10 - \frac{1}{8}x^2 - \frac{3}{4}x^2$$

$$0 = 10 - \frac{7}{8}x^2.$$

Solving for x, we get $x = \sqrt{80/7} \approx 3.381$. Since $x = t$ in the parametrization, we have the upper bound for t as $\sqrt{80/7}$. Furthermore, $\mathbf{r}'(t) = \langle 1, 3 \rangle$ so that $|\mathbf{r}'(t)| = \sqrt{1^2 + 3^2} = \sqrt{10}$, which gives $ds = |\mathbf{r}'(t)| \, dt = \sqrt{10} \, dt$. The integral is

$$\int_C \left(10 - \frac{1}{8}x^2 - \frac{1}{12}y^2\right) ds = \int_0^{\sqrt{80/7}} \left(10 - \frac{1}{8}(t)^2 - \frac{1}{12}(3t)^2\right) \sqrt{10} \, dt$$

$$= \sqrt{10} \int_0^{\sqrt{80/7}} \left(10 - \frac{7}{8}t^2\right) dt$$

$$= \sqrt{10} \left[10t - \frac{7}{24}t^3\right]_0^{\sqrt{80/7}}$$

$$= \sqrt{10}\left(10\left(\sqrt{80/7}\right) - \frac{7}{24}\left(\sqrt{80/7}\right)^3\right).$$

Using a calculator, the area of the wall is about 71.27 meters².

Example 44.5: Find $\int_C (16 - x^2 - y^2)\, ds$, where C is the line from $(0,0)$ to $(4,0)$.

Solution: Path C is parametrized:

$$\mathbf{r}(t) = \langle t, 0 \rangle, \quad \text{for } 0 \leq t \leq 4.$$

We have $\mathbf{r}'(t) = \langle 1, 0 \rangle$ and $|\mathbf{r}'(t)| = \sqrt{1^2 + (0)^2} = 1$. Since $ds = |\mathbf{r}'(t)|\, dt$, we have $ds = 1\, dt$, or simply $ds = dt$. Thus,

$$\int_C (16 - x^2 - y^2)\, ds = \int_0^4 (16 - t^2)\, dt = \frac{128}{3}$$

This example illustrates that the single-variable integrals along the x-axis are a special case of the scalar line integral, where the path is a line and the endpoints lie along the x-axis. The same would be true for a single-variable integral along the y-axis (x and y being dummy variables in this context).

Suppose that we parameterized the line C from $(0,0)$ to $(4,0)$ as $\mathbf{r}(t) = \langle 4t, 0 \rangle$ for $0 \leq t \leq 1$. Thus, we have $\mathbf{r}'(t) = \langle 4, 0 \rangle$ and $|\mathbf{r}'(t)| = \sqrt{4^2 + (0)^2} = 4$. The line integral is now

$$\int_C (16 - (4t)^2 - (0)^2)\, ds = \int_0^1 (16 - 16t^2)\, 4\, dt$$

$$= 64 \int_0^1 (1 - t^2)\, dt$$

$$= 64 \left[t - \frac{1}{3}t^3 \right]_0^1 = 64 \left(\frac{2}{3}\right) = \frac{128}{3}.$$

As expected, we get the same result. Roughly speaking, the first parametrization, $\mathbf{r}(t) = \langle t, 0 \rangle$ where $0 \leq t \leq 4$, had a "speed" of $|\mathbf{r}'(t)| = 1$, while the second parametrization, with $\mathbf{r}(t) = \langle 4t, 0 \rangle$ and $0 \leq t \leq 1$ (an interval one-fourth the length of the original interval), covered the same path with a "speed" of $|\mathbf{r}'(t)| = 4$.

Example 44.6: A glass rod of consistent thickness is in the shape of a quarter-circle of radius 3, modeled by $r(t) = \langle 3\cos t, 3\sin t\rangle$, where $0 \le t \le \frac{\pi}{2}$. However, it is not uniformly dense. Suppose that its density x centimeters left and right of the origin, and y centimeters above and below the origin, is given by $d(x,y) = x + y + xy$, where d is grams per centimeter. Find the mass of this glass rod.

Solution: This is equivalent to evaluating the scalar line integral

$$\int_C (x + y + xy)\, ds.$$

Differentiating, we have $r'(t) = \langle -3\sin t, 3\cos t\rangle$, so that $|r'(t)| = \sqrt{(-3\sin t)^2 + (3\cos t)^2} = 3$. From $r(t) = \langle 3\cos t, 3\sin t\rangle$, we have $x(t) = 3\cos t$ and $y(t) = 3\sin t$. Substitutions are made:

$$\int_C (x + y + xy)\, ds = \int_0^{\pi/2} \left((3\cos t) + (3\sin t) + (3\cos t)(3\sin t)\right)\, 3\, dt$$

$$= 9\int_0^{\pi/2} (\cos t + \sin t + 3\cos t \sin t)\, dt$$

$$= 9\left[\sin t - \cos t + \frac{3}{2}\sin^2 t\right]_0^{\pi/2}$$

The evaluation is simplified since $\sin 0 = 0$, $\cos 0 = 1$, $\sin\frac{\pi}{2} = 1$ and $\cos\frac{\pi}{2} = 0$:

$$9\left(\sin\left(\frac{\pi}{2}\right) - \cos\left(\frac{\pi}{2}\right) + \frac{3}{2}\sin^2\left(\frac{\pi}{2}\right)\right) - 9\left(\sin(0) - \cos(0) + \frac{3}{2}\sin^2(0)\right)$$

$$= 9\left(1 + \frac{3}{2}\right) - 9(-1) = \frac{63}{2}\text{ grams.}$$

45. Vector Line Integrals: Work & Circulation

Let $\mathbf{F}(x,y,z) = \langle M(x,y,z), N(x,y,z), P(x,y,z)\rangle$ represent a vector field in R^3, and let C be a *directed* path in R^3 parametrized by $\mathbf{r}(t) = \langle x(t), y(t), z(t)\rangle$ for $a \le t \le b$. The word "directed" means that the path must be traversed in a specified direction.

The **vector line integral** of $\mathbf{F}$ along C is given by

$$\int_C \mathbf{F} \cdot d\mathbf{r},$$

where $d\mathbf{r} = \mathbf{r}'(t) = \frac{d}{dt}\mathbf{r}(t)$. A line integral of this form is also defined in R^2, where the vector field is $\mathbf{F}(x,y) = \langle M(x,y), N(x,y)\rangle$ and C is parametrized by $\mathbf{r}(t) = \langle x(t), y(t)\rangle$.

A common "descriptive" way to describe this line integral is

$$\int_C \mathbf{F} \cdot \mathbf{T} \, ds.$$

As the particle moves along the path C, the vector field either "helps" or "hinders" this particle. In order to remove the particle's speed from consideration, the path is segmented into equally-sized sub-segments using the ds segmentation, where $ds = |\mathbf{r}'(t)|\, dt$. This effectively forces the particle to maintain a constant speed, and without loss of generality, we can use the unit tangent vector, $\mathbf{T}(t) = \frac{\mathbf{r}'(t)}{|\mathbf{r}'(t)|}$, to represent the constant speed. Thus, at any position along the path, one of three situations occurs:

- The vector $\mathbf{F}$ at this position points in the same direction as $\mathbf{T}$. That is, $\mathbf{F}$ and $\mathbf{T}$ are acute, and $\mathbf{F} \cdot \mathbf{T} > 0$. Vector $\mathbf{F}$ is "helping" the particle as though it was pushing it from behind.

- The vector $\mathbf{F}$ at this position points in an opposing direction as $\mathbf{T}$. That is, $\mathbf{F}$ and $\mathbf{T}$ are obtuse, and $\mathbf{F} \cdot \mathbf{T} < 0$. Vector $\mathbf{F}$ is hindering the particle's forward movement, as though it were pushing from the front.

- The vector $\mathbf{F}$ at this position is orthogonal direction as $\mathbf{T}$, and $\mathbf{F} \cdot \mathbf{T} = 0$. Vector $\mathbf{F}$ has no effect on the particle's forward movement.

The integral then sums (in the sense of integration) all of the dot products along the path. If the result of the line integral is positive, then the vector field **F** had a net positive effect on the particle's movement. If the line integral is negative, then the vector field **F** had a net negative effect on the particle's movement. If the line integral is 0, then the vector field **F** had a net-zero effect on the particle's movement.

We take the descriptive form of the line integral and make substitutions:

$$\int_C \mathbf{F} \cdot \mathbf{T} \, ds = \int_C \mathbf{F} \cdot \frac{\mathbf{r}'(t)}{|\mathbf{r}'(t)|} |\mathbf{r}'(t)| \, dt.$$

Note that $|\mathbf{r}'(t)|$ cancels, so we have

$$\int_C \mathbf{F} \cdot \mathbf{r}'(t) \, dt = \int_a^b \mathbf{F} \cdot d\mathbf{r},$$

where $d\mathbf{r}$ is shorthand for $\mathbf{r}'(t) \, dt$, and $a \le t \le b$.

The usual process to determine a line integral is the following:

1) Parameterize the path C in variable t. This will give $\mathbf{r}(t) = \langle x(t), y(t), z(t) \rangle$. It will also give the bounds of integration a and b.
2) Find $\mathbf{r}'(t) = \frac{d}{dt} \mathbf{r}(t)$, which will give $\langle x'(t), y'(t), z'(t) \rangle$.
3) Substitute $x(t)$, $y(t)$ and $z(t)$ (from Step 1) into $\mathbf{F}(x, y, z)$. This will give **F** in terms of t.
4) Find $\mathbf{F} \cdot d\mathbf{r}$, which will be a function in terms of t.
5) Integrate the result from Step 4 with respect to t and evaluate at the bounds a and b.

Note: the integral below is a common alternative way to express a line integral:

$$\int_C M(x, y, z) \, dx + N(x, y, z) \, dy + P(x, y, z) \, dz.$$

In this form, the expression $\mathbf{F} \cdot d\mathbf{r}$ has been expanded, where $d\mathbf{r}$ is denoted as $\langle dx, dy, dz \rangle$. It's important to remember that this is equivalent to $\int_C \mathbf{F} \cdot d\mathbf{r}$ and is a single integral in variable t.

The integrals

$$\int_C \mathbf{F} \cdot \mathbf{T} \, ds, \quad \int_C \mathbf{F} \cdot d\mathbf{r} \quad \& \quad \int_C M(x, y, z) \, dx + N(x, y, z) \, dy + P(x, y, z) \, dz$$

are all equivalent. These line integrals are used to show the **work** done by a vector field on a particle. If the path is a loop, the movement of a particle along the loop is called **circulation**.

◆ • ◆ • ◆ • ◆ • ◆

Example 45.1: Find $\int_C \mathbf{F} \cdot d\mathbf{r}$, where $\mathbf{F}(x,y) = \langle -y, x \rangle$ and C is the line segment from $P_0 = (4,0)$ to $P_1 = (0,4)$.

Solution: A sketch of the path C (in bold-black, with its direction shown by an arrow) with the vectors of $\mathbf{F}$ show that the vector field generally points in the same direction as the direction of movement along C. Thus, we expect that the line integral will be positive.

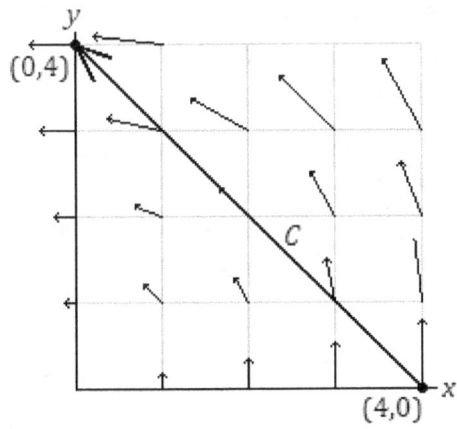

To find $\int_C \mathbf{F} \cdot d\mathbf{r}$, we follow the steps listed above.

1) Parameterize the path C in variable t and note the bounds:

$$\mathbf{r}(t) = \langle 4,0 \rangle + t\langle -4,4 \rangle = \langle 4 - 4t, 4t \rangle, \qquad 0 \leq t \leq 1.$$

2) Find $\mathbf{r}'(t) = \frac{d}{dt}\mathbf{r}(t)$:

$$\frac{d}{dt}\mathbf{r}(t) = \frac{d}{dt}\langle 4 - 4t, 4t \rangle = \langle -4, 4 \rangle.$$

3) Substitute $x(t)$ and $y(t)$ (from Step 1) into $\mathbf{F}(x,y)$:

$$\mathbf{F}(x(t), y(t)) = \langle -4t, 4 - 4t \rangle.$$

4) Find **F** · d**r**:

F · d**r** = ⟨−4t, 4 − 4t⟩ · ⟨−4,4⟩ = (−4t)(−4) + (4 − 4t)(4) = 16.

5) Integrate the result from Step 4 with respect to t:

$$\int_C \mathbf{F} \cdot d\mathbf{r} = \int_0^1 16 \, dt = [16t]_0^1 = 16.$$

The positive quantity of the line integral suggests that particle is "helped" by the vector field as it moves along the path C.

◆ • ◆ • ◆ • ◆ • ◆

Example 45.2: Find $\int_C \mathbf{F} \cdot d\mathbf{r}$, where $\mathbf{F}(x,y) = \langle xy, 1-x \rangle$ and C is the portion of the unit circle that begins at $P_0 = (1,0)$ and continues counterclockwise to $P_1 = (-1,0)$.

Solution: The path C is sketched and its direction noted. Note that the vectors in **F** generally point against (or orthogonal) to the direction of C. Thus, we expect that the line integral will be negative.

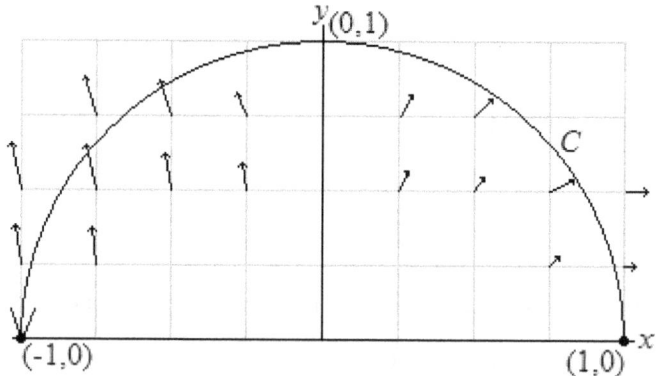

(Only the vectors of **F** "near" C are shown)

Parametrize the path C by the usual parameterization of the unit circle. Note that this is half a circle, so the upper bound of t is π:

$$\mathbf{r}(t) = \langle \cos t, \sin t \rangle, \quad \text{for } 0 \le t \le \pi.$$

As a result, $d\mathbf{r} = \langle -\sin t, \cos t\rangle$ and $\mathbf{F}(x(t), y(t)) = \langle \cos t \sin t, 1 - \cos t\rangle$. We then have

$$\mathbf{F} \cdot d\mathbf{r} = \langle -\sin t, \cos t\rangle \cdot \langle \cos t \sin t, 1 - \cos t\rangle$$
$$= -\sin^2 t \cos t + \cos t - \cos^2 t.$$

This is integrated, using u-du substitution for the first term and an identity for the third term: The identity $\cos^2 t = \frac{1}{2} + \frac{1}{2}\cos 2t$ is used for the last term.

$$\int_C \mathbf{F} \cdot d\mathbf{r} = \int_0^\pi (-\sin^2 t \cos t + \cos t - \cos^2 t)\, dt$$

$$= \int_0^\pi \left(-\sin^2 t \cos t + \cos t - \left(\frac{1}{2} + \frac{1}{2}\cos 2t\right)\right) dt$$

$$= \left[-\frac{1}{3}\sin^3 t + \sin t - \frac{1}{2}t - \frac{1}{4}\sin 2t\right]_0^\pi$$

Evaluated at the bounds, we have

$$\left(-\frac{1}{3}\sin^3 \pi + \sin \pi - \frac{1}{2}\pi - \frac{1}{4}\sin 2\pi\right)$$
$$- \left(-\frac{1}{3}\sin^3 0 + \sin 0 - \frac{1}{2}(0) - \frac{1}{4}\sin 2(0)\right) = -\frac{\pi}{2}.$$

The particle is "hindered" by the vector field as it moves along the path C.

◆ • ◆ • • • ◆ • •

Example 45.3: Find $\int_C \mathbf{F} \cdot d\mathbf{r}$, where $\mathbf{F}(x, y, z) = \langle x, xy, y + z^2\rangle$ and C is the line segment from $P_0 = (1, 2, -4)$ to $P_1 = (3, 5, 1)$.

Solution: The line segment C is parameterized as

$$\mathbf{r}(t) = \langle 1 + 2t, 2 + 3t, -4 + 5t\rangle, \qquad 0 \le t \le 1.$$

Now, find $\mathbf{r}'(t) = \frac{d}{dt}\mathbf{r}(t)$:

$$\frac{d}{dt}\mathbf{r}(t) = \frac{d}{dt}\langle 1 + 2t, 2 + 3t, -4 + 5t\rangle = \langle 2, 3, 5\rangle.$$

Substitute $x(t)$, $y(t)$ and $z(t)$ into $\mathbf{F}(x,y,z)$:

$$\mathbf{F}(x,y,z) = \langle x, xy, y+z^2 \rangle$$
$$\mathbf{F}(x(t),y(t),z(t)) = \langle 1+2t, (1+2t)(2+3t), (2+3t)+(-4+5t)^2 \rangle.$$

Simplified, we have

$$\mathbf{F}(t) = \mathbf{F}(x(t),y(t),z(t)) = \langle 1+2t, 6t^2+7t+2, 25t^2-37t+18 \rangle.$$

Find $\mathbf{F} \cdot d\mathbf{r}$:

$$\mathbf{F} \cdot d\mathbf{r} = \langle 1+2t, 6t^2+7t+2, 25t^2-37t+18 \rangle \cdot \langle 2,3,5 \rangle$$
$$= 2(1+2t) + 3(6t^2+7t+2) + 5(25t^2-37t+18)$$
$$= 143t^2 - 160t + 98.$$

Now, integrate with respect to t:

$$\int_0^1 (143t^2 - 160t + 98)\, dt = \left[\frac{143}{3}t^3 - 80t^2 + 98t\right]_0^1$$
$$= \frac{143}{3} - 80 + 98$$
$$= \frac{197}{3}.$$

Thus, $\int_C \mathbf{F} \cdot d\mathbf{r} = \frac{197}{3}$, a positive quantity, indicating that the vector field $\mathbf{F}$ "helped" the particle as it moved from $P_0 = (1,2,-4)$ to $P_1 = (3,5,1)$.

Example 45.4: Evaluate $\int_C z\, dx + (1-x)\, dy + 2y\, dz$, where C is the helix $\mathbf{r}(t) = (\cos t)\mathbf{i} + (\sin t)\mathbf{j} + t\mathbf{k}$, $0 \leq t \leq 2\pi$.

Solution: From the integral, we can infer the vector field $\mathbf{F}$. We see that

$$M(x,y,z) = z, \quad N(x,y,z) = 1-x \quad \text{and} \quad P(x,y,z) = 2y,$$

so that $\mathbf{F}(x,y,z) = \langle z, 1-x, 2y \rangle$.

From the helix, we have $\mathbf{r}(t) = (\cos t)\mathbf{i} + (\sin t)\mathbf{j} + t\mathbf{k}$, so that $x(t) = \cos t$, $y(t) = \sin t$ and $z(t) = t$. Thus, $\mathbf{F}(t) = \langle z, 1-x, 2y \rangle = \langle t, 1-\cos t, 2\sin t \rangle$.

Also from the helix, we have $d\mathbf{r} = \frac{d}{dt}\langle \cos t, \sin t, t\rangle = \langle -\sin t, \cos t, 1\rangle$. The line integral is now evaluated:

$$\int_C \mathbf{F} \cdot d\mathbf{r} = \int_C \langle t, 1 - \cos t, 2\sin t\rangle \cdot \langle -\sin t, \cos t, 1\rangle \, dt$$

$$= \int_0^{2\pi} (-t\sin t + (1 - \cos t)\cos t + 2\sin t) \, dt$$

$$= \int_0^{2\pi} (-t\sin t + \cos t - \cos^2 t + 2\sin t) \, dt.$$

The antiderivative of $-t\sin t$ is found by integration-by-parts and is $t\cos t - \sin t$, while the antiderivative of $-\cos^2 t$ is found by utilizing the identity $\cos^2 t = \frac{1}{2} + \frac{1}{2}\cos 2t$. When simplified, we have

$$\int_0^{2\pi} (-t\sin t + \cos t - \cos^2 t + 2\sin t) \, dt$$

$$= \left[t\cos t - \sin t - \frac{1}{2}t - \frac{1}{4}\sin 2t - 2\cos t\right]_0^{2\pi}$$

Evaluated at the bounds, we have $(2\pi - \pi - 2) - (-2) = \pi$.

Example 45.5: Evaluate $\int_C xy\,dx + x^2\,dy$, where C is the arc of the parabola $y = x^2$ from $(0,0)$ to $(2,4)$, followed by a straight line from $(2,4)$ back to $(0,0)$.

Solution: From the integral form, we see that $\mathbf{F}(x,y) = \langle xy, x^2 \rangle$. The path C is composed of two smaller paths. Let C_1 be the parabolic arc, and C_2 be the line. Thus, the parametrizations are:

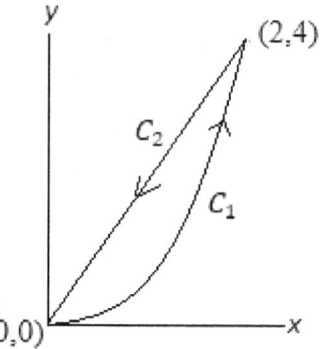

C_1: $\mathbf{r}_1(t) = \langle t, t^2 \rangle$, $\quad 0 \le t \le 2$,
C_2: $\mathbf{r}_2(t) = \langle 2 - 2t, 4 - 4t \rangle$,
$\quad 0 \le t \le 1$.

In such cases, the entire path C is the union of its sub-paths, so that

$$\int_C \mathbf{F}\cdot d\mathbf{r} = \int_{C_1 \cup C_2} \mathbf{F}\cdot d\mathbf{r} = \int_{C_1} \mathbf{F}\cdot d\mathbf{r}_1 + \int_{C_2} \mathbf{F}\cdot d\mathbf{r}_2.$$

For the parabolic arc, we have $d\mathbf{r}_1 = \langle 1, 2t \rangle$ and $\mathbf{F}(t) = \langle t^3, t^2 \rangle$. Thus,

$$\int_{C_1} \mathbf{F}\cdot d\mathbf{r}_1 = \int_{C_1} \langle t^3, t^2 \rangle \cdot \langle 1, 2t \rangle\,dt$$

$$= \int_0^2 3t^3\,dt$$

$$= \left[\frac{3}{4}t^4\right]_0^2 = 12.$$

For the line, $d\mathbf{r}_2 = \langle -2, -4 \rangle$ and $\mathbf{F}(t) = \langle 8t^2 - 16t + 8, 4t^2 - 8t + 4 \rangle$.

$$\int_{C_2} \mathbf{F}\cdot d\mathbf{r}_2 = \int_{C_2} \langle 8t^2 - 16t + 8, 4t^2 - 8t + 4 \rangle \cdot \langle -2, -4 \rangle\,dt$$

$$= \int_0^1 (-32t^2 + 64t - 32)\,dt$$

$$= \left[-\frac{32}{3}t^3 + 32t^2 - 32t\right]_0^1 = -\frac{32}{3}.$$

Therefore, $\int_C xy\,dx + x^2\,dy = 12 + \left(-\frac{32}{3}\right) = \frac{4}{3}$.

46. Vector Line Integrals: Flux

A second form of a line integral can be defined to describe the flow of a medium *through* a permeable membrane. Let $\mathbf{F}(x,y) = \langle M(x,y), N(x,y) \rangle$ be a vector field in R^2, representing the flow of the medium, and let C be a directed path, representing the permeable membrane. The **flux** (flow) of $\mathbf{F}$ through C is given by the flux line integral

$$\int_C \mathbf{F} \cdot \mathbf{n}\, ds.$$

Here, $\mathbf{n}$ represents a unit normal vector to C. The orientation of $\mathbf{n}$ is important, since it will define the "positive" direction of the flow. There are two cases:

- If C is a path with different starting and ending points, then $\mathbf{n}$ will point orthogonally to $\mathbf{T}$, the unit tangent vector, "to the right"; that is, as one moves along C in the direction of $\mathbf{T}$, then $\mathbf{n}$ points to the right of $\mathbf{T}$. Another way to describe this is that $\mathbf{n} \times \mathbf{T}$ would be in the direction of positive z, or out of the sheet of paper toward the viewer.

- If C is a simple closed loop (same starting and ending point, does not cross itself) that is traversed counterclockwise, then $\mathbf{n}$ points outward. Note that this also maintains the "pointing to the right" rule.

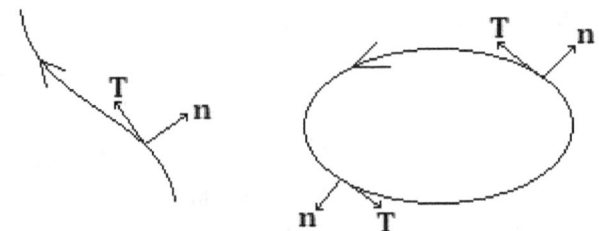

Left: Path C with a direction defined. Unit tangent $\mathbf{T}$ points in the forward direction, and $\mathbf{n}$ is orthogonal to $\mathbf{T}$ to $\mathbf{T}$'s right. **Right:** a simple closed loop path C in the counterclockwise orientation. $\mathbf{T}$ points in the forward direction, and $\mathbf{n}$ is orthogonal to $\mathbf{T}$. Note that in all cases, $\mathbf{n} \times \mathbf{T}$ points up from this page.

The plan is to compare the direction of $\mathbf{F}$ with $\mathbf{n}$ at each point along the path C, which is segmented into equally-sized sub-segments for the moment. If $\mathbf{F}$ and $\mathbf{n}$ point in the same direction, then their dot product $\mathbf{F} \cdot \mathbf{n}$ is positive. If the two vectors point in opposite directions, then $\mathbf{F} \cdot \mathbf{n}$ is negative, and if the two vectors are orthogonal to one another, then $\mathbf{F} \cdot \mathbf{n}$ is zero. The integral then "sums" all such possible dot products, resulting in a value that represents positive net flow (if the value is positive), negative net flow (if the value is negative), or no net flow (if the value is zero).

If C is parameterized as $\mathbf{r}(t) = \langle x(t), y(t) \rangle$, then recall that

$$\mathbf{T}(t) = \frac{\mathbf{r}'(t)}{|\mathbf{r}'(t)|} = \frac{\langle x'(t), y'(t) \rangle}{\sqrt{(x'(t))^2 + (y'(t))^2}}.$$

Since $\mathbf{n}$ is a right-angle turn clockwise from $\mathbf{T}$, then

$$\mathbf{n}(t) = \frac{\langle y'(t), -x'(t) \rangle}{\sqrt{(x'(t))^2 + (y'(t))^2}}.$$

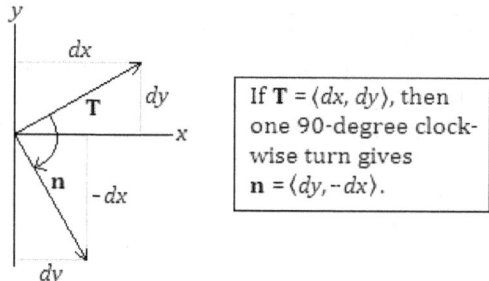

If $\mathbf{T} = \langle dx, dy \rangle$, then one 90-degree clockwise turn gives $\mathbf{n} = \langle dy, -dx \rangle$.

Thus, the flux line integral can be rewritten in terms of t:

$$\int_C \mathbf{F} \cdot \mathbf{n}\, ds = \int_C \mathbf{F}(x(t), y(t)) \cdot \frac{\langle y'(t), -x'(t) \rangle}{\sqrt{(x'(t))^2 + (y'(t))^2}} \sqrt{(x'(t))^2 + (y'(t))^2}\, dt$$

The radicals cancel, and after taking the dot product, we have the short form

$$\int_C \mathbf{F} \cdot \mathbf{n}\, ds = \int_a^b \left(M \frac{dy}{dt} - N \frac{dx}{dt} \right) dt = \int_a^b M\, dy - N\, dx.$$

Remember, despite the notation, this is an integral in variable t.

Example 46.1: Find the flux of $\mathbf{F}(x,y) = \langle 2,0 \rangle$ through the line segment from $(3,0)$ to $(0,3)$.

Solution: The line segment C is parameterized first:

$$\mathbf{r}(t) = \langle 3 - 3t, 3t \rangle \quad \text{for} \quad 0 \le t \le 1. \quad \text{(See Example 12.2)}$$

Thus, $\mathbf{r}'(t) = \langle -3, 3 \rangle$. From this, we have $\frac{dx}{dt} = -3$ and $\frac{dy}{dt} = 3$. Since $\mathbf{F}$ is a constant vector field, no substitutions need to be made. The flux of $\mathbf{F}(x,y) = \langle M, N \rangle = \langle 2, 0 \rangle$ through C is given by

$$\int_a^b M \, dy - N \, dx = \int_0^1 ((2)(3) - (0)(-3)) \, dt = \int_0^1 6 \, dt = 6.$$

In one unit of time, 6 units of mass flow through C. This can be seen graphically as the area of the shaded region below:

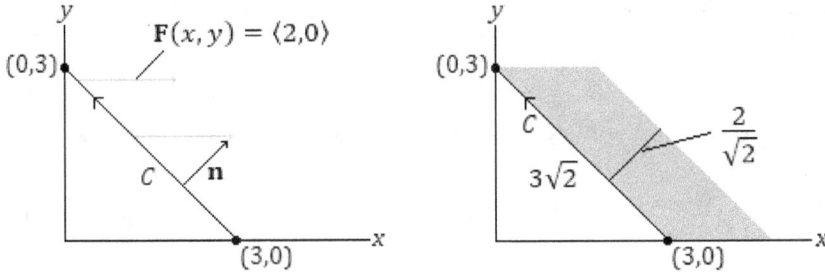

Left: Path C is shown, along with some vectors $\mathbf{F}(x,y) = \langle 2,0 \rangle$ in gray. Each of these vectors has a magnitude of 2 units. Note that the vectors $\mathbf{F}$ cross C at a 45-degree angle, so that the component of these vectors in the direction of $\mathbf{n}$ is $2/\sqrt{2}$. **Right:** The total flow is represented by the gray parallelogram, whose area is the base (length of C) multiplied by the height (the component of $\mathbf{F}$ in the direction of $\mathbf{n}$). We have $(3\sqrt{2})(2/\sqrt{2}) = 6$.

♦ • • ♦ • • ♦ • • ♦

Example 46.2: Find the flux of $\mathbf{F}(x,y) = \langle 3xy, x - y \rangle$ through the parabolic arc $y = x^2$ between $(-1,1)$ and $(4,16)$.

Solution: The parabolic arc is parameterized as

$$\mathbf{r}(t) = \langle t, t^2 \rangle, \quad \text{for} \quad -1 \le t \le 4.$$

Thus, $\mathbf{r}'(t) = \langle 1, 2t \rangle$. Furthermore,

$$\mathbf{F}(t) = \mathbf{F}(x(t), y(t)) = \langle 3(t)(t^2), (t) - (t^2) \rangle = \langle 3t^3, t - t^2 \rangle,$$

where $3xy = 3(t)(t^2) = 3t^3$ and $x - y = (t) - (t^2) = t - t^2$.

Therefore, the flux of $\mathbf{F}(x,y) = \langle 3xy, x-y \rangle$ through the parabolic arc is

$$\int_{-1}^{4} (3t^3)(2t) - (t-t^2)(1) dt = \int_{-1}^{4} (6t^4 + t^2 - t) dt$$

$$= \left[\frac{6}{5}t^5 + \frac{1}{3}t^3 - \frac{1}{2}t^2 \right]_{-1}^{4}$$

This simplifies to

$$= \left(\frac{6}{5}(4)^5 + \frac{1}{3}(4)^3 - \frac{1}{2}(4)^2 \right) - \left(\frac{6}{5}(-1)^5 + \frac{1}{3}(-1)^3 - \frac{1}{2}(-1)^2 \right) = \frac{7465}{6}.$$

About 1,244.167 units of mass flow through this membrane per unit of time.

♦ • ♦ • ♦ • ♦ • ♦

Example 46.3: Find the flux of $\mathbf{F}(x,y) = \langle x,y \rangle$ through the line connecting $(0,0)$ to (a,b).

Solution: The line is parameterized as $\mathbf{r}(t) = \langle at, bt \rangle$ for $0 \leq t \leq 1$, and so $\mathbf{r}'(t) = \langle a,b \rangle$. Furthermore, $\mathbf{F}(t) = \mathbf{F}(x(t), y(t)) = \langle at, bt \rangle$. Thus, the flux is

$$\left. \begin{array}{ll} M = at & dy = b \\ N = bt & dx = a \end{array} \right\} \quad \int_0^1 (at)(b) - (bt)(a) \, dt = \int_0^1 (abt - abt) \, dt = 0.$$

This result is not surprising: any line connecting $(0,0)$ to (a,b) is parallel to the stream-lines formed by the vector field $\mathbf{F}(x,y) = \langle x,y \rangle$. At no time (or place) does $\mathbf{F}$ ever pass through such a line.

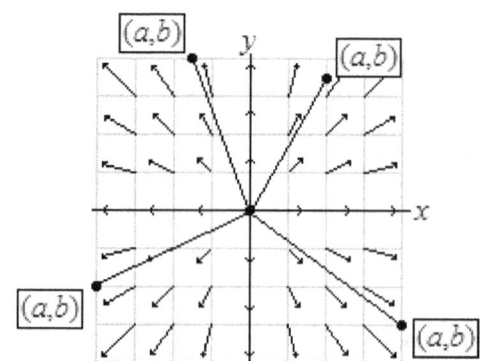

Example 46.4: Find the flux of $F(x, y) = \langle 3x, 5y \rangle$ through the circle $x^2 + y^2 = 1$, traversed counterclockwise.

Solution: The circle is parameterized as $r(t) = \langle \cos t, \sin t \rangle$ for $0 \leq t \leq 2\pi$. Thus, $r'(t) = \langle -\sin t, \cos t \rangle$. The vector field is written in terms of t:

$$F(t) = F(x(t), y(t)) = \langle 3 \cos t, 5 \sin t \rangle.$$

Thus, we have $M = 3 \cos t$, $N = 5 \sin t$, $\frac{dy}{dt} = \cos t$ and $\frac{dx}{dt} = -\sin t$:

$$\int_0^{2\pi} ((3 \cos t)(\cos t) - (5 \sin t)(-\sin t)) \, dt = \int_0^{2\pi} (3 \cos^2 t + 5 \sin^2 t) \, dt.$$

We use the identities $\cos^2 t = \frac{1}{2} + \frac{1}{2}\cos 2t$ and $\sin^2 t = \frac{1}{2} - \frac{1}{2}\cos 2t$ to simplify the integrand:

$$\int_0^{2\pi} (3 \cos^2 t + 5 \sin^2 t) \, dt = \int_0^{2\pi} \left(3 \left(\frac{1}{2} + \frac{1}{2} \cos 2t \right) + 5 \left(\frac{1}{2} - \frac{1}{2} \cos 2t \right) \right) dt$$

$$= \int_0^{2\pi} (4 - \cos 2t) \, dt$$

$$= \left[4t - \frac{1}{2} \sin 2t \right]_0^{2\pi} = 8\pi.$$

♦ ♦ ♦ • • ♦ • • ♦

The flux of $F = \langle M, N \rangle$ through a simple closed loop C traversed counterclockwise, such as in Example 46.4, can be calculated using the following formula:

$$\int_C F \cdot n \, ds = \iint_R \left(\frac{\partial M}{\partial x} + \frac{\partial N}{\partial y} \right) dA,$$

where R is the region enclosed by C. This is called the **Divergence Theorem** in R^2. The general Divergence Theorem is discussed later in Section 54.

Example 46.5: Use the Divergence Theorem to find the flux of $F(x,y) = \langle 5y^2, \sqrt{e^x} \rangle$ through the triangle traversed from vertices $(1,1)$, $(5,1)$ and $(3,5)$, back to $(1,1)$, in that order.

Solution: Note that $\frac{\partial M}{\partial x} + \frac{\partial N}{\partial y} = 0$. Thus, the net flux is 0. This means that equal amounts of mass are entering and exiting through the boundaries per unit of time.

◆ • ◆ • ◆ • ◆ • ◆

Example 46.6: Use the Divergence Theorem to find the flux of $F(x,y) = \langle 3x, 5y \rangle$ through the circle $x^2 + y^2 = 1$, traversed counterclockwise. (This is a repeat of Example 46.4)

Solution: Since $M(x,y) = 3x$, then $\frac{\partial M}{\partial x} = 3$, and since $N(x,y) = 5y$, then $\frac{\partial N}{\partial y} = 5$. Thus, we have

$$\iint_R \left(\frac{\partial M}{\partial x} + \frac{\partial N}{\partial y} \right) dA = \iint_R (3+5)\, dA = \iint_R 8\, dA = 8 \iint_R dA.$$

Note that $\iint_R dA$ represents the area of R. Since R is a circle of radius 1, its area is $\pi(1)^2 = \pi$. Thus, the flux of $F(x,y) = \langle 3x, 5y \rangle$ through the circle $x^2 + y^2 = 1$ is

$$8 \iint_R dA = 8\pi.$$

◆ • ◆ • ◆ • ◆ • ◆

Example 46.7: Use the Divergence Theorem to find the flux of $F(x,y) = \langle xy, 3 \rangle$ through the rectangle traversed from $(0,0)$ to $(0,3)$ to $(6,3)$ to $(6,0)$ to $(0,0)$.

Solution: The path C is a rectangle traced clockwise, not counterclockwise as is required by the Divergence Theorem. We can proceed, but must negate the final result to account for the clockwise movement. We have

$$\iint_R \left(\frac{\partial M}{\partial x} + \frac{\partial N}{\partial y} \right) dA = \int_0^6 \int_0^3 y\, dy\, dx.$$

The inside integral is

$$\int_0^3 y\, dy = \left[\frac{1}{2}y^2\right]_0^3 = \frac{9}{2}.$$

The outside integral is

$$\int_0^6 \left(\frac{9}{2}\right) dx = \left(\frac{9}{2}\right)(6) = 27.$$

Since the path around the rectangle is traced clockwise, the result is negated:

$$\int_C \mathbf{F} \cdot \mathbf{n}\, ds = -27.$$

◆ • ◆ • ◆ • ◆ • ◆

47. Conservative Vector Fields

Given a function $z = f(x, y)$, its gradient is $\nabla f = \langle f_x, f_y \rangle$. Thus, ∇f is a **gradient** (or **conservative**) **vector field**, and the function f is called a **potential function**.

Suppose we are given the vector field first, in the form $\mathbf{F}(x, y) = \langle M(x, y), N(x, y) \rangle$. Can we show that this is a conservative vector field? Recall that $f_{xy} = f_{yx}$ is true by Clairaut's Theorem. Assuming such a function f exists, we infer that $f_x = M$ and that $f_y = N$, and observing that $f_{xy} = f_{yx}$, this is equivalent to showing that $M_y = N_x$. In other words, if f exists, then $M_y = N_x$, and if $M_y = N_x$ is true, then f exists. (The exceptions to this property are rare and not relevant to this discussion).

To summarize, if given a vector field $\mathbf{F}(x, y) = \langle M(x, y), N(x, y) \rangle$, then two cases result:

- If $M_y = N_x$, then $\mathbf{F}$ is conservative, and there exists a potential function ϕ.
- If $M_y \neq N_x$, then $\mathbf{F}$ is not conservative and no such potential function ϕ exists.

Example 47.1: Determine if $F(x, y) = \langle 3x^2y^2, 2x^3y \rangle$ is conservative. If it is, find a potential function $f(x, y)$ such that $\nabla f = F$.

Solution: From F, we have $M(x, y) = 3x^2y^2$ and $N(x, y) = 2x^3y$. We find M_y and N_x:

$$M_y = 6x^2y \quad \text{and} \quad N_x = 6x^2y.$$

Since $M_y = N_x$, then F is conservative, and there exists a function $f(x, y)$ such that $f_x = 3x^2y^2$ and $f_y = 2x^3y$. Since we assume that $f_x = 3x^2y^2$, we integrate it with respect to x:

$$\int 3x^2y^2 \, dx = x^3y^2 + g(y).$$

Here, $g(y)$ represents a possible term in variable y, noting that under differentiation with respect to x, $\frac{\partial}{\partial x} g(y) = 0$. Now differentiate this result with respect to y:

$$\frac{\partial}{\partial y}(x^3y^2 + g(y)) = 2x^3y + g'(y).$$

This is compared to $N(x, y) = 2x^3y$:

$$2x^3y + g'(y) = 2x^3y.$$

This suggests $g'(y) = 0$ so that integrating, $g(y) = k$, a constant. Thus, the potential function has the form $f(x, y) = x^3y^2 + k$. In such cases, any constants are set to 0. This leaves

$$f(x, y) = x^3y^2$$

as a potential function of F. This is easily checked by showing that $f_x = 3x^2y^2$ and $f_y = 2x^3y$.

◆ • ◆ • ◆ • ◆ • ◆

Example 47.2: Determine if $F(x, y) = \langle xy, 1 - x^2 \rangle$ is conservative. If it is, find the potential function $f(x, y)$ such that $\nabla f = F$.

Solution: We have $M(x, y) = xy$ and $N(x, y) = 1 - x^2$. Observe that $M_y = x$ and $N_x = -2x$. Since $M_y \neq N_x$, vector field F is not conservative, and there does not exist a function whose gradient is F.

Example 47.3: Determine if $\mathbf{F}(x,y) = \langle y - 3, x + 2 \rangle$ is conservative. If it is, find a potential function f.

Solution: We have $M(x,y) = y - 3$ and $N(x,y) = x + 2$. Observe that $M_y = 1$ and $N_x = 1$. Since $M_y = N_x$, the vector field $\mathbf{F}$ is conservative. To determine $f(x,y)$, we first integrate $M(x,y)$ with respect to x:

$$\int (y - 3)\, dx = xy - 3x + g(y).$$

Differentiating this result with respect to y, we have

$$\frac{\partial}{\partial y}(xy - 3x + g(y)) = x + g'(y).$$

This is compared to $N(x,y) = x + 2$:

$$x + g'(y) = x + 2.$$

Thus, $g'(y) = 2$, so that $g(y) = 2y + k$ (any constants of integration can be set to 0). A potential function is

$$f(x,y) = xy - 3x + 2y,$$

which we check by showing that $\nabla f(x,y) = \mathbf{F}(x,y)$:

$$f_x(x,y) = y - 3, \qquad f_y(x,y) = x + 2.$$

These are precisely the components of $\mathbf{F}$, so $\nabla f(x,y) = \mathbf{F}(x,y)$.

> Given a conservative vector field $\mathbf{F}(x,y) = \langle M(x,y), N(x,y) \rangle$, a "shortcut" to find a potential function $f(x,y)$ is to integrate $M(x,y)$ with respect to x, and $N(x,y)$ with respect to y, and to form the union of the terms in each antiderivative. However, check that the alleged potential function's partial derivatives with respect to x, and respect to y, do give M and N, respectively.

Example 47.4: Given the conservative vector field $F(x,y) = \langle 3x^2y^2, 2x^3y \rangle$, find a potential function, $f(x,y)$.

Solution: Integrate $M(x,y) = 3x^2y^2$ with respect to x, and $N(x,y) = 2x^3y$ with respect to y:

$$\int 3x^2y^2 \, dx = x^3y^2 \quad \text{and} \quad \int 2x^3y \, dy = x^3y^2.$$

Observing the two antiderivatives, we infer that $f(x,y) = x^3y^2$ may be a potential function. A check that $f_x = M(x,y) = 3x^2y^2$ and $f_y = N(x,y) = 2x^3y$ shows that this function is a correct potential function. Any constants of integration can be ignored.

♦ • ♦ • ♦ • ♦ • ♦

Example 47.5: Given the conservative vector field

$$F(x,y) = \langle 3x^2 + 2y, 2x - 2y \rangle,$$

find the potential function, $f(x,y)$.

Solution: Integrate $3x^2 + 2y$ with respect to x, and $2x - 2y$ with respect to y:

$$\int (3x^2 + 2y) \, dx = x^3 + 2xy \quad \text{and} \quad \int (2x - 2y) \, dy = 2xy - y^2.$$

The union of terms from these two antiderivatives is $f(x,y) = x^3 + 2xy - y^2$. We check that this is actually a potential function: $f_x = 3x^2 + 2y = M(x,y)$, and $f_y = 2x - 2y = N(x,y)$. Thus, this is a correct potential function.

♦ • ♦ • ♦ • ♦ • ♦

Example 47.6: A student is given the vector field $F(x,y) = \langle x^2, xy \rangle$. He then integrates x^2 with respect to x, getting $\int x^2 \, dx = \frac{1}{3}x^3$, and integrates xy with respect to y, getting $\int xy \, dy = \frac{1}{2}xy^2$. He concludes that the potential function is $f(x,y) = \frac{1}{3}x^3 + \frac{1}{2}xy^2$. Explain the error.

Solution: Vector field F is *not* conservative since $M_y \neq N_x$. Thus, there is no potential function that generates F. Note that the alleged potential function, $f(x,y) = \frac{1}{3}x^3 + \frac{1}{2}xy^2$, does not generate F since $f_x = x^2 + \frac{1}{2}y^2$, which is not equal to x^2. In other words, $\nabla f \neq F$.

Vector fields in R^3 can also be conservative, where $w = f(x, y, z)$ is a potential function of a vector field $\nabla f = \mathbf{F}(x, y, z) = \langle f_x, f_y, f_z \rangle$. However, showing that a vector field $\mathbf{F}$ in R^3 is conservative is found by showing that curl $\mathbf{F} = \mathbf{0}$. The **curl** of a vector field is discussed in Section 52.

♦ • ♦ • ♦ • ♦ • ♦

48. Fundamental Theorem of Line Integrals

If $\mathbf{F}$ is a conservative vector field in R^2 with $f(x, y)$ as its potential function, and C is a directed path with endpoints $a = (x_0, y_0)$ and $b = (x_1, y_1)$, then

$$\int_C \mathbf{F} \cdot d\mathbf{r} = [f(x,y)]_a^b = f(b) - f(a) = f(x_1, y_1) - f(x_0, y_0).$$

This is called the **Fundamental Theorem of Line Integrals** (FTLI). In this case, there is no need to parametrize the path, as the value of the line integral depends only on the potential function evaluated at the endpoints, then subtracted in the usual manner of integration. A couple of corollaries follow:

- Line integrals in a conservative vector field are *path independent*, meaning that any path from a to b will result in the same value of the line integral.

- If the path C is a simple loop, meaning it starts and ends at the same point and does not cross itself, and $\mathbf{F}$ is a conservative vector field, then the line integral is 0.

♦ • ♦ • ♦ • ♦ • ♦

Example 48.1: Evaluate $\int_C \mathbf{F} \cdot d\mathbf{r}$, where $\mathbf{F}(x, y) = \langle 3x^2 y^2, 2x^3 y \rangle$ and C is the line segment from $a = (1,2)$ to $b = (4, -3)$.

Solution: From a previous example, we showed that $\mathbf{F}$ is conservative, and that a potential function is $f(x, y) = x^3 y^2$. Therefore,

$$\int_C \mathbf{F} \cdot d\mathbf{r} = [x^3 y^2]_{(1,2)}^{(4,-3)}$$
$$= (4)^3(-3)^2 - (1)^3(2)^2$$
$$= 576 - 4 = 572.$$

Note that we did not parametrize the line segment to solve this line integral.

Example 48.2: Evaluate $\int_C \mathbf{F} \cdot d\mathbf{r}$, where $\mathbf{F}(x,y) = \langle 2x, 3y \rangle$ and C is any path from $a = (1,0)$ to $b = (0,1)$.

Solution: Let's try a few common paths. Suppose C is a line from a to b. We have $\mathbf{r}(t) = \langle 1-t, t \rangle$, where $0 \le t \le 1$. Thus, $\mathbf{r}'(t) = \langle -1, 1 \rangle$ and $\mathbf{F}(t) = \langle 2(1-t), 3t \rangle$. The line integral is

$$\int_C \mathbf{F} \cdot d\mathbf{r} = \int_0^1 \langle 2(1-t), 3t \rangle \cdot \langle -1, 1 \rangle \, dt$$

$$= \int_0^1 (5t - 2) \, dt$$

$$= \left[\frac{5}{2} t^2 - 2t \right]_0^1$$

$$= \frac{5}{2} - 2$$

$$= \frac{1}{2}.$$

Now suppose C is a quarter circle, centered at the origin, with radius 1. It is parametrized as $\mathbf{r}(t) = \langle \cos t, \sin t \rangle$, where $0 \le t \le \frac{\pi}{2}$. As a result, $\mathbf{r}'(t) = \langle -\sin t, \cos t \rangle$ and $\mathbf{F}(t) = \langle 2\cos t, 3\sin t \rangle$. The line integral is

$$\int_C \mathbf{F} \cdot d\mathbf{r} = \int_0^{\pi/2} \langle 2\cos t, 3\sin t \rangle \cdot \langle -\sin t, \cos t \rangle \, dt$$

$$= \int_0^{\pi/2} \sin t \cos t \, dt$$

$$= \left[\frac{1}{2} \sin^2 t \right]_0^{\pi/2} \quad \begin{cases} u = \sin t \\ du = \cos t \, dt. \end{cases}$$

$$= \frac{1}{2}.$$

Suppose C is a parabola $x = 1 - y^2$. It is parametrized as $\mathbf{r}(t) = \langle 1 - t^2, t \rangle$, where $0 \leq t \leq 1$. Thus, $\mathbf{r}'(t) = \langle -2t, 1 \rangle$ and $\mathbf{F}(t) = \langle 2(1 - t^2), 3t \rangle$. The line integral is

$$\int_C \mathbf{F} \cdot d\mathbf{r} = \int_0^1 \langle 2(1 - t^2), 3t \rangle \cdot \langle -2t, 1 \rangle \, dt$$

$$= \int_0^1 (4t^3 - t) \, dt$$

$$= \left[t^4 - \frac{1}{2} t^2 \right]_0^1$$

$$= 1 - \frac{1}{2}$$

$$= \frac{1}{2}.$$

It appears that regardless the path, the line integral is $\int_C \mathbf{F} \cdot d\mathbf{r} = \frac{1}{2}$. Although three examples are not a "proof" that this assertion is true, it suggests that it might be worth approaching the problem from a different perspective.

Observe that the vector field $\mathbf{F}$ is conservative: $M(x, y) = 2x$, so that $M_y = 0$, and $N(x, y) = 3y$, so that $N_x = 0$. A potential function is $f(x, y) = x^2 + \frac{3}{2} y^2$ (You should verify this). Thus, using the Fundamental Theorem of Line Integrals, we have

$$\int_C \mathbf{F} \cdot d\mathbf{r} = \left[x^2 + \frac{3}{2} y^2 \right]_{(1,0)}^{(0,1)}$$

$$= \left((0)^2 + \frac{3}{2}(1)^2 \right) - \left((1)^2 + \frac{3}{2}(0)^2 \right)$$

$$= \frac{3}{2} - 1$$

$$= \frac{1}{2}.$$

This example illustrates that in a conservative vector field, the line integral along any path between two fixed endpoints will always give the same result. Rather than try many different paths, it's easier to first check whether $\mathbf{F}$ is conservative. If it is, then skip the parametrization step entirely, and proceed to finding a potential function and using the Fundamental Theorem of Line Integrals.

Example 48.3: Evaluate $\int_C \mathbf{F} \cdot d\mathbf{r}$, where $\mathbf{F}(x,y) = \langle y, x+2y \rangle$ and C is a sequence of line segments from (1,3) to (2,7) to (−4,0) to (8,2).

Solution: We check first to see if $\mathbf{F}$ is conservative: $M_y = 1$ and $N_x = 1$. Since $M_y = N_x$, then $\mathbf{F}$ is conservative, and it is not necessary to parametrize the sequence of line segments. Instead, we find f and evaluate it by using the Fundamental Theorem of Line Integrals. We need a potential function. Note that

$$\int y\, dx = xy \quad \text{and} \quad \int (x+2y)\, dy = xy + y^2.$$

Thus, $f(x,y) = xy + y^2$ is a potential function. We check by finding ∇f: $f_x = y$ and $f_y = x + 2y$. These are M and N, respectively, so $f(x,y) = xy + y^2$ is a correct potential function. Therefore,

$$\int_C \mathbf{F} \cdot d\mathbf{r} = [xy + y^2]_{(1,3)}^{(8,2)}$$

$$= \left((8)(2) + (2)^2\right) - \left((1)(3) + (3)^2\right)$$

$$= 20 - 12$$

$$= 8.$$

The intermediate points were ignored. Only the starting and ending points of the path are needed.

♦ • ♦ • • ♦ • • ♦ • ♦

Example 48.4: Evaluate $\int_C \mathbf{F} \cdot d\mathbf{r}$, where $\mathbf{F}(x,y) = \langle 2x, 3y^2 \rangle$ and C is given by $\mathbf{r}(t) = \langle t^2, 5t \rangle$ for $-1 \le t \le 3$.

Solution: Note that $M_y = 0$ and that $N_x = 0$. Since $M_y = N_x$, then $\mathbf{F}$ is conservative, and that $f(x,y) = x^2 + y^3$ is the potential function. Since $\mathbf{F}$ is conservative, the actual path of C is not relevant. We just need its two endpoints. When $t = -1$, we have $\mathbf{r}(-1) = \langle(-1)^2, 5(-1)\rangle = \langle 1, -5 \rangle$, and when $t = 3$, we have $\mathbf{r}(3) = \langle(3)^2, 5(3)\rangle = \langle 9, 15 \rangle$. Note that $\langle 1, -5 \rangle$ and $\langle 9, 15 \rangle$ are vectors, but if their feet are placed at the origin, then their heads point to the ordered pairs (1, −5) and (9,15). In this way, the point as ordered pairs can be inferred from a vector.

Therefore, the line integral is

$$\int_C \mathbf{F} \cdot d\mathbf{r} = [x^2 + y^3]_{(1,-5)}^{(9,15)} = ((9)^2 + (15)^3) - ((1)^2 + (-5)^3) = 3580.$$

Example 48.5: The contour map of $z = f(x,y)$ is below, for $-4 \leq x \leq 4$ and $-4 \leq y \leq 4$. Suppose that vector field $\mathbf{F}(x,y) = \nabla f(x,y)$.

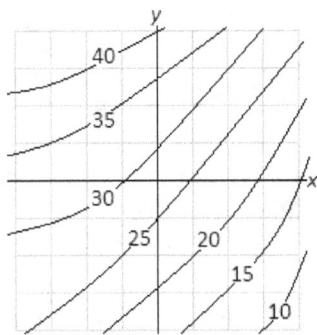

Evaluate the following:

a) $\int_C \mathbf{F} \cdot d\mathbf{r}$, where C is any path from $(2,-1)$ to $(-3,1)$.

b) $\int_C \mathbf{F} \cdot d\mathbf{r}$, where C is any path from $(-1,0)$ to $(5,3)$, then to $(3,-2)$.

c) $\int_C \mathbf{F} \cdot d\mathbf{r}$, where C is a circle of radius 2, centered at the origin.

Solution:

a) Since $\mathbf{F}(x,y) = \nabla f(x,y)$, then f is a potential function of the vector field $\mathbf{F}$, and $\mathbf{F}$ is conservative. Thus, $\mathbf{F}$ is path-independent, and only the starting and ending points of C are relevant. Note that from the contour map, we have $z = f(2,-1) = 20$ as the starting point, and $z = f(-3,1) = 35$ as the ending point. By the Fundamental Theorem of Line Integrals, we have

$$\int_C \mathbf{F} \cdot d\mathbf{r} = [f(x,y)]_{(2,-1)}^{(-3,1)}$$

$$= f(-3,1) - f(2,-1)$$

$$= 35 - 20$$

$$= 15.$$

b) Because $\mathbf{F}$ is conservative, only the starting and ending points of the path are relevant. Note that $f(-1,0) = 30$ and that $f(3,-2) = 15$. Thus, $\int_C \mathbf{F} \cdot d\mathbf{r} = 15 - 30 = -15$.

c) Since $\mathbf{F}$ is a conservative vector field and C is a closed simple loop, then $\int_C \mathbf{F} \cdot d\mathbf{r} = 0$.

49. Green's Theorem

Let $\mathbf{F}(x,y) = \langle M(x,y), N(x,y) \rangle$ be a vector field in R^2, and C is a path that starts and ends at the same point such that it does not cross itself. Such a path is called a *simple closed loop*, and it will enclose a region R. Assume M and N and its first partial derivatives are defined within R including its boundary C. Furthermore, the path is to be traversed (circulated) in a counterclockwise direction, called the *positive orientation*. If these conditions are met, then the line integral around the simple loop path may be evaluated by a double integral. This is called **Green's Theorem**, and is written

$$\int_C \mathbf{F} \cdot d\mathbf{r} = \iint_R (N_x - M_y)\, dA.$$

The expression $N_x - M_y$ is the *curl* of $\mathbf{F}$ in R^2.

If $\mathbf{F}$ is a conservative vector field, then $M_y = N_x$, so that $N_x - M_y = 0$. Thus, in a conservative vector field, all line integrals along a simple closed loop path evaluate to 0. In a physical sense, there is no net circulation around the loop, and a conservative vector field is often called a *rotation-free* (or *irrotational*) vector field.

When calculating a line integral, you should check two things:

- Is the vector field conservative?
- Is the path a simple closed loop?

The following table will help plan the calculation accordingly.

	F is conservative	F is not conservative
C is a simple closed loop	0	Use Green's Theorem
C is not a loop of any kind (it has different start and end points).	Find the potential function $\varphi(x,y)$ and calculate the line integral by the Fundamental Theorem of Line Integrals (The FTLI)	Parameterize the path(s) in variable t and calculate the line integral directly.

Example 49.1: Evaluate $\int_C \mathbf{F} \cdot d\mathbf{r}$, where $\mathbf{F}(x, y) = \langle y, 4x \rangle$ and C is a triangle, traversed from (0,0) to (2,0) to (2,4) back to (0,0).

Solution: Sketch C and observe that it is a simple closed loop that is traversed counterclockwise:

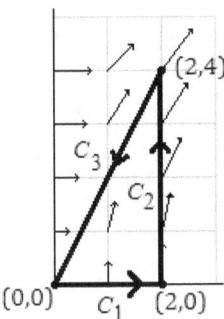

To evaluate $\int_C \mathbf{F} \cdot d\mathbf{r}$ as a sequence of line integrals, the path must be divided into three smaller paths: C_1 being the line from (0,0) to (2,0), C_2 being the line from (2,0) to (2,4), and C_3 being the line from (2,4) to (0,0).

- For C_1, the path is parameterized $\mathbf{r}_1(t) = \langle 2t, 0 \rangle$ with $0 \leq t \leq 1$, so that $\mathbf{r}_1'(t) = \langle 2, 0 \rangle$ and $\mathbf{F}(t) = \langle 0, 8t \rangle$. Thus, $\mathbf{F} \cdot d\mathbf{r}_1 = \langle 0, 8t \rangle \cdot \langle 2, 0 \rangle = 0$, so that $\int_{C_1} \mathbf{F} \cdot d\mathbf{r}_1 = 0$. In the above image, note that the vector field elements are orthogonal to the segment C_1.

- For C_2, the path is parameterized $\mathbf{r}_2(t) = \langle 2, 4t \rangle$ with $0 \leq t \leq 1$, so that $\mathbf{r}_2'(t) = \langle 0, 4 \rangle$ and $\mathbf{F}(t) = \langle 4t, 8 \rangle$. Thus, $\mathbf{F} \cdot d\mathbf{r}_2 = \langle 4t, 8 \rangle \cdot \langle 0, 4 \rangle = 32$, so that $\int_{C_2} \mathbf{F} \cdot d\mathbf{r}_2 = \int_0^1 32 \, dt = [32t]_0^1 = 32$. The vector field elements agree with the direction of C_2.

- For C_3, the path is parameterized $\mathbf{r}_3(t) = \langle 2 - 2t, 4 - 4t \rangle$ with $0 \leq t \leq 1$, so that $\mathbf{r}_3'(t) = \langle -2, -4 \rangle$ and $\mathbf{F}(t) = \langle 4 - 4t, 8 - 8t \rangle$. Thus, $\mathbf{F} \cdot d\mathbf{r}_3 = \langle 4 - 4t, 8 - 8t \rangle \cdot \langle -2, -4 \rangle = 40t - 40$, which gives $\int_{C_3} \mathbf{F} \cdot d\mathbf{r}_3 = \int_0^1 (40t - 40) \, dt = [20t^2 - 40t]_0^1 = -20$. The vector field elements disagree (point against) the direction of C_3.

Since

$$\int_C \mathbf{F} \cdot d\mathbf{r} = \int_{C_1} \mathbf{F} \cdot d\mathbf{r}_1 + \int_{C_2} \mathbf{F} \cdot d\mathbf{r}_2 + \int_{C_3} \mathbf{F} \cdot d\mathbf{r}_3,$$

the line integral is

$$\int_C \mathbf{F} \cdot d\mathbf{r} = 0 + 32 - 20 = 12.$$

Using Green's Theorem, the problem is much shorter. The curl is $N_x - M_y = 4 - 1 = 3$, so that

$$\int_C \mathbf{F} \cdot d\mathbf{r} = \iint_R 3 \, dA$$
$$= 3 \iint_R dA$$
$$= 3(4)$$
$$= 12.$$

The constant integrand was moved to the front, leaving $\iint_R dA$, which is the area of region R. Using geometry, the area of R, a triangle with base 2 and height 4, is $\frac{1}{2}(2)(4) = 4$. Which method was faster?

◆ • ◆ ◆ • • ◆ • • ◆

Example 49.2: Evaluate $\int_C \mathbf{F} \cdot d\mathbf{r}$, where $\mathbf{F}(x, y) = \langle 2xy, x \rangle$ and C traverses from (2,0) to (−2,0) along a semi-circle of radius 2, centered at the origin, in the counter-clockwise direction, then from (−2,0) back to (2,0) along a straight line.

Solution: Path C is a simple closed loop traversed in a counterclockwise direction, as shown below.

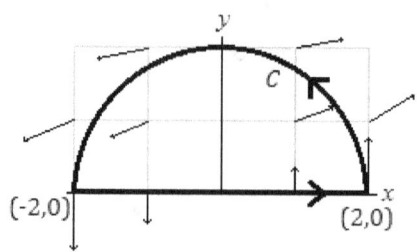

To find $\int_C \mathbf{F} \cdot d\mathbf{r}$, we use Green's Theorem. The curl is $N_x - M_y = 1 - 2x$. Since the region R is a semicircle of radius 2, the double integral is best evaluated using polar coordinates.

$$\int_C \mathbf{F} \cdot d\mathbf{r} = \iint_R (N_x - M_y)\, dA$$

$$= \iint_R (1 - 2x)\, dA$$

$$= \int_0^\pi \int_0^2 (1 - 2r\cos\theta)\, r\, dr\, d\theta$$

$$= \int_0^\pi \int_0^2 (r - 2r^2 \cos\theta)\, dr\, d\theta.$$

The inside integral is evaluated with respect to r:

$$\int_0^2 (r - 2r^2 \cos\theta)\, dr = \left[\frac{1}{2}r^2 - \frac{2}{3}r^3 \cos\theta\right]_0^2 = 2 - \frac{16}{3}\cos\theta.$$

This is then integrated with respect to θ:

$$\int_0^\pi \left(2 - \frac{16}{3}\cos\theta\right) d\theta = \left[2\theta - \frac{16}{3}\sin\theta\right]_0^\pi = 2\pi.$$

The line integral along C is $\int_C \mathbf{F} \cdot d\mathbf{r} = 2\pi$. There is positive circulation along this path induced by the vector field.

◆ ◆ ◆ ◆ ◆ ◆ ◆ ◆ ◆

Example 49.3: Evaluate $\int_C \mathbf{F} \cdot d\mathbf{r}$, where $\mathbf{F}(x, y) = \langle 3y, -x + y \rangle$ and C traverses a rectangle from $(1,1)$ to $(1,6)$ to $(7,6)$ to $(7,1)$ back to $(1,1)$.

Solution: A sketch of the path C shows it to be a simple closed loop traversed in a *clockwise* direction. To use Green's Theorem, the path needs to be traversed it in the counterclockwise direction, which is equivalent to traversing each segment in its opposite direction. This means that the result will be multiplied by -1 to account for this "opposite" direction.

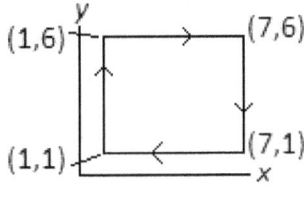

Using Green's Theorem, the curl is $N_x - M_y = -1 + (-3) = -4$:

$$\iint_R (N_x - M_y)\, dA = \iint_R (-4)\, dA = -4 \iint_R dA.$$

The double integral $\iint_R dA$ is the area of the rectangle, which is $(6)(5) = 30$. Thus,

$$\iint_R (N_x - M_y)\, dA = -4(30) = -120.$$

However, since C was traversed in the opposite direction, the result is negated:

$$\int_C \mathbf{F} \cdot d\mathbf{r} = 120.$$

♦ ♦ ♦ ♦ ♦ ♦ ♦ ♦ ♦ ♦

Example 49.4: Evaluate $\int_C \mathbf{F} \cdot d\mathbf{r}$, where $\mathbf{F}(x, y) = \langle 5x^4 + y^2, 2yx \rangle$ and C is an ellipse with major axis of 12 along the x-axis, and minor axis of 8 along the y-axis, in a counterclockwise direction.

Solution: Green's Theorem gives

$$\iint_R (N_x - M_y)\, dA = \iint_R (2y - 2y)\, dA = \iint_R 0\, dA = 0.$$

Note that $\mathbf{F}$ is conservative, since $M_y = N_x$ (equivalently, the curl is 0). There is no need to parameterize the ellipse.

♦ ♦ ♦ ♦ ♦ ♦ ♦ ♦ ♦ ♦

Green's Theorem can be used to find the line integral of a non-loop path. We "close off" the path forming a loop, as this next example shows:

Example 49.5: Evaluate $\int_C \mathbf{F} \cdot d\mathbf{r}$, where $\mathbf{F}(x, y) = \langle 2y, x^2 \rangle$ and C is a sequence of line segments from (0,0) to (3,0) to (3,4) to (–4,4).

Solution: The path C is not a simple closed loop nor is the vector field conservative. Thus, we would have to parametrize each line segment one at a time and determine the value of each line integral individually. Instead, we can add in the final line segment, that from (–4,4) to (0,0), creating a simple closed loop traversed counter-clockwise, and use Green's Theorem.

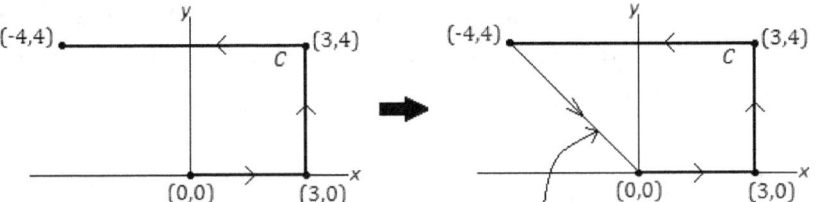

The path C is not a closed loop.　　Add in the final segment to make a closed loop.

Using Green's Theorem, we have

$$\iint_R (N_x - M_y)\, dA = \int_0^4 \int_{-y}^3 (2x - 2)\, dx\, dy = 2\int_0^4 \int_{-y}^3 (x - 1)\, dx\, dy.$$

The inside integral is evaluated:

$$\int_{-y}^3 (x - 1)\, dx = \left[\frac{1}{2}x^2 - x\right]_{-y}^3$$
$$= \left(\frac{9}{2} - 3\right) - \left(\frac{1}{2}y^2 + y\right)$$
$$= -\frac{1}{2}y^2 - y + \frac{3}{2}.$$

This result is then integrated with respect to y:

$$2\int_0^4 \left(-\frac{1}{2}y^2 - y + \frac{3}{2}\right) dy = \left[-\frac{1}{3}y^3 - y^2 + 3y\right]_0^4 = -\frac{76}{3}.$$

We now evaluate the line integral from $(-4,4)$ to $(0,0)$, the segment that was added in to form the closed loop.

The line is parameterized by $\mathbf{r}(t) = \langle -4 + 4t, 4 - 4t \rangle$, where $0 \le t \le 1$. Thus, $\mathbf{r}'(t) = \langle 4, -4 \rangle$ and $\mathbf{F}(t) = \langle 2(4 - 4t), (-4 + 4t)^2 \rangle = \langle 8 - 8t, 16t^2 - 32t + 16 \rangle$. Along this path segment, the line integral is

$$\int_C \mathbf{F} \cdot d\mathbf{r} = \int_0^1 \langle 8 - 8t, 16t^2 - 32t + 16 \rangle \cdot \langle 4, -4 \rangle\, dt$$
$$= \int_0^1 (-64t^2 + 96t - 32)\, dt$$
$$= \left[-\frac{64}{3}t^3 + 48t^2 - 32t\right]_0^1$$
$$= -\frac{16}{3}.$$

Therefore, the line integral from (0,0) to (3,0) to (3,4) to (−4,4) is the value found from Green's Theorem, $-\frac{76}{3}$, subtracted by the value of the line integral along the segment used to "close off" the region, $-\frac{16}{3}$.

This is $\int_C \mathbf{F} \cdot d\mathbf{r} = -\frac{76}{3} - \left(-\frac{16}{3}\right) = -\frac{60}{3} = -20.$

As a check, here are the individual line integrals along the three line segments:

- From (0,0) to (3,0): the line is parameterized by $\mathbf{r}(t) = \langle 3t, 0 \rangle$ with bounds $0 \le t \le 1$. Thus, $\mathbf{r}'(t) = \langle 3, 0 \rangle$ and $\mathbf{F}(t) = \langle 0, 9t^2 \rangle$, so $\int_C \mathbf{F} \cdot d\mathbf{r} = \int_0^1 \langle 0, 9t^2 \rangle \cdot \langle 3, 0 \rangle \, dt = \int_0^1 0 \, dt = 0.$

- From (3,0) to (3,4): the line is parameterized by $\mathbf{r}(t) = \langle 3, 4t \rangle$, with bounds $0 \le t \le 1$. Thus, $\mathbf{r}'(t) = \langle 0, 4 \rangle$ and $\mathbf{F}(t) = \langle 8t, 9 \rangle$, so $\int_C \mathbf{F} \cdot d\mathbf{r} = \int_0^1 \langle 8t, 9 \rangle \cdot \langle 0, 4 \rangle \, dt = \int_0^1 36 \, dt = [36t]_0^1 = 36.$

- From (3,4) to (−4,4): the line is parameterized by $\mathbf{r}(t) = \langle 3 - 7t, 4 \rangle$, with bounds $0 \le t \le 1$. Thus, $\mathbf{r}'(t) = \langle -7, 0 \rangle$ and $\mathbf{F}(t) = \langle 8, (3-7t)^2 \rangle$, so, $\int_C \mathbf{F} \cdot d\mathbf{r} = \int_0^1 \langle 8, (3-7t)^2 \rangle \cdot \langle -7, 0 \rangle \, dt = \int_0^1 -56 \, dt = -56.$

The sum is $0 + 36 - 56 = -20$, which agrees with the earlier answer.

◆ ◆ ◆ ◆ ◆ ◆ ◆ ◆ ◆

If C is a simple closed loop, then the region R bounded by C is **simply connected**. All of the regions in the preceding examples in this section are simply connected. A simply connected region in the plane has no holes.

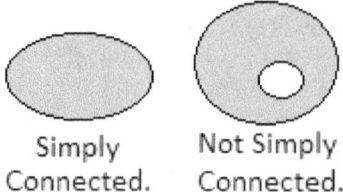

Simply Connected.   Not Simply Connected.

Green's Theorem requires a simply connected region R. However, a non-simply connected region can be made into two (or more) simply connected regions by dividing the region carefully.

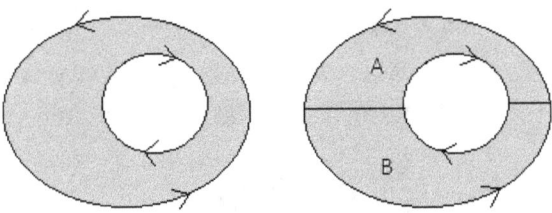

In the above image, a non-simply connected region is strategically divided into two subregions, A and B, that are each simply connected. Notice that the counterclockwise circulation is preserved in both cases. The line integrals along the two "cuts" will cancel, since the flow is in opposite directions depending on whether A or B is being considered. Green's Theorem can then be applied to each subregion, and often combined into one double integral covering the entire region.

◆ • ◆ • ◆ • ◆ • ◆

Example 49.6: Evaluate $\int_C \mathbf{F} \cdot d\mathbf{r}$, where $\mathbf{F}(x,y) = \langle e^x + 2y, 7x - \sin y \rangle$ and C is the boundary of a region R enclosed by two concentric circles, centered at the origin, one of radius 5 and the other of radius 3. Assume the circulation in the outer circle is counterclockwise, and that the circulation on the inner circle is clockwise.

Solution: The region R and its boundary C are shown below.

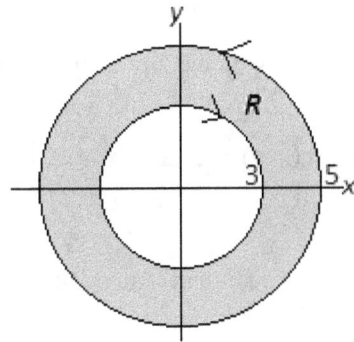

Using Green's Theorem, the curl is $N_x - M_y = 7 - 2 = 5$. Thus,

$$\int_C \mathbf{F} \cdot d\mathbf{r} = \iint_R 5 \, dA$$

$$= 5 \iint_R dA$$

$$= 5(\text{area of the ring})$$

$$= 5(25\pi - 9\pi) = 80\pi.$$

Note that $\iint_R dA$ is the area of R represented as a double integral, so we can verify using geometry. The area inside a circle of radius 5 is 25π, and the area inside a circle of radius 3 is 9π.

♦ ♦ ♦ ♦ ♦ ♦ ♦ ♦ ♦ ♦

50. Surface Area Integrals

Let $\mathbf{r}(u, v) = \langle x(u, v), y(u, v), z(u, v) \rangle$ parametrically describe a surface S in R^3. Then the surface area of S over a region of integration R is given by

$$\iint_S dS = \iint_R |\mathbf{r}_u \times \mathbf{r}_v| \, dA.$$

If the surface is defined explicitly in the form $z = f(x, y)$, then the surface can be parametrized as

$$\mathbf{r}(x, y) = \langle x, y, f(x, y) \rangle.$$

Its partial derivatives are

$$\mathbf{r}_x = \langle 1, 0, f_x(x, y) \rangle \quad \text{and} \quad \mathbf{r}_y = \langle 0, 1, f_y(x, y) \rangle.$$

The cross product is

$$\mathbf{r}_x \times \mathbf{r}_y = \langle -f_x(x, y), -f_y(x, y), 1 \rangle,$$

and the magnitude of this cross product is

$$|\mathbf{r}_x \times \mathbf{r}_y| = \sqrt{(-f_x(x,y))^2 + (-f_y(x,y))^2 + 1^2}$$
$$= \sqrt{(f_x(x,y))^2 + (f_y(x,y))^2 + 1}$$

Thus, in the case of a surface being described by an explicitly-defined function, the area of the surface S over a region of integration R is

$$\iint_S dS = \iint_R \sqrt{(f_x(x,y))^2 + (f_y(x,y))^2 + 1} \; dx \; dy.$$

◆ ◆ ◆ ◆ ◆ ◆ ◆ ◆ ◆

Example 50.1: Find the surface area of the plane with intercepts $(6,0,0)$, $(0,4,0)$ and $(0,0,10)$ that is in the first octant.

Solution: The plane's equation is $\frac{x}{6} + \frac{y}{4} + \frac{z}{10} = 1$, or $10x + 15y + 6z = 60$. Below is a sketch of the surface S, the plane in the first octant, and its region of integration R in the xy-plane:

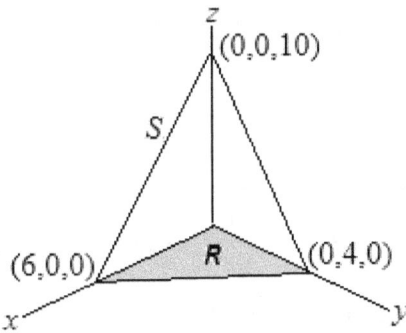

Solving for z, we have $z = 10 - \frac{5}{3}x - \frac{5}{2}y$. Therefore, the plane can be written parametrically:

$$\mathbf{r}(x,y) = \left(x, y, 10 - \frac{5}{3}x - \frac{5}{2}y\right).$$

Its partial derivatives are $\mathbf{r}_x = \left(1, 0, -\frac{5}{3}\right)$ and $\mathbf{r}_y = \left(0, 1, -\frac{5}{2}\right)$, and the cross product is

$$\mathbf{r}_x \times \mathbf{r}_y = \left(\frac{5}{3}, \frac{5}{2}, 1\right).$$

Therefore, the magnitude is

$$|\mathbf{r}_x \times \mathbf{r}_y| = \sqrt{\left(\frac{5}{3}\right)^2 + \left(\frac{5}{2}\right)^2 + 1^2} = \sqrt{\frac{361}{36}} = \frac{19}{6}.$$

The surface area is

$$\iint_S dS = \iint_R |\mathbf{r}_x \times \mathbf{r}_y|\, dA = \frac{19}{6} \iint_R dA.$$

Note that $\iint_R dA$ is the area of the region of integration R. Using the formula for area of a triangle, the area of R is $\frac{1}{2}(6)(4) = 12$. Thus, the surface area of the plane $z = 10 - \frac{5}{3}x - \frac{5}{2}y$ in the first octant is $\frac{19}{6}(12) = 38$ square units.

♦ • ♦ • ♦ • ♦ • ♦

Example 50.2: Find the surface area of the paraboloid $z = 9 - x^2 - y^2$ that extends above the xy-plane.

Solution: Parametrically, the paraboloid is $\mathbf{r}(x, y) = \langle x, y, 9 - x^2 - y^2 \rangle$, and its partial derivatives are $\mathbf{r}_x = \langle 1, 0, -2x \rangle$ and $\mathbf{r}_y = \langle 0, 1, -2y \rangle$. Therefore, their cross product is

$$\mathbf{r}_x \times \mathbf{r}_y = \langle 2x, 2y, 1 \rangle,$$

and the magnitude of the cross product is

$$|\mathbf{r}_x \times \mathbf{r}_y| = \sqrt{(2x)^2 + (2y)^2 + 1^2} = \sqrt{4x^2 + 4y^2 + 1}.$$

The paraboloid intersects the xy-plane ($z = 0$) at a circle of radius 3, centered at the origin, so that the region of integration R is given by $x^2 + y^2 \leq 9$. Therefore, the surface area of the paraboloid $z = 9 - x^2 - y^2$ that extends above the xy-plane is given by

$$\iint_S dS = \iint_R \sqrt{4x^2 + 4y^2 + 1}\, dA.$$

In rectangular coordinates, this is a difficult integrand to integrate. Instead, we use polar coordinates to rewrite this surface-area integral in terms of r and θ:

$$\iint_R \sqrt{4x^2 + 4y^2 + 1}\, dA = \int_0^{2\pi} \int_0^3 \sqrt{4r^2 + 1}\, r\, dr\, d\theta.$$

The inside integral is evaluated first:

$$\int_0^3 \sqrt{4r^2+1}\, r\, dr = \left[\frac{1}{12}(4r^2+1)^{3/2}\right]_0^3 = \frac{1}{12}(37^{3/2}-1).$$

Then, the outside integral is evaluated to find the surface area:

$$\frac{1}{12}(37^{3/2}-1)\int_0^{2\pi} d\theta = \frac{\pi}{6}(37^{3/2}-1), \text{ or about } 117.32 \text{ units}^2.$$

Example 50.3: Find the surface area of the hemisphere $x^2 + y^2 + z^2 = 25$ such that $x \geq 0$.

Solution: We can write this explicitly by solving for x:

$$x = f(y,z) = \sqrt{25 - y^2 - z^2}.$$

Thus, the hemisphere is parameterized as

$$\mathbf{r}(y,z) = \langle \sqrt{25 - y^2 - z^2}, y, z \rangle.$$

The partial derivatives are found first:

$$\mathbf{r}_y = \left(-\frac{y}{\sqrt{25 - y^2 - z^2}}, 1, 0\right) \text{ and } \mathbf{r}_z = \left(-\frac{z}{\sqrt{25 - y^2 - z^2}}, 0, 1\right).$$

The cross product is determined:

$$\mathbf{r}_y \times \mathbf{r}_z = \left(1, \frac{y}{\sqrt{25 - y^2 - z^2}}, \frac{z}{\sqrt{25 - y^2 - z^2}}\right).$$

Then the magnitude of the cross product is determined and simplified:

$$|\mathbf{r}_y \times \mathbf{r}_z| = \sqrt{1^2 + \left(\frac{y}{\sqrt{25-y^2-z^2}}\right)^2 + \left(\frac{z}{\sqrt{25-y^2-z^2}}\right)^2}$$

$$= \sqrt{1 + \frac{y^2}{25-y^2-z^2} + \frac{z^2}{25-y^2-z^2}}$$

$$= \sqrt{\frac{25-y^2-z^2}{25-y^2-z^2} + \frac{y^2}{25-y^2-z^2} + \frac{z^2}{25-y^2-z^2}}$$

$$= \sqrt{\frac{25 - y^2 - z^2 + y^2 + z^2}{25 - y^2 - z^2}}$$

$$= \frac{5}{\sqrt{25 - y^2 - z^2}}.$$

Thus, the surface area of the hemisphere is

$$\iint_R \frac{5}{\sqrt{25 - y^2 - z^2}} \, dA,$$

where R is the region of integration on the yz-plane, a circle of radius 5 centered at the origin. Rewrite this integral in terms of r and θ:

$$5 \int_0^{2\pi} \int_0^5 \frac{1}{\sqrt{25 - r^2}} \, r \, dr \, d\theta.$$

The inside integral is evaluated using u-du substitution:

$$\int_0^5 \frac{1}{\sqrt{25 - r^2}} r \, dr = \left[-\sqrt{25 - r^2} \right]_0^5 = 5.$$

Then the outer integral is evaluated:

$$5(5) \int_0^{2\pi} d\theta = 25(2\pi) = 50\pi \text{ units}^2.$$

Note that the surface area of a sphere of radius r is $A = 4\pi r^2$. Thus, the surface area of a hemisphere of radius 5 is $\frac{1}{2}(4\pi(5)^2) = 50\pi$. An alternative method of this example using spherical coordinates is presented next.

Example 50.4: Use spherical coordinates to find the surface area of the hemisphere $x^2 + y^2 + z^2 = 25$ where $x \geq 0$.

Solution: Since the hemisphere lies "above" the yz-plane. Thus, when describing this hemisphere in spherical coordinates, the variable ϕ will be reckoned from the positive x-axis, such that $\phi = 0$ is the positive x-axis and $\phi = \frac{\pi}{2}$ is the yz-plane. The radius is fixed, so $\rho = 5$. The conversions are:

$$x = 5\cos\phi, \quad y = 5\sin\phi \cos\theta, \quad z = 5\sin\phi \sin\theta.$$

We can describe the parameterize the hemisphere using variables ϕ and θ:

$$\mathbf{r}(\phi, \theta) = \langle 5 \cos \phi, 5 \sin \phi \cos \theta, 5 \sin \phi \sin \theta \rangle,$$
$$\text{where} \quad 0 \le \phi \le \frac{\pi}{2} \quad \text{and} \quad 0 \le \theta \le 2\pi.$$

The partial derivatives are

$$\mathbf{r}_\phi = \langle -5 \sin \phi, 5 \cos \phi \cos \theta, 5 \cos \phi \sin \theta \rangle,$$
$$\mathbf{r}_\theta = \langle 0, -5 \sin \phi \sin \theta, 5 \sin \phi \cos \theta \rangle.$$

The cross product looks intimidating, but trigonometric identities will help simplify it:

$$\mathbf{r}_\phi \times \mathbf{r}_\theta = \begin{vmatrix} \mathbf{i} & \mathbf{j} & \mathbf{k} \\ -5 \sin \phi & 5 \cos \phi \cos \theta & 5 \cos \phi \sin \theta \\ 0 & -5 \sin \phi \sin \theta & 5 \sin \phi \cos \theta \end{vmatrix}$$

$$= \begin{vmatrix} 5 \cos \phi \cos \theta & 5 \cos \phi \sin \theta \\ -5 \sin \phi \sin \theta & 5 \sin \phi \cos \theta \end{vmatrix} \mathbf{i} - \begin{vmatrix} -5 \sin \phi & 5 \cos \phi \sin \theta \\ 0 & 5 \sin \phi \cos \theta \end{vmatrix} \mathbf{j}$$

$$+ \begin{vmatrix} -5 \sin \phi & 5 \cos \phi \cos \theta \\ 0 & -5 \sin \phi \sin \theta \end{vmatrix} \mathbf{k}$$

$$= (25 \cos \phi \sin \phi \cos^2 \theta + 25 \cos \phi \sin \phi \sin^2 \theta) \mathbf{i}$$
$$- (-25 \sin^2 \phi \cos \theta) \mathbf{j} + (25 \sin^2 \phi \sin \theta) \mathbf{k}$$
$$= (25 \cos \phi \sin \phi) \mathbf{i} + (25 \sin^2 \phi \cos \theta) \mathbf{j} + (25 \sin^2 \phi \sin \theta) \mathbf{k}$$

The magnitude is found next. Note that by certain factoring steps, the Pythagorean identity appears twice (=1), greatly simplifying the result:

$$|\mathbf{r}_\phi \times \mathbf{r}_\theta| = \sqrt{(25 \cos \phi \sin \phi)^2 + (25 \sin^2 \phi \cos \theta)^2 + (25 \sin^2 \phi \sin \theta)^2}$$
$$= \sqrt{625(\cos^2 \phi \sin^2 \phi + \sin^4 \phi \cos^2 \theta + \sin^4 \phi \sin^2 \theta)}$$
$$= 25\sqrt{\cos^2 \phi \sin^2 \phi + \sin^4 \phi (\cos^2 \theta + \sin^2 \theta)}$$
$$= 25\sqrt{\cos^2 \phi \sin^2 \phi + \sin^4 \phi}$$
$$= 25\sqrt{\sin^2 \phi (\cos^2 \phi + \sin^2 \phi)}$$
$$= 25\sqrt{\sin^2 \phi}$$
$$= 25 \sin \phi.$$

Therefore, the surface area of the hemisphere is

$$\int_0^{2\pi} \int_0^{\pi/2} 25 \sin \phi \, d\phi \, d\theta.$$

The inside integral is evaluated:

$$\int_0^{\pi/2} 25 \sin\phi \, d\phi = [-25 \cos\phi]_0^{\pi/2}$$
$$= -25\left(\cos\frac{\pi}{2} - \cos 0\right)$$
$$= -25(0 - 1)$$
$$= 25.$$

Then, the integral with respect to θ is evaluated:

$$\int_0^{2\pi} 25 \, d\theta = 25(2\pi) = 50\pi.$$

◆ • ◆ • ◆ • ◆ • ◆

Example 50.5: A circular cylinder $x^2 + y^2 = 36$ intersects the plane $x + z = 10$. Find the surface area of this plane that is cut off by the cylinder, and then find the surface area of the cylinder that is bounded below by the xy-plane and above by the plane $x + z = 10$.

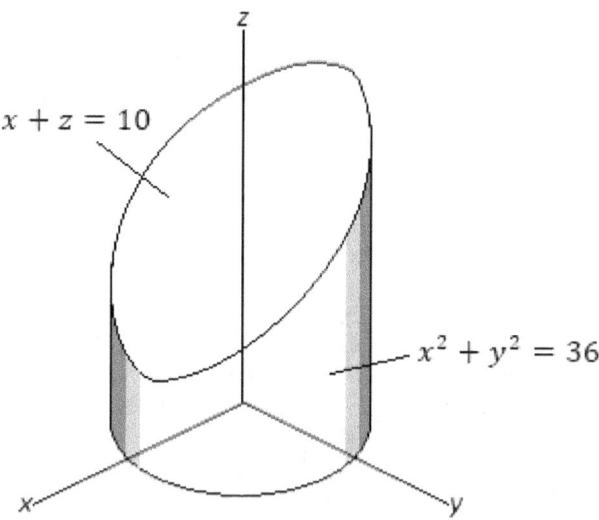

Solution: For the plane $x + z = 10$, we solve for z, getting $z = 10 - x$. Thus, the plane is parametrized by

$$\mathbf{r}(x, y) = \langle x, y, 10 - x \rangle.$$

Note that we use y as a parameter since the plane does extend into the y direction, even though values of y do not govern the values of z. (If it helps, think of the plane as $x + 0y + z = 10$).

The partial derivatives are $\mathbf{r}_x = \langle 1, 0, -1 \rangle$ and $\mathbf{r}_y = \langle 0, 1, 0 \rangle$, and their cross product is

$$\mathbf{r}_x \times \mathbf{r}_y = \langle 1, 0, 1 \rangle,$$

with magnitude $|\mathbf{r}_x \times \mathbf{r}_y| = \sqrt{2}$.

Thus, the surface area of the plane is

$$\iint_R |\mathbf{r}_x \times \mathbf{r}_y|\, dA = \sqrt{2} \iint_R dA,$$

where $\iint_R dA$ is the area of the region of integration R. Since R is a circle of radius 6, we have $\iint_R dA = 36\pi$, so that the surface area of the plane is $36\pi\sqrt{2}$ units².

The surface area of the cylinder bounded by the xy-plane ($z = 0$) and the plane $x + z = 10$ (written $z = 10 - x = 10 - 6\cos\theta$) is found in a similar manner. First, we parametrize the cylinder:

$$\mathbf{r}(\theta, z) = \langle 6\cos\theta, 6\sin\theta, z \rangle,$$
where $0 \le \theta \le 2\pi$ and $0 \le z \le 10 - 6\cos\theta$.

The partial derivatives are $\mathbf{r}_\theta = \langle -6\sin\theta, 6\cos\theta, 0 \rangle$ and $\mathbf{r}_z = \langle 0, 0, 1 \rangle$, and their cross product is

$$\mathbf{r}_\theta \times \mathbf{r}_z = \langle 6\cos\theta, 6\sin\theta, 0 \rangle.$$

The magnitude of the cross product is

$$\begin{aligned}|\mathbf{r}_\theta \times \mathbf{r}_z| &= \sqrt{(6\cos\theta)^2 + (6\sin\theta)^2 + 0^2} \\ &= \sqrt{36(\cos^2\theta + \sin^2\theta)} \\ &= \sqrt{36} \\ &= 6.\end{aligned}$$

The surface area of the cylinder is

$$\int_0^{2\pi} \int_0^{10-6\cos\theta} 6 \, dz \, d\theta = 6 \int_0^{2\pi} \int_0^{10-6\cos\theta} dz \, d\theta$$

$$= 6 \int_0^{2\pi} (10 - 6\cos\theta) \, d\theta$$

$$= 6[10\theta - 6\sin\theta]_0^{2\pi}$$

$$= 120\pi \text{ units}^2.$$

◆ ◆ ◆ ◆ ◆ ◆ ◆ ◆ ◆

Example 50.6: Find the surface area of the side of the cone $z = \sqrt{x^2 + y^2}$ where $1 \leq z \leq 4$. This is called a *band*. The solid formed by removing the apex from any conical or pyramidal object is called a *frustum*.

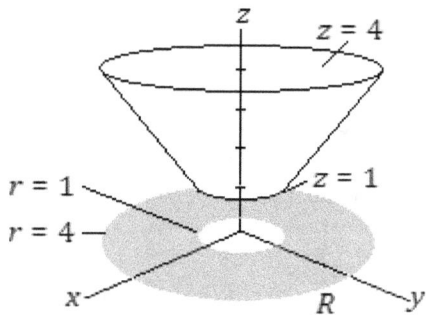

Solution: We have

$$\mathbf{r}(x, y) = \langle x, y, \sqrt{x^2 + y^2} \rangle.$$

We will determine the bounds of integration in a moment.

The partial derivatives are

$$\mathbf{r}_x = \langle 1, 0, \frac{x}{\sqrt{x^2 + y^2}} \rangle \quad \text{and} \quad \mathbf{r}_y = \langle 0, 1, \frac{y}{\sqrt{x^2 + y^2}} \rangle.$$

The cross product is

$$\mathbf{r}_x \times \mathbf{r}_y = \begin{vmatrix} \mathbf{i} & \mathbf{j} & \mathbf{k} \\ 1 & 0 & \dfrac{x}{\sqrt{x^2+y^2}} \\ 0 & 1 & \dfrac{y}{\sqrt{x^2+y^2}} \end{vmatrix}$$

$$= \begin{vmatrix} 0 & \dfrac{x}{\sqrt{x^2+y^2}} \\ 1 & \dfrac{y}{\sqrt{x^2+y^2}} \end{vmatrix} \mathbf{i} - \begin{vmatrix} 1 & \dfrac{x}{\sqrt{x^2+y^2}} \\ 0 & \dfrac{y}{\sqrt{x^2+y^2}} \end{vmatrix} \mathbf{j} + \begin{vmatrix} 1 & 0 \\ 0 & 1 \end{vmatrix} \mathbf{k}$$

$$= -\dfrac{x}{\sqrt{x^2+y^2}} \mathbf{i} - \dfrac{y}{\sqrt{x^2+y^2}} \mathbf{j} + \mathbf{k}.$$

The magnitude of the cross product is

$$|\mathbf{r}_x \times \mathbf{r}_y| = \sqrt{\left(-\dfrac{x}{\sqrt{x^2+y^2}}\right)^2 + \left(-\dfrac{y}{\sqrt{x^2+y^2}}\right)^2 + 1^2}$$

$$= \sqrt{\dfrac{x^2}{x^2+y^2} + \dfrac{y^2}{x^2+y^2} + 1}$$

$$= \sqrt{\dfrac{x^2}{x^2+y^2} + \dfrac{y^2}{x^2+y^2} + \dfrac{x^2+y^2}{x^2+y^2}}$$

$$= \sqrt{\dfrac{x^2+y^2+x^2+y^2}{x^2+y^2}}$$

$$= \sqrt{\dfrac{2(x^2+y^2)}{x^2+y^2}} = \sqrt{2}.$$

The region of integration R is the area between two concentric circles, one of radius 1 and the other of radius 4. This is the "shadow" cast by the side of the conical band onto the xy-plane. Thus, the surface area of the band on the cone $z = \sqrt{x^2+y^2}$ where $1 \leq z \leq 4$ is given by

$$\iint_R \sqrt{2}\, dA = \sqrt{2} \iint_R dA$$

$$= \sqrt{2}(\pi(4)^2 - \pi(1)^2)$$

$$= 15\pi\sqrt{2} \text{ units}^2.$$

We used geometry to determine the area between the two circles, represented by $\iint_R dA$.

In the following example, this problem is evaluated again using cylindrical coordinates.

◆ • ◆ • ◆ • • ◆ • •

Example 50.7: Use cylindrical coordinates to find the surface area of the cone $z = \sqrt{x^2 + y^2}$ where $1 \leq z \leq 4$.

Solution: In rectangular coordinates, the cone is parameterized as

$$\mathbf{r}(x, y) = \langle x, y, \sqrt{x^2 + y^2}\rangle.$$

Letting $x = r\cos\theta$ and $y = r\sin\theta$, we have

$$\mathbf{r}(r, \theta) = \langle r\cos\theta, r\sin\theta, \sqrt{(r\cos\theta)^2 + (r\sin\theta)^2}\rangle = \langle r\cos\theta, r\sin\theta, r\rangle.$$

The bounds are $1 \leq r \leq 4$ and $0 \leq \theta \leq 2\pi$.

The partial derivatives are

$$\mathbf{r}_\theta = \langle -r\sin\theta, r\cos\theta, 0\rangle \quad \text{and} \quad \mathbf{r}_r = \langle \cos\theta, \sin\theta, 1\rangle.$$

The cross product is

$$\mathbf{r}_\theta \times \mathbf{r}_r = \begin{vmatrix} \mathbf{i} & \mathbf{j} & \mathbf{k} \\ -r\sin\theta & r\cos\theta & 0 \\ \cos\theta & \sin\theta & 1 \end{vmatrix}$$
$$= \langle r\cos\theta, r\sin\theta, -r\sin^2\theta - r\cos^2\theta\rangle$$
$$= \langle r\cos\theta, r\sin\theta, -r\rangle.$$

The magnitude is

$$|\mathbf{r}_\theta \times \mathbf{r}_r| = \sqrt{(r\cos\theta)^2 + (r\sin\theta)^2 + (-r)^2}$$
$$= \sqrt{r^2(\cos^2\theta + \sin^2\theta) + r^2}$$
$$= \sqrt{r^2 + r^2}$$
$$= r\sqrt{2}.$$

The surface area is given by

$$\iint_R |\mathbf{r}_\theta \times \mathbf{r}_r|\, dA = \int_0^{2\pi} \int_1^4 r\sqrt{2}\, dr\, d\theta$$

$$= \sqrt{2} \int_0^{2\pi} \int_1^4 r\, dr\, d\theta$$

$$= \sqrt{2} \int_0^{2\pi} \left[\frac{r^2}{2}\right]_1^4 d\theta$$

$$= \sqrt{2} \int_0^{2\pi} \frac{15}{2}\, d\theta$$

$$= \frac{15}{2}\sqrt{2} \int_0^{2\pi} d\theta$$

$$= \frac{15}{2}\sqrt{2}(2\pi)$$

$$= 15\pi\sqrt{2} \text{ units}^2.$$

♦ ♦ ♦ ♦ ♦ ♦ ♦ ♦ ♦

Example 50.8: Find the surface area of $\mathbf{r}(u,v) = \langle u+v, u-v, 2uv \rangle$ over the circular region $u^2 + v^2 \leq 9$.

Solution: Taking partial derivatives of **r**, we have

$$\mathbf{r}_u(u,v) = \langle 1,1,2v \rangle \quad \text{and} \quad \mathbf{r}_v(u,v) = \langle 1,-1,2u \rangle.$$

Thus, $\mathbf{r}_u \times \mathbf{r}_v = \langle 2u+2v, 2v-2u, -2 \rangle$, and its magnitude is

$$|\mathbf{r}_u \times \mathbf{r}_v| = \sqrt{(2u+2v)^2 + (2v-2u)^2 + (-2)^2}$$

$$= \sqrt{4u^2 + 8uv + 4v^2 + 4v^2 - 8uv + 4u^2 + 4}$$

$$= \sqrt{4 + 8u^2 + 8v^2}.$$

The surface area is given by

$$\iint_R |\mathbf{r}_u \times \mathbf{r}_v|\, dA = \iint_R \sqrt{4 + 8u^2 + 8v^2}\, du\, dv.$$

Converting to polar coordinates, and noting that the region of integration is inside a circle of radius 3, we have

$$\iint_R \sqrt{4 + 8u^2 + 8v^2}\, du\, dv = \int_0^{2\pi} \int_0^3 \sqrt{4 + 8r^2}\, r\, dr\, d\theta.$$

The integral with respect to r is evaluated:

$$\int_0^3 \sqrt{4 + 8r^2}\, r\, dr = \left[\frac{1}{24}(4 + 8r^2)^{3/2}\right]_0^3 = \frac{1}{24}(76^{3/2} - 8).$$

The outside integral is then evaluated:

$$\int_0^{2\pi} \left(\frac{1}{24}(76^{3/2} - 8)\right) d\theta = \frac{1}{24}(76^{3/2} - 8) \int_0^{2\pi} d\theta$$

$$= \frac{\pi}{12}(76^{3/2} - 8)$$

$$\approx 171.36 \text{ units}^2.$$

♦ • ♦ • ♦ • ♦ • ♦ • ♦

The generic form of the surface-area integral (in parameters u and v), $\iint_R |\mathbf{r}_u \times \mathbf{r}_v|\, dA$, does not distinguish whether u and v are rectangular, polar, cylindrical or spherical coordinates. Thus, the area differential is always $dA = du\, dv$.

In some examples, we used a non-rectangular coordinate system to set up the integral. In such a case, we do *not* write in the usual Jacobian associated with that system. See Examples 50.4 and 50.7 for two such cases.

However, if after setting it up in a particular coordinate system, we decide to integrate it in a different coordinate system, then we must make all the necessary substitutions and then include the Jacobian. See Examples 50.2 and 50.8 for two such cases where we did include the Jacobian.

51. General Surface Integrals

The area of a surface S in R^3 defined parametrically by $\mathbf{r}(u,v) = \langle x(u,v), y(u,v), z(u,v) \rangle$ over a region of integration R in the input-variable (uv) plane is given by

$$\iint_S dS = \iint_R |\mathbf{r}_u \times \mathbf{r}_v|\, dA.$$

Let $w = f(x, y, z)$ be a function defined over this surface. The **surface integral**, where $f(\mathbf{r}(u, v)) = f(x(u, v), y(u, v), z(u, v))$, is given by

$$\iint_S f(\mathbf{r}(u,v))\, dS = \iint_R f(\mathbf{r}(u,v))|\mathbf{r}_u \times \mathbf{r}_v|\, dA.$$

When the surface S is defined explicitly by a function $z = g(x, y)$, then $\mathbf{r}(x, y) = \langle x, y, g(x, y) \rangle$, and the surface integral can be rewritten

$$\iint_S f(x,y,z)\, dS = \iint_R f(x,y,g(x,y))\sqrt{(g_x(x,y))^2 + (g_y(x,y))^2 + 1}\, dA,$$

Where $dS = |\mathbf{r}_u \times \mathbf{r}_v|\, dA = \sqrt{(g_x(x,y))^2 + (g_y(x,y))^2 + 1}\, dA.$

Surface area integrals are a special case of surface integrals, where $f(x, y, z) = 1$. Surface integrals can be interpreted in many ways. Some examples are discussed at the end of this section.

◆ • ◆ • ◆ • ◆ • ◆

Example 51.1: Let S be the surface $z = 12 - 4x - 3y$ contained in the first quadrant. Find $\iint_S (x + yz)\, dS$.

Solution: Here, $z = g(x,y) = 12 - 4x - 3y$, so that $g_x = -4$ and $g_y = -3$. Thus, dS is

$$dS = \sqrt{(-4)^2 + (-3)^2 + 1} = \sqrt{26}\, dA.$$

The integrand is written in terms of x and y, with $z = 12 - 4x - 3y$:

$$x + yz = x + y(12 - 4x - 3y) = x + 12y - 4xy - 3y^2$$

The region of integration R is the footprint of the surface S projected onto the xy-plane. Below is a sketch of S and its region of integration R. Letting $dA = dy\, dx$, we have $0 \le y \le -\frac{4}{3}x + 4$ and $0 \le x \le 3$ as the bounds of R:

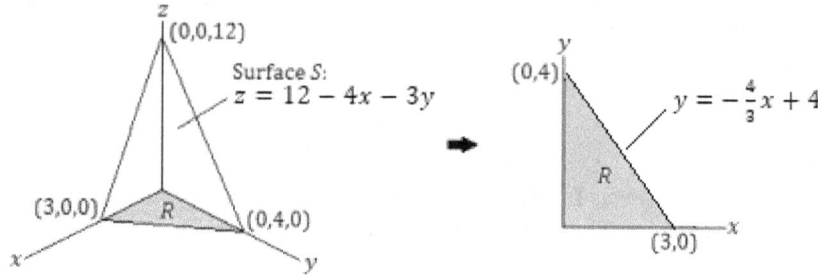

The surface integral is now

$$\iint_S (x + yz)\, dS = \iint_R (x + 12y - 4xy - 3y^2)\sqrt{26}\, dA$$

$$= \sqrt{26} \int_0^3 \int_0^{-(4/3)x+4} (x + 12y - 4xy - 3y^2)\, dy\, dx.$$

The inside integral is

$$\int_0^{-(4/3)x+4} (x + 12y - 4xy - 3y^2)\, dy = [xy + (6 - 2x)y^2 - y^3]_0^{-(4/3)x+4}$$

Note that the two middle terms, $12y - 4xy$, can be written $(12 - 4x)y$, which gives $(6 - 2x)y^2$ after integration with respect to y. Substituting and simplifying, we obtain

$$x\left(-\frac{4}{3}x + 4\right) + (6 - 2x)\left(-\frac{4}{3}x + 4\right)^2 - \left(-\frac{4}{3}x + 4\right)^3$$

$$= -\frac{32}{27}x^3 + \frac{28}{3}x^2 - 28x + 32.$$

This is now integrated with respect to x:

$$\sqrt{26} \int_0^3 \left(-\frac{32}{27}x^3 + \frac{28}{3}x^2 - 28x + 32\right) dx$$

$$= \sqrt{26}\left[-\frac{8}{27}x^4 + \frac{28}{9}x^3 - 14x^2 + 32x\right]_0^3 = 30\sqrt{26}.$$

Example 51.2: Find $\iint_S x^2 \, dS$, where S is the portion of sphere of radius 4, centered at the origin, such that $x \geq 0$ and $z \geq 0$.

Solution: The surface is a quarter-sphere bounded by the xy and yz planes. Sketch S and from it, infer the region of integration R:

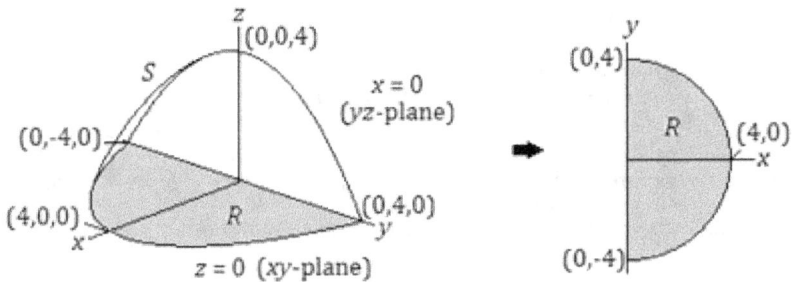

In rectangular coordinates, the hemisphere is $x^2 + y^2 + z^2 = 16$, so that $z = g(x,y) = \sqrt{16 - x^2 - y^2}$. From this, we obtain partial derivatives

$$z_x = \frac{-x}{\sqrt{16 - x^2 - y^2}} \quad \text{and} \quad z_y = \frac{-y}{\sqrt{16 - x^2 - y^2}}.$$

Thus,

$$dS = \sqrt{\left(\frac{-x}{\sqrt{16 - x^2 - y^2}}\right)^2 + \left(\frac{-y}{\sqrt{16 - x^2 - y^2}}\right)^2 + 1} \, dA$$

$$= \frac{4}{\sqrt{16 - x^2 - y^2}} \, dA.$$

(See Example 50.3 for a similar example with all steps shown.)

Using polar coordinates, where $x = r\cos\theta$, $y = r\sin\theta$, and $r^2 = x^2 + y^2$, the region R is described by $0 \leq r \leq 4$ and $-\frac{\pi}{2} \leq \theta \leq \frac{\pi}{2}$. Due to the change of variables, the differential element is now $dA = r \, dr \, d\theta$. The surface integrals now is written

$$\iint_S x^2 \, dS = \int_{-\pi/2}^{\pi/2} \int_0^4 (r\cos\theta)^2 \frac{4}{\sqrt{16 - r^2}} r \, dr \, d\theta.$$

This simplifies to

$$4 \int_{-\pi/2}^{\pi/2} \int_0^4 \left(\frac{r^3 \cos^2\theta}{\sqrt{16 - r^2}}\right) dr \, d\theta.$$

315

To antidifferentiate $\frac{r^3}{\sqrt{16-r^2}}$, we use the form taken from a table of integrals:

$$\int \frac{r^3}{\sqrt{a^2-r^2}}\, dr = -\frac{1}{3}(2a^2+r^2)\sqrt{a^2-r^2}.$$

Thus,

$$\int_0^4 \frac{r^3}{\sqrt{16-r^2}}\, dr = \left[-\frac{1}{3}(32+r^2)\sqrt{16-r^2}\right]_0^4 = \frac{128}{3}.$$

We now evaluate the outer integral, using the identity $\cos^2\theta = \frac{1}{2}+\frac{1}{2}\cos(2\theta)$. Note that the constant $\frac{128}{3}$ moves to the front of the integral:

$$4\left(\frac{128}{3}\right)\int_{-\pi/2}^{\pi/2} \cos^2\theta\, d\theta = \frac{512}{3}\int_{-\pi/2}^{\pi/2}\left(\frac{1}{2}+\frac{1}{2}\cos 2\theta\right) d\theta$$

$$= \frac{256}{3}\int_{-\pi/2}^{\pi/2}(1+\cos 2\theta)\, d\theta$$

$$= \frac{256}{3}\left[\theta+\frac{1}{2}\sin 2\theta\right]_{-\pi/2}^{\pi/2}$$

$$= \frac{256}{3}\left[\left(\frac{\pi}{2}+\frac{1}{2}\sin(\pi)\right)-\left(-\frac{\pi}{2}+\frac{1}{2}\sin(-\pi)\right)\right]$$

$$= \frac{256}{3}\pi.$$

◆ • ◆ • ◆ • ◆ • ◆

In the next example, we revisit the previous example using spherical coordinates.

Example 51.3: Find $\iint_S x^2\, dS$, where S is the portion of sphere of radius 4, centered at the origin, such that $x \geq 0$ and $z \geq 0$. Parameterize S using spherical coordinates.

Solution: Using spherical coordinates ϕ and θ, and the "usual" parameterization of a sphere with a fixed radius (in this case, $\rho = 4$), we have

$$\mathbf{r}(\phi,\theta) = \langle 4\sin\phi\cos\theta,\, 4\sin\phi\sin\theta,\, 4\cos\phi\rangle,$$
$$\text{where } -\frac{\pi}{2}\leq \theta \leq \frac{\pi}{2} \text{ and } 0\leq \phi \leq \frac{\pi}{2}.$$

From this, we determine $|\mathbf{r}_\phi \times \mathbf{r}_\theta|$. The derivation is lengthy but not difficult, as many trigonometric identities can be used to simplify. See Example 50.4 for one such example. We have

$$|\mathbf{r}_\phi \times \mathbf{r}_\theta| = 16 \sin \phi.$$

Thus, the surface integral is

$$\iint_S x^2 \, dS = \int_{-\pi/2}^{\pi/2} \int_0^{\pi/2} (4 \sin \phi \cos \theta)^2 \, 16 \sin \phi \, d\phi \, d\theta.$$

where $x^2 = (4 \sin \phi \cos \theta)^2$ and $dS = |\mathbf{r}_\phi \times \mathbf{r}_\theta| \, dA = 16 \sin \phi \, d\phi \, d\theta$. This simplifies to

$$256 \int_{-\pi/2}^{\pi/2} \int_0^{\pi/2} \sin^3 \phi \cos^2 \theta \, d\phi \, d\theta.$$

Because the bounds are constant and the integrand held by multiplication, we can rewrite the integral as

$$256 \int_{-\pi/2}^{\pi/2} \int_0^{\pi/2} \sin^3 \phi \cos^2 \theta \, d\phi \, d\theta$$

$$= 256 \left(\int_0^{\pi/2} \sin^3 \phi \, d\phi \right) \left(\int_{-\pi/2}^{\pi/2} \cos^2 \theta \, d\theta \right).$$

Both require some techniques of trigonometric integration. For the integrand $\cos^2 \theta$, we use the identity $\cos^2 \theta = \frac{1}{2} + \frac{1}{2} \cos(2\theta)$:

$$\int_{-\pi/2}^{\pi/2} \cos^2 \theta \, d\theta = \int_{-\pi/2}^{\pi/2} \left(\frac{1}{2} + \frac{1}{2} \cos(2\theta) \right) d\theta$$

$$= \left[\frac{1}{2} \theta + \frac{1}{4} \sin(2\theta) \right]_{-\pi/2}^{\pi/2}$$

$$= \left(\frac{\pi}{4} + 0 \right) - \left(-\frac{\pi}{4} + 0 \right)$$

$$= \frac{\pi}{2}.$$

For the integrand, $\sin^3 \phi$ is rewritten as $\sin^2 \phi \sin \phi = (1 - \cos^2 \phi) \sin \phi$ and u-du substitution is used:

$$\int_0^{\pi/2} \sin^3 \phi \, d\phi = \int_0^{\pi/2} \sin^2 \phi \sin \phi \, d\phi$$

$$= \int_0^{\pi/2} (1 - \cos^2 \phi) \sin \phi \, d\phi$$

$$= \int_0^{\pi/2} \sin \phi - \cos^2 \phi \sin \phi \, d\phi$$

$$= \left[-\cos \phi + \frac{1}{3} \cos^3 \phi \right]_0^{\pi/2}$$

$$= 0 - \left(-1 + \frac{1}{3} \right)$$

$$= \frac{2}{3}.$$

Assembling this information together, we have

$$\iint_S x^2 \, dS = 256 \left(\int_0^{\pi/2} \sin^3 \phi \, d\phi \right) \left(\int_{-\pi/2}^{\pi/2} \cos^2 \theta \, d\theta \right)$$

$$= 256 \left(\frac{2}{3} \right) \left(\frac{\pi}{2} \right)$$

$$= \frac{256}{3} \pi.$$

You can decide if this method is more efficient than using rectangular coordinates. Clearly, both methods work.

◆ • ◆ • ◆ • ◆ • ◆

Note that we did *not* include the Jacobian $\rho^2 \sin \phi$ when we developed the integral in the previous example (Example 51.3) in variables ϕ and θ. This is because we originally parameterized the surface in ϕ and θ, in which case, the area differential elements will always be $dA = du \, dv$, or $dA = d\phi \, d\theta$ in this case. The derivation of $dS = |\mathbf{r}_\phi \times \mathbf{r}_\theta| \, dA$ does not "know" whether the variables represent rectangular, spherical or cylindrical coordinate systems.

However, if we parameterize the surface in generic variables u and v, and then midway through the problem decide to integrate with respect to a different coordinate system, then we *must* include the Jacobian when we convert the area differential element. We saw this in the example prior to the last one (Example 51.2).

Setting up a surface integral is usually not difficult. However, in many cases, the integrands can be difficult to antidifferentiate. A computer, tables of integrals or numerical methods may need to be used. This is shown in the next example.

Example 51.4: Evaluate the integral $\iint_S x^2 z \, dS$, where S is the paraboloid $z = g(x, y) = 1 - x^2 - y^2$ over the xy-plane.

Solution: From the surface S, we have partial derivatives $g_x = -2x$ and $g_y = -2y$. Thus,

$$dS = \sqrt{(-2x)^2 + (-2y)^2 + 1} \, dA = \sqrt{4x^2 + 4y^2 + 1} \, dA.$$

The surface integral is

$$\iint_S x^2 z \, dS = \iint_R x^2(1 - x^2 - y^2)\sqrt{4x^2 + 4y^2 + 1} \, dA.$$

This is a difficult integral to evaluate if we remain in rectangular coordinates. Thus, we convert to polar coordinates, where the region of integration R is a circle of radius 1, centered at the origin on the xy-plane:

$$\iint_R x^2(1 - x^2 - y^2)\sqrt{4x^2 + 4y^2 + 1} \, dA$$

$$= \int_0^{2\pi} \int_0^1 (r \cos \theta)^2 (1 - r^2)\sqrt{4r^2 + 1}\, r \, dr \, d\theta$$

$$= \int_0^{2\pi} \int_0^1 \cos^2 \theta \, (r^3 - r^5)\sqrt{4r^2 + 1} \, dr \, d\theta,$$

This integrand is not much easier to antidifferentiate either. Using a computer or numerical methods,

$$\int_0^1 (r^3 - r^5)\sqrt{4r^2 + 1} \, dr \approx 0.143,$$

Meanwhile, $\int_0^{2\pi} \cos^2 \theta \, d\theta = \pi$ (using a trigonometric identity such as in the previous example). Thus,

$$\iint_S x^2 z \, dS \approx 0.143\pi.$$

Applications of Surface Integrals

There are a handful of common applications of surface integrals that may help one intuitively understand them better. For example, if $z = g(x,y)$ is a surface S with uniform thickness, and $f(x,y,z)$ represents the density at each point (x,y,z) on the surface, then $\iint_S f(x,y,z)\, dS$ can be interpreted as the *total mass* of S.

From Example 51.1, we had $\iint_S (x+yz)\, dS$, where S was the surface $z = 12 - 4x - 3y$ contained in the first quadrant. If x, y and z are measured in meters, and $f(x,y,z) = x + yz$ is the density of the object at the point (x,y,z) in kilograms per square meter, then $\iint_S (x+yz)\, dS$ is the total mass of the surface, in kilograms. We could interpret the result by claiming that this surface has a total mass of $30\sqrt{26}$ kilograms.

In Example 51.2, suppose that the integrand $f(x,y,z) = x^2$ represents the density of a population of bacteria (in thousands per square centimeter) on the surface at any given point (x,y,z), where the variables are measured in centimeters. Then, the total population would be given by $\iint_S x^2\, dS = \frac{256}{3}\pi \approx 268$, or about 268,000 bacteria.

Furthermore, the *average density* of the object represented by the surface would be its total mass divided by the surface's area:

$$\text{Average density} = \frac{\iint_S f(x,y,z)\, dS}{\iint_S dS}.$$

In Example 51.1, the surface area of S is $\iint_S dS = 6\sqrt{26}$ square meters. Thus, the average density of the object is $\frac{30\sqrt{26}}{6\sqrt{26}} = 5$ kilograms per square meter. In Example 51.2, the surface area of the quarter-sphere is $\iint_S dS = \frac{1}{4}\left(\frac{4}{3}\pi(4)^3\right) = \frac{64}{3}\pi$, so the average density of the bacteria on this surface is $\frac{(256/3)\pi}{(64/3)\pi} = 4$, or 4,000 bacteria per square centimeter.

Surface integrals are also used to find the flow of material through a surface, discussed in Section 53, Flux Integrals.

52. The Del Operator: Divergence and Curl

Let $\mathbf{F}(x,y,z) = \langle M(x,y,z), N(x,y,z), P(x,y,z) \rangle$ be a vector field in R^3. The **del operator** is represented by the symbol ∇, and is written

$$\nabla = \left\langle \frac{\partial}{\partial x}, \frac{\partial}{\partial y}, \frac{\partial}{\partial z} \right\rangle, \quad \text{or} \quad \nabla = \langle \partial_x, \partial_y, \partial_z \rangle.$$

By itself, the del operator is meaningless. It must be combined with a vector field **F** via a dot product or cross product to be meaningful. For example, the del operator can be combined with a vector field **F** as a dot product:

$$\nabla \cdot \mathbf{F} = \left\langle \frac{\partial}{\partial x}, \frac{\partial}{\partial y}, \frac{\partial}{\partial z} \right\rangle \cdot \mathbf{F}(x,y,z)$$

$$= \left\langle \frac{\partial}{\partial x}, \frac{\partial}{\partial y}, \frac{\partial}{\partial z} \right\rangle \cdot \langle M(x,y,z), N(x,y,z), P(x,y,z) \rangle$$

$$= \frac{\partial}{\partial x} M(x,y,z) + \frac{\partial}{\partial y} N(x,y,z) + \frac{\partial}{\partial z} P(x,y,z)$$

This is called the **divergence** of **F** and is written shorthand as div $\mathbf{F} = \nabla \cdot \mathbf{F} = M_x + N_y + P_z$. In R^2, we have div $\mathbf{F} = \nabla \cdot \mathbf{F} = M_x + N_y$.

> div **F** is a scalar function.

The del operator can also be combined with **F** as a cross product:

$$\nabla \times \mathbf{F} = \begin{vmatrix} \mathbf{i} & \mathbf{j} & \mathbf{k} \\ \partial_x & \partial_y & \partial_z \\ M & N & P \end{vmatrix} = (P_y - N_z)\mathbf{i} - (P_x - M_z)\mathbf{j} + (N_x - M_y)\mathbf{k}.$$

This is called the **curl** of **F**, and is written shorthand as

$$\text{curl } \mathbf{F} = \nabla \times \mathbf{F} = \langle P_y - N_z, M_z - P_x, N_x - M_y \rangle.$$

> curl **F** is a vector field.

The second component, $M_z - P_x$, is simplified slightly by distributing the leading negative. Also note that the third component, $N_x - M_y$, is the integrand for Green's Theorem. Thus, we will see that Green's Theorem is a special case of a higher-dimension analog called Stokes' Theorem that uses curl **F**.

Example 52.1: Given $\mathbf{F}(x,y,z) = \langle xy^2, x^2yz^2, 2xz^4 \rangle$, find div **F** and curl **F**.

Solution: For div **F**, we have

$$\text{div } \mathbf{F} = \nabla \cdot \mathbf{F}$$
$$= M_x + N_y + P_z$$
$$= y^2 + x^2z^2 + 8xz^3.$$

For curl **F**, we have

$$\text{curl } \mathbf{F} = \nabla \times \mathbf{F}$$
$$= \langle P_y - N_z, M_z - P_x, N_x - M_y \rangle$$
$$= \langle 0 - 2x^2yz, 0 - 2z^4, 2xyz^2 - y^2 \rangle$$
$$= \langle -2x^2yz, -2z^4, 2xyz^2 - y^2 \rangle.$$

◆ • ◆ • ◆ • ◆ • ◆

Example 52.2: Given $\mathbf{F}(x,y,z) = \langle 2y, xz, x + 2y \rangle$, find div **F** and curl **F**.

Solution: For div **F**, we have

$$\text{div } \mathbf{F} = \nabla \cdot \mathbf{F}$$
$$= M_x + N_y + P_z$$
$$= 0 + 0 + 0 = 0.$$

For curl **F**, we have

$$\text{curl } \mathbf{F} = \nabla \times \mathbf{F}$$
$$= \langle P_y - N_z, M_z - P_x, N_x - M_y \rangle$$
$$= \langle 2 - x, -1, z - 2 \rangle.$$

When div **F** = 0, the vector field is **incompressible**.

◆ • ◆ • ◆ • ◆ • ◆

Be careful with the syntax when using the symbol ∇. If f is a scalar function, then ∇f is the gradient of f. If **F** is a vector field, then $\nabla \cdot \mathbf{F}$ is the divergence of **F**, and $\nabla \times \mathbf{F}$ is the curl of **F**. However, statements like $\nabla \mathbf{F}$ and $\nabla \cdot f$ have no meaning. On the other hand, statements like $\nabla \cdot \nabla f$, $\nabla \times \nabla f$ and $\nabla \cdot (\nabla \times \mathbf{F})$ are well-defined.

Example 52.3: Given $\mathbf{F}(x, y, z) = \langle 2xyz^3,\ x^2z^3,\ 3x^2yz^2 \rangle$, find div **F** and curl **F**.

Solution: For div **F**, we have

$$\begin{aligned} \text{div } \mathbf{F} &= \nabla \cdot \mathbf{F} \\ &= M_x + N_y + P_z \\ &= 2yz^3 + 6x^2yz. \quad \text{(Note that } N_y = 0\text{)} \end{aligned}$$

For curl **F**, we have

$$\begin{aligned} \text{curl } \mathbf{F} &= \nabla \times \mathbf{F} \\ &= \langle P_y - N_z, M_z - P_x, N_x - M_y \rangle \\ &= \langle 3x^2z^2 - 3x^2z^2, 6xyz^2 - 6xyz^2, 2xz^3 - 2xz^3 \rangle \\ &= \langle 0,0,0 \rangle = \mathbf{0}. \end{aligned}$$

When curl **F** = **0**, the vector field is **irrotational**.

♦ • ♦ • ♦ • ♦ • ♦

Example 52.4: Given $\mathbf{F}(x, y, z) = \langle 2, 1, -4 \rangle$, find div **F** and curl **F**.

Solution: For div **F**, we have

$$\begin{aligned} \text{div } \mathbf{F} &= \nabla \cdot \mathbf{F} \\ &= M_x + N_y + P_z \\ &= 0. \end{aligned}$$

For curl **F**, we have

$$\begin{aligned} \text{curl } \mathbf{F} &= \nabla \times \mathbf{F} \\ &= \langle P_y - N_z, M_z - P_x, N_x - M_y \rangle \\ &= \langle 0 - 0, 0 - 0, 0 - 0 \rangle \\ &= \langle 0,0,0 \rangle = \mathbf{0}. \end{aligned}$$

A constant vector field is both incompressible (div **F** = **0**) and irrotational (curl **F** = **0**).

The divergence operator is used to show (quantify) how a vector field flows through a region bounded by permeable membranes. The region can then be made as small as we desire, down to a point. Thus, divergence can show the existences of a **source** (where, roughly speaking, the flow radiates away from the point), or a **sink** (where a flow collects into a point).

If the divergence of a vector field **F** is 0, then there are no sources nor sinks in **F**. If a certain amount of mass flows into a region, then the same amount must flow away from the region in order to maintain the balance, and thus, the flow is incompressible. The flow of fluid, as modeled by a vector field **F**, is a good example of an incompressible field. It is not possible to compress an idealized fluid. On the other hand, heat or gasses can be compressed, allowing for sources and/or sinks.

The curl operator is used to show (quantify) the tendency for the vector field **F** to create "spin", and this spin is defined around a vector representing the axis of spin, at any given point. Thus, in a vector field **F**, there is super-imposed another vector field, curl **F**, which consists of vectors that serve as axes of rotation for any possible "spinning" within **F**. In a physical sense, "spin" creates circulation, and curl **F** is often used to show how a vector field might induce a current through a wire or loop immersed within that field. If curl **F** = **0**, then the vector field **F** induces no spin (or circulation).

Curl can be defined on a vector field within R^2, as shown below:

◆ • ◆ ◆ • • ◆ • • ◆

Example 52.5: Given $\mathbf{F}(x, y) = \langle xy, 2x^2 \rangle$. Find curl **F**.

Solution: Rewrite **F** to include a third component of 0:

$$\mathbf{F}(x, y, z) = \langle xy, 2x^2, 0 \rangle.$$

Thus,

$$\begin{aligned} \text{curl } \mathbf{F} &= \nabla \times \mathbf{F} \\ &= \langle P_y - N_z, M_z - P_x, N_x - M_y \rangle \\ &= \langle 0, 0, N_x - M_y \rangle \\ &= \langle 0, 0, 3x \rangle. \end{aligned}$$

A vector field **F** confined to the xy-plane (R^2) may induce a spin, and if so, all axes of rotations point into the third dimension, orthogonal to the xy-plane. At each point within some bounded region in R^2, there may be a spin. While some spins may cancel others, the net result will be evident at the boundary, where such spins then induce a current around that boundary.

Curl **F** may also be used to show if **F** is conservative. In general, if curl **F** = **0**, then **F** is (usually) conservative. We then find a possible potential function $f(x, y, z)$ such that $\nabla f = \mathbf{F}$.

Example 52.6: Given $\mathbf{F}(x, y, z) = \langle 2xyz^3, x^2z^3, 3x^2yz^2 \rangle$, show that curl **F** = **0**, and find a potential function $f(x, y, z)$ such that $\nabla f = \mathbf{F}$.

Solution: From an earlier example, we showed that curl **F** = **0**:

$$\begin{aligned}\text{curl } \mathbf{F} &= \nabla \times \mathbf{F} \\ &= \langle P_y - N_z, M_z - P_x, N_x - M_y \rangle \\ &= \langle 3x^2z^2 - 3x^2z^2, 6xyz^2 - 6xyz^2, 2xz^3 - 2xz^3 \rangle \\ &= \langle 0,0,0 \rangle = \mathbf{0}.\end{aligned}$$

This suggests that **F** is probably conservative. We seek a potential function by antidifferentiating M with respect to x, N with respect to y, and P with respect to z, and examining the results:

$$\int 2xyz^3 \, dx = x^2yz^3; \quad \int x^2z^3 \, dy = x^2yz^3; \quad \int 3x^2yz^2 \, dz = x^2yz^3.$$

Observe that all three antiderivatives result in x^2yz^3. We check by showing that $\nabla x^2yz^3 = \mathbf{F}$. It is, and thus, $f(x, y, z) = x^2yz^3$ is a potential function of **F**, so that **F** is a conservative vector field in R^3. In this case, **F** is also called a **gradient vector field**.

> In general, if a function $f(x, y, z)$ has continuous second-order derivatives over the relevant domain, then ∇f is a gradient vector field, and curl $\nabla f = \nabla \times \nabla f = \mathbf{0}$.
>
> Furthermore, if given $\mathbf{F}(x, y, z) = \langle M(x, y, z), N(x, y, z), P(x, y, z) \rangle$ and assuming M, N and P have continuous first-ordered partial derivatives, then div curl $\mathbf{F} = \nabla \cdot (\nabla \times \mathbf{F}) = 0$.

◆ ◆ ◆ ◆ ◆ ◆ ◆ ◆ ◆

Example 52.7: Given $\mathbf{F}(x, y, z) = \langle xy^2, x^2yz^2, 2xz^4 \rangle$, verify that div curl **F** = 0.

Solution: From an earlier example, curl $\mathbf{F} = \langle -2x^2yz, -2z^4, 2xyz^2 - y^2 \rangle$. Thus,

$$\begin{aligned}\nabla \cdot (\nabla \times \mathbf{F}) &= \frac{\partial}{\partial x}(-2x^2yz) + \frac{\partial}{\partial y}(-2z^4) + \frac{\partial}{\partial z}(2xyz^2 - y^2) \\ &= -4xyz + 0 + 4xzy \\ &= 0.\end{aligned}$$

Example 52.8: Find $\int_C \mathbf{F} \cdot d\mathbf{r}$, where

$$\mathbf{F}(x, y, z) = \left\langle \frac{2x}{z}, \frac{1}{z}, -\frac{x^2 + y}{z^2} \right\rangle,$$

and C is is a line segment from $(2,1,3)$ to $(4,4,4)$, then another line segment from $(4,4,4)$ to $(5,7,6)$.

Solution: It is possible that $\mathbf{F}$ is a conservative (gradient) vector field. We find curl $\mathbf{F}$:

$$\text{curl } \mathbf{F} = \langle P_y - N_z, M_z - P_x, N_x - M_y \rangle$$
$$= \left\langle -\frac{1}{z^2} - \left(-\frac{1}{z^2}\right), -\frac{2x}{z^2} - \left(-\frac{2x}{z^2}\right), 0 - 0 \right\rangle$$
$$= \langle 0, 0, 0 \rangle = \mathbf{0}.$$

Since curl $\mathbf{F} = \mathbf{0}$, then $\mathbf{F}$ is likely conservative. We now find $f(x, y, z)$ such that $\nabla f = \mathbf{F}$. We antidifferentiate M with respect to x, N with respect to y, and P with respect to z, and examine the results:

$$\int \frac{2x}{z} \, dx = \frac{x^2}{z}; \quad \int \frac{1}{z} \, dy = \frac{y}{z}; \quad \int \left(-\frac{x^2 + y}{z^2}\right) dz = \frac{x^2 + y}{z}.$$

The union of these terms is

$$f(x, y, z) = \frac{x^2 + y}{z}.$$

This is a potential function since $f_x = M = \frac{2x}{z}$, $f_y = N = \frac{1}{z}$, and $f_z = P = -\frac{x^2 + y}{z^2}$.

Thus, the line integral can be determined by using the Fundamental Theorem of Line Integrals, and avoiding the need to parameterize the line segments:

$$\int_C \mathbf{F} \cdot d\mathbf{r} = \left[\frac{x^2 + y}{z}\right]_{(2,1,3)}^{(5,7,6)}$$
$$= \left(\frac{(5)^2 + (7)}{(6)}\right) - \left(\frac{(2)^2 + (1)}{(3)}\right)$$
$$= \frac{32}{6} - \frac{5}{3} = \frac{11}{3}.$$

Visualizing Divergence

To see divergence at a point, visualize a small box around that point, then infer whether more mass is entering into this box than leaving, more is leaving the box than entering, or equal amounts are flowing into and out of the box. For example, suppose a point P, shown below, is located within a vector field represented by arrows:

A box is drawn around P, and we see that the vectors "entering" from the left are the same magnitude as those "leaving" to the right. If we scale the box down and assume a similar behavior of the vector field for these smaller boxes, then it is reasonable to infer that in this case, there are equal amounts of material entering as leaving. Thus, there is no divergence at P.

In the image below, the arrows differ in magnitude, but it is still evident that there are equal amounts of material entering as leaving. There is no divergence at P.

In the next image, it appears more material is entering than is leaving. Thus, at P, there is negative divergence, and P is a sink.

In this image, more material is leaving than is entering, so at P, there is positive divergence and P is a source:

Visualizing Curl

Curl is the tendency of a vector field to cause a spin at a point, the spin rotating around an axis of revolution. However, when viewing a vector field, "seeing" curl is not as obvious. It should not be confused with any apparent "curviness" of a vector field. A fluid may flow along a non-straight-line path yet have no curl.

To see evidence of curl at a point P, look for vectors that seem to shear (face opposite directions) near P, or look for any concentric behavior of the flow lines. However, even this won't strongly indicate curl.

For example, in the image below, there is probably a curl vector at P. Note the opposing directions of the vector field.

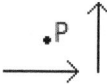

In the next image, there is probably curl at P as well:

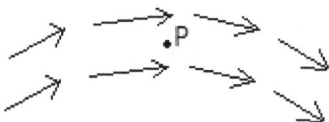

However, in the next image, there may not be any curl at P:

The formula for curl **F** allows us to quantify the curl at any given point, which is helpful since trying to infer it from an image of a vector field may be difficult.

The following are examples of vector fields and their divergence and curl:

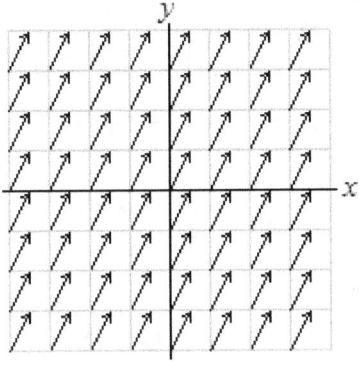

$$F(x, y) = \langle 1, 2 \rangle$$
div **F** = 0
curl **F** = **0**.

Constant vector fields have no divergence and no curl.

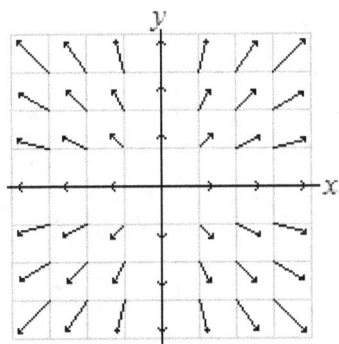

$$F(x, y) = \langle x, y \rangle$$
div **F** = 2
curl **F** = **0**.

All vectors emanate away from the origin and grow in magnitude. Draw a small box anywhere and note that more mass is moving "out" than entering "in". All points in the plane are considered "sources". Divergence is positive. There is no rotation, so curl is 0.

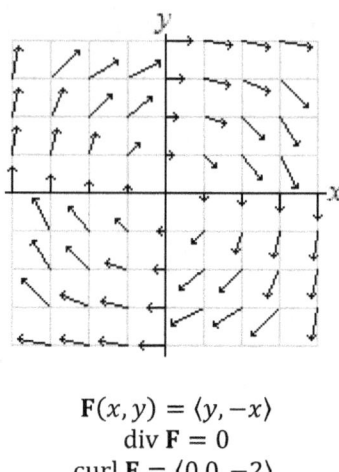

$$\mathbf{F}(x,y) = \langle y, -x \rangle$$
$$\text{div } \mathbf{F} = 0$$
$$\text{curl } \mathbf{F} = \langle 0, 0, -2 \rangle.$$

This field has no divergence, but it does have curl. Since curl is negative, the spin is clockwise, and the curl vectors point in the negative z direction ("into" the page).

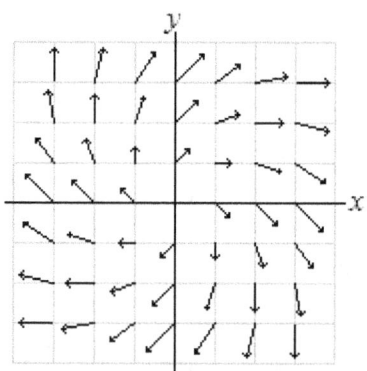

$$\mathbf{F}(x,y) = \langle x+y, y-x \rangle$$
$$\text{div } \mathbf{F} = 2$$
$$\text{curl } \mathbf{F} = \langle 0, 0, -2 \rangle.$$

There is divergence at all points: draw a small box anywhere and note that more mass is moving "out" than entering "in". There is also curl: note the general clockwise "spiral" nature of the vector field (hence, the -2 in the z-component of curl). A point anywhere in the plane would be compelled to spin.

53. Flux Integrals

Let S be an *orientable* surface within R^3. In this context, an orientable surface is one with two distinct sides. At any point on an orientable surface, there exists two normal vectors, one pointing in the opposite direction of the other. Most surfaces, especially those defined explicitly by $z = f(x, y)$, are orientable. An example of a non-orientable surface is the Moebius Strip. However, these odd surfaces will not play a role in the following discussion.

Let $\mathbf{F}(x, y, z) = \langle M(x, y, z), N(x, y, z), P(x, y, z) \rangle$ be a vector field in R^3. Suppose $\mathbf{F}$ represents the flow of some medium, *e.g.* heat or fluid, through R^3. The question that arises is: how much flow, as defined by $\mathbf{F}$, passes through the surface S in a given unit of time? We make the reasonable assumption that S is completely permeable.

At each point on the surface S, there exists two vectors: one being $\mathbf{F}$ representing the flow, and a unit normal vector $\mathbf{n}$, representing the positive direction. If $\mathbf{F}$ and $\mathbf{n}$ point in the same direction (their angle is acute), then their dot product $\mathbf{F} \cdot \mathbf{n}$ is positive, and at this point we say the flow is positive. Similarly, if $\mathbf{F}$ and $\mathbf{n}$ point in opposite directions, their dot product is negative, and we say that there is negative flow at this point. It is possible that $\mathbf{F} \cdot \mathbf{n}$ is 0, in which case there is no flow through the surface at the point. Since $\mathbf{F}$ can vary in length, the values given by the dot products can vary in size too.

To gain a rough sense of the total net flow, or **flux**, of a vector field $\mathbf{F}$ through a surface S, we sum all such dot products $\mathbf{F} \cdot \mathbf{n}$. To sum "all" of the dot products at every point on the surface means to take an integral. The flux of a vector field $\mathbf{F}$ through a surface S is given by

$$\iint_R \mathbf{F} \cdot \mathbf{n} \, dS.$$

Here, R is the region over which the double integral is evaluated. Region R is the "shadow" surface S makes on the xy-plane.

A *closed* surface is one that encloses a finite-volume subregion of R^3 in such a way that there is an "inside" and "outside". Examples of closed surfaces are cubes, spheres, ellipsoids, and so on.

> **Comment:** the notions of "positive" and "negative", and of "up" and "down", vary depending on the context. For a typical surface, "positive" direction of flow is usually arbitrarily chosen. For closed surfaces, positive flow is always taken to be from the inside to the outside. That is, normal vectors $\mathbf{n}$ point "away" from the interior of the subregion.

Setting up a flux integral requires a number of steps. First, a surface S must be given. If the surface is defined explicitly such as $z = f(x,y)$, then its parameterization is

$$\mathbf{r}(x,y) = \langle x, y, f(x,y) \rangle.$$

From this, we can find *unit* normal vectors $\mathbf{n}$ by using the formulas

$$\mathbf{n} = \frac{\langle f_x(x,y), f_y(x,y), -1 \rangle}{\sqrt{f_x^2(x,y) + f_y^2(x,y) + 1}} \quad \text{or} \quad \mathbf{n} = \frac{\langle -f_x(x,y), -f_y(x,y), 1 \rangle}{\sqrt{f_x^2(x,y) + f_y^2(x,y) + 1}}.$$

Recall from the discussion of surface area integrals that

$$dS = \sqrt{f_x^2(x,y) + f_y^2(x,y) + 1} \, dA.$$

Thus, substitutions can be made into the flux integral:

$$\iint_R \mathbf{F} \cdot \mathbf{n} \, dS$$

$$= \iint_R \langle M, N, P \rangle \cdot \frac{\langle f_x(x,y), f_y(x,y), -1 \rangle}{\sqrt{f_x^2(x,y) + f_y^2(x,y) + 1}} \sqrt{f_x^2(x,y) + f_y^2(x,y) + 1} \, dA.$$

Note that the ratio of expressions $\sqrt{f_x^2(x,y) + f_y^2(x,y) + 1}$ is 1 (that is, they cancel). The flux integral is now

$$\iint_R \langle M, N, P \rangle \cdot \langle f_x(x,y), f_y(x,y), -1 \rangle \, dA,$$

$$\text{or} \quad \iint_R \langle M, N, P \rangle \cdot \langle -f_x(x,y), -f_y(x,y), 1 \rangle \, dA.$$

After taking the dot product, the integrand is a function in variables x and y, and normal techniques are used to evaluate the double integrals.

In the examples that follow, we abuse the notation slightly: the vector $\mathbf{n}$ may not be a unit vector. As long as the normal vector is derived carefully and has the appearance shown above, it will be sufficient.

Example 53.1: Find the flux of the vector field $\mathbf{F}(x, y, z) = \langle 1,2,3 \rangle$ through the square S in the xy-plane with vertices $(0,0)$, $(1,0)$, $(0,1)$ and $(1,1)$, where positive flow is the positive z direction. (Since the surface S lies in the xy-plane, it is identical to R in this case).

Solution: Since positive flow is in the direction of positive z, and the surface S is on the xy-plane itself, then a unit normal to R is $\mathbf{n} = \langle 0,0,1 \rangle$. Thus, $\mathbf{F} \cdot \mathbf{n} = \langle 1,2,3 \rangle \cdot \langle 0,0,1 \rangle = 3$, and the flux is given by

$$\iint_{R(=S)} 3 \, dA = 3 \iint_R dA = 3(\text{Area of } R) = 3(1) = 3.$$

In any unit of time, a total flow of 3 units of mass per unit of time will flow through S. In this example, the answer could be reasoned without performing the actual integration. Note that from each vector $\langle 1,2,3 \rangle$, only the z-component 3 is relevant. That is, in the positive z direction, the fluid flows at a rate of 3 units of mass per unit of time.

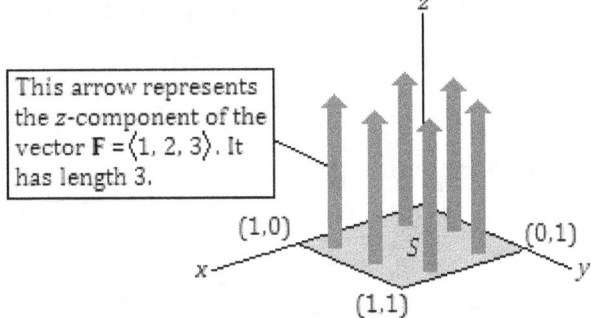

At each point in the square S, a vector $\langle 1,2,3 \rangle$ is drawn, so it stands to reason that the flux can be viewed as the volume of a box with the square S as its base, and a height of 3; thus, the volume = $(1)(1)(3) = 3$, the flux. If the flow was of water, this box holds the water that flowed through the square S in one unit of time in the direction of positive z.

The other components in $\mathbf{F}$ indicate that the flow travels in a direction that is not orthogonal to S. But the fact does remain that regardless how far the fluid may travel in the x or y directions, after 1 unit of time, 3 units of mass will have flowed in the z-direction, and that is exactly what we are seeking to determine.

This geometrical phenomenon is known as *Cavalieri's Principle*. You may have "seen" this principle when stacking coins. A perfectly vertical stack will appear as a cylinder and its volume can be easily determined. If the stack is disturbed so that it leans but does not fall over, the vertical height has not changed, nor has the volume. The x or y offset in the lean has no effect on the volume of the stack.

Example 53.2: Find the flux of the vector field $\mathbf{F}(x, y, z) = \langle x, y, -z \rangle$ through the portion of the plane in the first octant with intercepts (4,0,0), (0,8,0) and (0,0,10), where positive flow is defined to be in the positive z direction.

Solution: First, we find an equation for the plane (this is surface S). Recall from Example 13.6 that a plane passing through $(a,0,0)$, $(0,b,0)$ and $(0,0,c)$ has the general form

$$\frac{x}{a} + \frac{y}{b} + \frac{z}{c} = 1.$$

Thus, the plane here is

$$\frac{x}{4} + \frac{y}{8} + \frac{z}{10} = 1.$$

Clearing fractions, we have $10x + 5y + 4z = 40$, or $z = 10 - \frac{5}{2}x - \frac{5}{4}y$.

From the plane, There are two normal vectors $\mathbf{n}$: $\left\langle -\frac{5}{2}, -\frac{5}{4}, -1 \right\rangle$ or $\left\langle \frac{5}{2}, \frac{5}{4}, 1 \right\rangle$. We choose $\mathbf{n} = \left\langle \frac{5}{2}, \frac{5}{4}, 1 \right\rangle$ since the problem defined positive flow to be in the positive z direction.

We now find $\mathbf{F} \cdot \mathbf{n}$. Note that since $z = 10 - \frac{5}{2}x - \frac{5}{4}y$, we write $\mathbf{F}$ in terms of x and y, where $\mathbf{F}(x, y) = \langle x, y, -z \rangle = \left\langle x, y, -\left(10 - \frac{5}{2}x - \frac{5}{4}y\right) \right\rangle$. Thus,

$$\mathbf{F} \cdot \mathbf{n} = \left\langle x, y, -10 + \frac{5}{2}x + \frac{5}{4}y \right\rangle \cdot \left\langle \frac{5}{2}, \frac{5}{4}, 1 \right\rangle$$

$$= \frac{5}{2}x + \frac{5}{4}y - 10 + \frac{5}{2}x + \frac{5}{4}y$$

$$= 5x + \frac{5}{2}y - 10.$$

This will be the integrand.

The bounds of integration lie in the xy-plane, the "footprint" of the surface S, $10x + 5y + 4z = 40$ as it is projected onto the xy-plane (this being R):

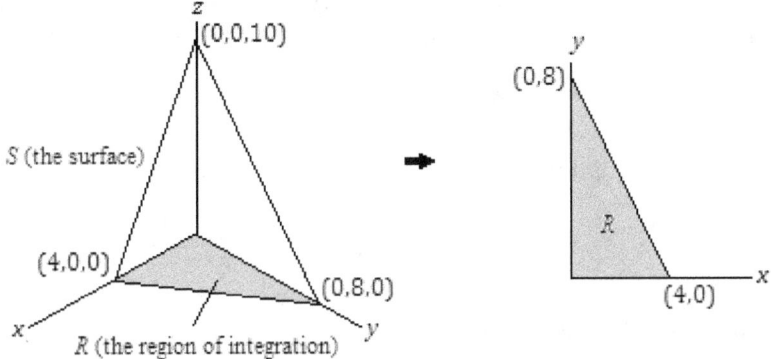

Choosing $dy\, dx$ as the order of integration, the bounds on y are $0 \le y \le 8 - 2x$, and the bounds on x are $0 \le x \le 4$. The flux of **F** through the surface S is

$$\int_0^4 \int_0^{8-2x} \left(5x + \frac{5}{2}y - 10\right) dy\, dx.$$

Evaluating the inside integral, we have

$$\int_0^{8-2x} \left(5x + \frac{5}{2}y - 10\right) dy = \left[5xy + \frac{5}{4}y^2 - 10y\right]_0^{8-2x}$$

$$= 5x(8 - 2x) + \frac{5}{4}(8 - 2x)^2 - 10(8 - 2x)$$

$$= 20x - 5x^2. \quad \text{(After simplification)}$$

This is integrated with respect to x:

$$\int_0^4 (20x - 5x^2)\, dx = \left[10x^2 - \frac{5}{3}x^3\right]_0^4 = 10(4)^2 - \frac{5}{3}(4)^3 = \frac{160}{3}.$$

The flux is positive, and we can say that in one unit of time, $\frac{160}{3}$ units of material flow through this surface.

◆ • ◆ • • ◆ • • ◆

It seems plausible that it should not matter in which direction we define to be positive flow. In the next example, we repeat this same problem but in a different "positive" direction.

Example 53.3: Find the flux of the vector field $\mathbf{F}(x, y, z) = \langle x, y, -z \rangle$ through the portion of the plane in the first octant with intercepts (4,0,0), (0,8,0) and (0,0,10), where positive flow is defined to be in the positive y direction.

Solution: From the equation of the plane $10x + 5y + 4z = 40$, solve for y, obtaining $y = 8 - 2x - \frac{4}{5}z$, and from this, two normal vectors are identified, $\langle -2, -1, -\frac{4}{5} \rangle$ or $\langle 2, 1, \frac{4}{5} \rangle$. Since the direction is positive y, choose $\mathbf{n} = \langle 2, 1, \frac{4}{5} \rangle$.

The vector field $\mathbf{F}$ is adjusted too. In place of y, we substitute $8 - 2x - \frac{4}{5}z$:

$$\mathbf{F}(x, z) = \langle x, y, -z \rangle = \langle x, 8 - 2x - \frac{4}{5}z, -z \rangle.$$

The dot product $\mathbf{F} \cdot \mathbf{n}$ is now determined:

$$\mathbf{F} \cdot \mathbf{n} = \langle x, 8 - 2x - \frac{4}{5}z, -z \rangle \cdot \langle 2, 1, \frac{4}{5} \rangle$$
$$= 2x + 8 - 2x - \frac{4}{5}z - \frac{4}{5}z$$
$$= 8 - \frac{8}{5}z.$$

For the bounds of integration, we look at the footprint of the surface S projected onto the xz plane:

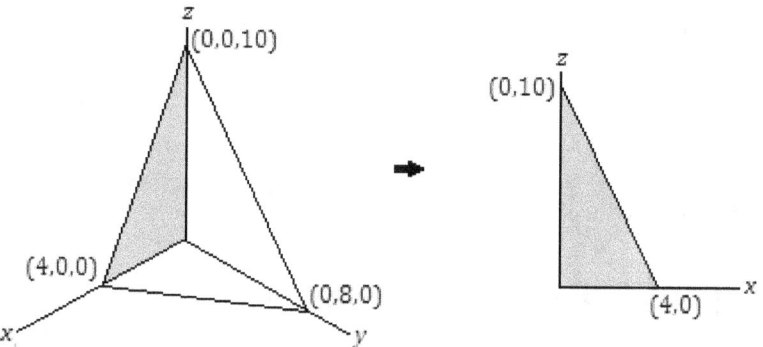

Choosing the $dz\,dx$ order of integration, the bounds are $0 \le z \le 10 - \frac{5}{2}x$ and $0 \le x \le 4$. The flux is given by the double integral

$$\int_0^4 \int_0^{10-(5/2)x} \left(8 - \frac{8}{5}z\right) dz\,dx.$$

The inside integral is evaluated:

$$\int_0^{10-(5/2)x} \left(8 - \frac{8}{5}z\right) dz = \left[8z - \frac{4}{5}z^2\right]_0^{10-(5/2)x}$$

$$= 8\left(10 - \frac{5}{2}x\right) - \frac{4}{5}\left(10 - \frac{5}{2}x\right)^2$$

$$= 20x - 5x^2. \quad \text{(after simplification)}$$

The outside integral is evaluated:

$$\int_0^4 (20x - 5x^2) \, dx = \left[10x^2 - \frac{5}{3}x^3\right]_0^4 = 10(4)^2 - \frac{5}{3}(4)^3 = \frac{160}{3}.$$

And we arrive at the same result. This should not be surprising. If you are feeling energetic, repeat the problem where the positive direction of flow is the positive x direction.

♦ • ♦ • ♦ • ♦ • ♦

Example 53.4: Find the flux of the vector field $\mathbf{F}(x, y, z) = \langle z, 2x, y \rangle$ through the portion of the paraboloid $z = 9 - x^2 - y^2$ above the xy-plane and confined to the first octant, where positive flow is in the positive z-direction.

Solution: Following the forms $\langle f_x, f_y, -1 \rangle$ or $\langle -f_x, -f_y, 1 \rangle$, the normal vectors to the surface are

$$\langle -2x, -2y, -1 \rangle \quad \text{or} \quad \langle 2x, 2y, 1 \rangle.$$

We use $\mathbf{n} = \langle 2x, 2y, 1 \rangle$ since it agrees with the positive z direction. Next, vector field $\mathbf{F}$ is written in terms of x and y only:

$$\mathbf{F}(x, y) = \langle 9 - x^2 - y^2, 2x, y \rangle.$$

Thus, the dot product $\mathbf{F} \cdot \mathbf{n}$ is:

$$\mathbf{F} \cdot \mathbf{n} = (9 - x^2 - y^2)(2x) + (2x)(2y) + (y)(1)$$
$$= (9 - x^2 - y^2)2x + 4xy + y.$$

The region of integration R is a filled-in quarter-circle on the xy-plane with radius 3, with its center at the origin. We use polar coordinates. Using the substitutions $x = r \cos \theta$ and $y = r \sin \theta$, the bounds of integration become $0 \le r \le 3$ and $0 \le \theta \le \frac{\pi}{2}$. Recall also that the area element $dA = r \, dr \, d\theta$.

The expression $(9 - x^2 - y^2)2x + 4xy + y$ is also written in terms of r and θ:

$$(9 - (r\cos\theta)^2 - (r\sin\theta)^2)2(r\cos\theta) + 4(r\cos\theta)(r\sin\theta) + (r\sin\theta).$$

Note that $(9 - (r\cos\theta)^2 - (r\sin\theta)^2) = 9 - r^2$. Thus, we have

$$(9 - r^2)2(r\cos\theta) + 4(r\cos\theta)(r\sin\theta) + (r\sin\theta).$$

This is further simplified:

$$18r\cos\theta - 2r^3\cos\theta + 4r^2\cos\theta\sin\theta + r\sin\theta.$$

Finally, the flux integral is

$$\int_0^{\pi/2}\int_0^3 (18r\cos\theta - 2r^3\cos\theta + 4r^2\cos\theta\sin\theta + r\sin\theta)\, r\, dr\, d\theta.$$

Distribute the r:

$$\int_0^{\pi/2}\int_0^3 (18r^2\cos\theta - 2r^4\cos\theta + 4r^3\cos\theta\sin\theta + r^2\sin\theta)\, dr\, d\theta.$$

The inside integral is evaluated:

$$\int_0^3 (18r^2\cos\theta - 2r^4\cos\theta + 4r^3\cos\theta\sin\theta + r^2\sin\theta)\, dr$$

$$= \left[6r^3\cos\theta - \frac{2}{5}r^5\cos\theta + r^4\cos\theta\sin\theta + \frac{1}{3}r^3\sin\theta\right]_0^3$$

$$= 162\cos\theta - \frac{486}{5}\cos\theta + 81\cos\theta\sin\theta + 9\sin\theta$$

$$= \frac{324}{5}\cos\theta + 81\cos\theta\sin\theta + 9\sin\theta. \quad \left\{162 - \frac{486}{5} = \frac{324}{5}\right\}$$

This is integrated with respect to θ:

$$\int_0^{\pi/2}\left(\frac{324}{5}\cos\theta + 81\cos\theta\sin\theta + 9\sin\theta\right)d\theta$$

$$= \left[\frac{324}{5}\sin\theta + \frac{81}{2}\sin^2\theta - 9\cos\theta\right]_0^{\pi/2}$$

Note that $\sin\left(\frac{\pi}{2}\right) = 1$, $\sin(0) = 0$, $\cos\left(\frac{\pi}{2}\right) = 0$ and $\cos(0) = 1$:

$$\left[\frac{324}{5}\sin\theta + \frac{81}{2}\sin^2\theta - 9\cos\theta\right]_0^{\pi/2}$$

$$= \left(\frac{324}{5}(1) + \frac{81}{2}(1) - 9(0)\right) - \left(\frac{324}{5}(0) + \frac{81}{2}(0) - 9(1)\right) = \frac{1143}{10}.$$

There is a lot of positive flow through this surface, as indicated by the result. As much work as this seemed to be, it took advantage of many of the nicer aspects of polar integration.

◆ ◆ ◆ ◆ ◆ ◆ ◆ ◆ ◆

The following three examples discuss flux through a closed surface. Thus, we must choose **n** to point away from the interior of the closed surface.

Example 53.5: Find the net flux of the vector field $\mathbf{F}(x, y, z) = \langle x, y, 1 \rangle$ through the closed surface of the hemisphere $x^2 + y^2 + z^2 = 4$ above the xy-plane, whose base is on the xy-plane.

Solution: For the hemisphere, the normal vectors will point in the direction of positive z, away from the interior. Isolate z in the equation $x^2 + y^2 + z^2 = 4$:

$$z = f(x, y) = \sqrt{4 - x^2 - y^2}.$$

Using the form $\mathbf{n} = \langle -f_x, -f_y, 1 \rangle$, we have

$$\mathbf{n} = \left\langle \frac{x}{\sqrt{4 - x^2 - y^2}}, \frac{y}{\sqrt{4 - x^2 - y^2}}, 1 \right\rangle.$$

There is no z-variable in **F**, so there is no need to make any substitutions. Vector field **F** is already in terms of x and y. The dot product $\mathbf{F} \cdot \mathbf{n}$ is

$$\mathbf{F} \cdot \mathbf{n} = \langle x, y, 1 \rangle \cdot \left\langle \frac{x}{\sqrt{4 - x^2 - y^2}}, \frac{y}{\sqrt{4 - x^2 - y^2}}, 1 \right\rangle$$

$$= \frac{x^2}{\sqrt{4 - x^2 - y^2}} + \frac{y^2}{\sqrt{4 - x^2 - y^2}} + 1$$

$$= \frac{x^2 + y^2}{\sqrt{4 - x^2 - y^2}} + 1.$$

This is the integrand of the flux integral. We will use polar coordinates to evaluate. The region of integration in the xy-plane is a circle of radius 2, centered at the origin. Using $x = r \cos \theta$ and $y = r \sin \theta$, with bounds $0 \le r \le 2$ and $0 \le \theta \le 2\pi$, the integrand becomes

$$\frac{r^2}{\sqrt{4-r^2}} + 1,$$

and the flux through the hemisphere is

$$\int_0^{2\pi} \int_0^2 \left(\frac{r^2}{\sqrt{4-r^2}} + 1\right) r \, dr \, d\theta.$$

The inside integral is evaluated. A table of integrals is used to determine the antiderivative:

$$\int_0^2 \left(\frac{r^2}{\sqrt{4-r^2}} + 1\right) r \, dr = \int_0^2 \left(\frac{r^3}{\sqrt{4-r^2}} + r\right) dr$$

$$= \left[-\frac{1}{3}(r^2+8)\sqrt{4-r^2} + \frac{1}{2}r^2\right]_0^2$$

$$= 2 - \left(-\frac{1}{3}(8)(2)\right) = \frac{22}{3}.$$

Then the outside integral is evaluated:

$$\int_0^{2\pi} \left(\frac{22}{3}\right) d\theta = \frac{22}{3} \int_0^{2\pi} d\theta = \frac{22}{3}(2\pi) = \frac{44\pi}{3}.$$

Now, we evaluate the flux through the base itself, the circle of radius 2 centered at the origin. Since this surface lies in the xy-plane, we will use $\mathbf{n} = \langle 0,0,-1 \rangle$ because the positive orientation of flow is *away* from the interior (note that $\langle 0,0,1 \rangle$ would point inside the object). The dot product $\mathbf{F} \cdot \mathbf{n}$ is

$$\mathbf{F} \cdot \mathbf{n} = \langle x, y, 1 \rangle \cdot \langle 0, 0, -1 \rangle = -1.$$

Thus, the flux through the circle of radius 2 is

$$\iint_R (-1) \, dA = -\iint_R dA = -4\pi.$$

Note that $\iint_R dA$ represents the area of the circle of radius 2, which is 4π. Adding the two flux values for the two surfaces that compose the closed surface, the total net flux through this hemisphere and its base is $\frac{44\pi}{3} + (-4\pi) = \frac{32\pi}{3}$.

Example 53.6: Find the net flux of $\mathbf{F}(x,y,z) = \langle 2x, y, 4z \rangle$ through a cube in the first octant with vertices (0,0,0), (1,0,0), (1,1,0), (0,1,0), (0,0,1), (1,0,1), (1,1,1) and (0,1,1).

Solution: To find the net flux, we need to find the flux through each of the box's six surfaces, then sum these values. The box is enclosed by the planes $x = 0$, $y = 0$, $z = 0$, $x = 1$, $y = 1$ and $z = 1$.

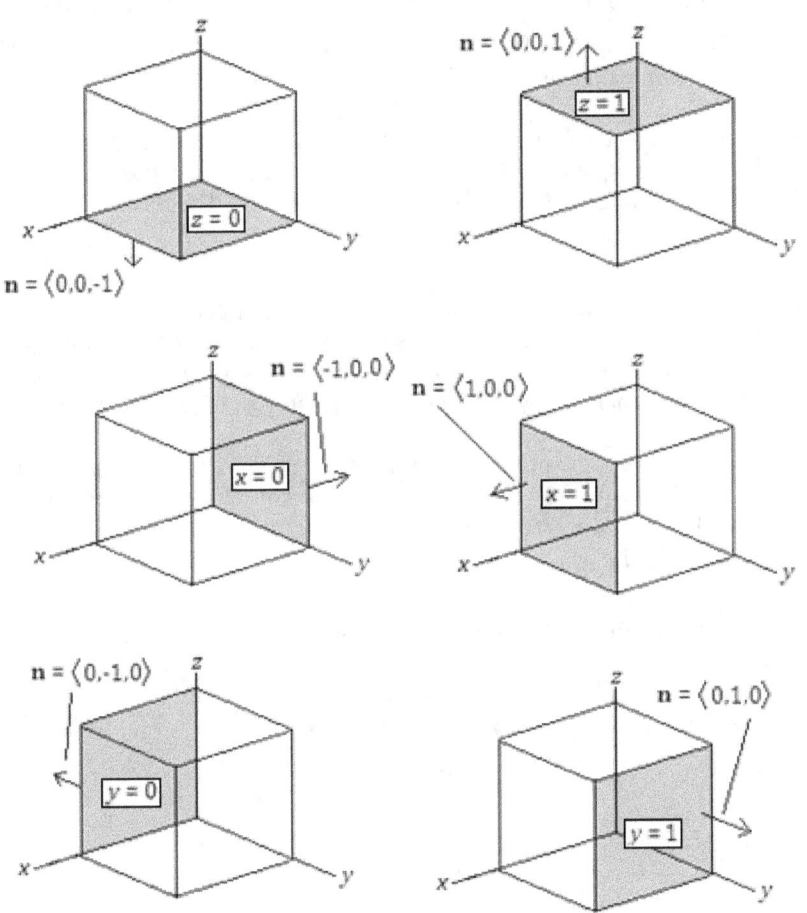

The six normal vectors **n** for each of the six surfaces of the box.
Note that each **n** points *away* from the interior of the cube.

For the surface $z = 0$, the normal vector points in the direction of negative z, so $\mathbf{n} = \langle 0, 0, -1 \rangle$. The equation $z = 0$ is substituted into **F**, so that $\mathbf{F}(x, y, 0) = \langle 2x, y, 0 \rangle$. Therefore, $\mathbf{F} \cdot \mathbf{n} = \langle 2x, y, 0 \rangle \cdot \langle 0, 0, -1 \rangle = 0$. The flux is zero—there is no flow generated by the vector field **F** through the surface $z = 0$.

For the surface $x = 0$, the normal vector points in the direction of negative x, so $\mathbf{n} = \langle -1,0,0 \rangle$. The equation $x = 0$ is substituted into $\mathbf{F}$, so that $\mathbf{F}(0, y, z) = \langle 0, y, 4z \rangle$. Therefore, $\mathbf{F} \cdot \mathbf{n} = \langle 0, y, 4z \rangle \cdot \langle -1,0,0 \rangle = 0$. The flux is zero—there is no flow generated by the vector field $\mathbf{F}$ through the surface $x = 0$.

For the surface $z = 1$, the normal vector points in the direction of positive z, so $\mathbf{n} = \langle 0,0,1 \rangle$. The equation $z = 1$ is substituted into $\mathbf{F}$, so that $\mathbf{F}(x, y, 1) = \langle 2x, y, 4(1) \rangle$. Therefore, $\mathbf{F} \cdot \mathbf{n} = \langle 2x, y, 4 \rangle \cdot \langle 0,0,1 \rangle = 4$. The flux through $z = 1$ is $\iint_R 4 \, dA = 4 \iint_R dA = 4(1) = 4$, where $\iint_R dA$ is the area of the surface, which is a square with side lengths 1.

For the surface $x = 1$, the normal vector points in the direction of positive x, so $\mathbf{n} = \langle 1,0,0 \rangle$. The equation $x = 1$ is substituted into $\mathbf{F}$, so that $\mathbf{F}(1, y, z) = \langle 2(1), y, 4z \rangle$. Therefore, $\mathbf{F} \cdot \mathbf{n} = \langle 2, y, 4z \rangle \cdot \langle 1,0,0 \rangle = 2$. The flux through $x = 1$ is given by $\iint_R 2 \, dA = 2 \iint_R dA = 2(1) = 2$.

For the surface $y = 0$, the normal vector points in the direction of negative y, so $\mathbf{n} = \langle 0, -1, 0 \rangle$. The equation $y = 0$ is substituted into $\mathbf{F}$, so that $\mathbf{F}(x, 0, z) = \langle 2x, 0, 4z \rangle$. Therefore, $\mathbf{F} \cdot \mathbf{n} = \langle 2x, 0, 4z \rangle \cdot \langle 0, -1, 0 \rangle = 0$. The flux is zero—there is no flow generated by the vector field $\mathbf{F}$ through the surface $y = 0$.

For the surface $y = 1$, the normal vector points in the direction of positive y, so $\mathbf{n} = \langle 0,1,0 \rangle$. The equation $y = 1$ is substituted into $\mathbf{F}$, so that $\mathbf{F}(x, 1, z) = \langle 2x, 1, 4z \rangle$. Therefore, $\mathbf{F} \cdot \mathbf{n} = \langle 2x, 1, 4z \rangle \cdot \langle 0,1,0 \rangle = 1$. The flux through $y = 1$ is given by $\iint_R 1 \, dA = 1 \iint_R dA = 1$.

Thus, the total net flux is the sum of these values: $0 + 0 + 0 + 4 + 2 + 1 = 7$ units of mass per unit of time.

◆ • ◆ • ◆ • ◆ • ◆

See an error? Have a suggestion?
Please see www.surgent.net/vcbook

Example 53.7: Find the net flux of $F(x, y, z) = \langle xy, -y, z\rangle$ through the closed surface composed of the cylinder $x^2 + y^2 = 4$ and the planes, $y = 0, z = 1$ and $z = 6$.

Solution: The object is shown below:

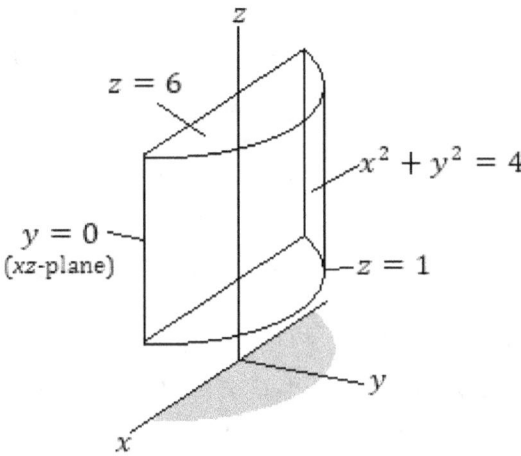

The flux through the planes can be found quickly.

For $z = 1$, we use $\mathbf{n} = \langle 0, 0, -1\rangle$ since the direction of positive flow will be away from the interior bounded by the surface. Also, since $z = 1$, we have $F(x, y, 1) = \langle xy, -y, 1\rangle$. Thus, we have $\mathbf{F} \cdot \mathbf{n} = \langle xy, -y, 1\rangle \cdot \langle 0, 0, -1\rangle = -1$. The region of integration R is a half-circle in the xy-plane of radius 2, its center at the origin (in gray, above). The flux through this plane is

$$\iint_R (-1) \, dA = -\iint_R dA = -\binom{\text{Area inside half of a}}{\text{circle of radius 2}} = -\frac{1}{2}\pi(2)^2 = -2\pi.$$

For $z = 6$, we use $\mathbf{n} = \langle 0, 0, 1\rangle$, which points upward, away from the interior bounded by the surface. Also, since $z = 6$, we have $F(x, y, 6) = \langle xy, -y, 6\rangle$. Thus, we have $\mathbf{F} \cdot \mathbf{n} = \langle xy, -y, 6\rangle \cdot \langle 0, 0, 1\rangle = 6$. The region of integration R is the same as above. The flux through this plane is

$$\iint_R (6) \, dA = 6\iint_R dA = 6\left(\frac{1}{2}\pi(2)^2\right) = 12\pi.$$

For $y = 0$, we use $\mathbf{n} = \langle 0, -1, 0\rangle$, keeping in mind we want the normal vector to point outward from the interior bounded by the surface. We have $F(x, 0, z) = \langle 0, 0, z\rangle$, so that $\mathbf{F} \cdot \mathbf{n} = 0$. Thus, the flux through this plane is 0.

For the cylinder $x^2 + y^2 = 4$, we parameterize it in cylindrical coordinates using two variables:

$$\mathbf{r}(u, v) = \langle 2\cos u, 2\sin u, v\rangle, \quad \text{where} \quad 0 \leq u \leq \pi \quad \text{and} \quad 1 \leq v \leq 6.$$

To find a normal vector $\mathbf{n}$ to this surface, find $\mathbf{r}_u$ and $\mathbf{r}_v$ and then find $\mathbf{r}_u \times \mathbf{r}_v$:

$$\mathbf{r}_u = \langle -2\sin u, 2\cos u, 0\rangle \quad \text{and} \quad \mathbf{r}_v = \langle 0,0,1\rangle;$$

$$\text{thus,} \quad \mathbf{n} = \mathbf{r}_u \times \mathbf{r}_v = \langle 2\cos u, 2\sin u, 0\rangle.$$

Vector field $\mathbf{F}$ is rewritten in terms of u and v, where $x = 2\cos u$ and $y = 2\sin u$:

$$\mathbf{F}(u,v) = \langle (2\cos u)(2\sin u), -(2\sin u), v\rangle = \langle 4\cos u \sin u, -2\sin u, v\rangle.$$

The dot product is

$$\mathbf{F} \cdot \mathbf{n} = \langle 4\cos u \sin u, -2\sin u, v\rangle \cdot \langle 2\cos u, 2\sin u, 0\rangle$$
$$= 8\cos^2 u \sin u - 4\sin^2 u.$$

The flux through the cylinder alone is

$$\int_1^6 \int_0^\pi (8\cos^2 u \sin u - 4\sin^2 u)\, du\, dv.$$

We use the identity $\sin^2 u = \frac{1}{2} - \frac{1}{2}\cos(2u)$, then simplify:

$$\int_1^6 \int_0^\pi \left(8\cos^2 u \sin u - 4\left(\frac{1}{2} - \frac{1}{2}\cos(2u)\right)\right) du\, dv$$
$$= \int_1^6 \int_0^{2\pi} (8\cos^2 u \sin u - 2 + 2\cos(2u))\, du\, dv.$$

The inside integral is evaluated:

$$\int_0^\pi (8\cos^2 u \sin u - 2 + 2\cos(2u))\, du = \left[-\frac{8}{3}\cos^3 u - 2u + \sin(2u)\right]_0^{2\pi}$$
$$= \frac{16}{3} - 2\pi.$$

The outside integral is evaluated:

$$\int_1^6 \left(\frac{16}{3} - 2\pi\right) dv = \left(\frac{16}{3} - 2\pi\right)\int_1^6 dv = \left(\frac{16}{3} - 2\pi\right)(5) = \frac{80}{3} - 10\pi.$$

The net flux through the closed surface is $-2\pi + 12\pi + \frac{80}{3} - 10\pi = \frac{80}{3}$.

♦ • • ♦ • • ♦ • • ♦

Determining the flux through a closed surfaces can be tedious since we usually must determine the flux through all surfaces of the object. However, there is a faster way to find the flux through such surfaces, using the divergence operator. This is called the *divergence theorem*.

54. The Divergence Theorem

Let S be a closed surface that encloses a subregion in R^3 in such a way that the surface creates a distinct inside and outside. Let $\mathbf{F}(x, y, z)$ be a vector field in R^3. To find the total flow of mass through S, we can use the **divergence theorem**:

$$\iint_R \mathbf{F} \cdot \mathbf{n}\, dS = \iiint_S \operatorname{div} \mathbf{F}\, dV.$$

We are slightly abusing the notation here. The subscript S in the triple integral is the surface, but the integral itself is evaluated over the region enclosed by the surface.

> If div $\mathbf{F}$ is a constant, then geometry may be used to determine $\iiint_S dV$.

♦ • • ♦ • • ♦ • • ♦

Example 54.1: Find the net flux of the vector field $\mathbf{F}(x, y, z) = \langle x, y, 1 \rangle$ through the closed surface of the hemisphere $x^2 + y^2 + z^2 = 4$ above the xy-plane, whose base is on the xy-plane. (This is a repeat of Example 53.5)

Solution: We use the divergence theorem. The divergence of $\mathbf{F}$ is

$$\operatorname{div} \mathbf{F} = \nabla \cdot \mathbf{F} = \langle \partial_x, \partial_y, \partial_z \rangle \cdot \langle x, y, 1 \rangle = 1 + 1 + 0 = 2.$$

Note that div $\mathbf{F}$ is constant, so we use geometry to solve for the flux:

$$\iiint_S \operatorname{div} \mathbf{F}\, dV = \iiint_S 2\, dV = 2 \iiint_S dV$$

$$= 2 \begin{pmatrix} \text{volume inside a hemisphere} \\ \text{with radius 2} \end{pmatrix}$$

$$= 2 \left(\frac{1}{2} \left(\frac{4}{3} \pi (2)^3 \right) \right) = \frac{32\pi}{3}.$$

Example 54.2: Find the net flux of $\mathbf{F}(x,y,z) = \langle 2x, y, 4z \rangle$ through a cube in the first octant with vertices $(0,0,0)$, $(1,0,0)$, $(1,1,0)$, $(0,1,0)$, $(0,0,1)$, $(1,0,1)$, $(1,1,1)$ and $(0,1,1)$.   (This is a repeat of Example 53.6)

Solution: Find the divergence of $\mathbf{F}$:

$$\text{div } \mathbf{F} = \nabla \cdot \mathbf{F} = \langle \partial_x, \partial_y, \partial_z \rangle \cdot \langle 2x, y, 4z \rangle$$
$$= 2 + 1 + 4$$
$$= 7.$$

Thus, the flux through the solid cube with side lengths 1 is

$$\iiint_S \text{div } \mathbf{F}\, dV = \iiint_S 7\, dV$$
$$= 7 \iiint_S dV$$
$$= 7(1)^3$$
$$= 7.$$

♦ • ♦ • ♦ • ♦ • ♦

Example 54.3: Find the net flux of $\mathbf{F}(x,y,z) = \langle xy, -y, z \rangle$ through the closed surface composed of the cylinder $x^2 + y^2 = 4$ and the planes, $y = 0$, $z = 1$ and $z = 6$.   (This is a repeat of Example 53.7)

Solution: The divergence of $\mathbf{F}$ is $\nabla \cdot \mathbf{F} = y$, and by the divergence theorem, the net flux through this closed surface is (using polar coordinates with $y = r\sin\theta$):

$$\iiint_S y\, dV = \int_1^6 \int_0^\pi \int_0^2 (r\sin\theta)\, r\, dr\, d\theta\, dz = \int_1^6 \int_0^\pi \int_0^2 r^2 \sin\theta\, dr\, d\theta\, dz.$$

Since the bounds are constants and the integrand is held by multiplication, we evaluate this triple integral as a product of three single-variable integrals:

$$\left(\int_1^6 dz\right)\left(\int_0^\pi \sin\theta\, d\theta\right)\left(\int_0^2 r^2\, dr\right) = (5)(2)\left(\frac{8}{3}\right) = \frac{80}{3}.$$

♦ • ♦ • ♦ • ♦ • ♦

Note the efficiency of the divergence theorem on closed surfaces by comparing the previous three examples with the earlier examples.

Example 54.4: Let $F(x,y,z) = \langle 5x + e^y, \sin(x^2) + 2y, \tan^{-1}(xy) - z \rangle$. Find the flux of F through the closed surface that is a right circular cone, including its base, with base radius of 3 on the yz-plane, and apex at $(4,0,0)$.

Solution: The divergence of F is $\nabla \cdot F = 5 + 2 - 1 = 6$ (you verify). The flux is given by

$$\iiint_S 6 \, dV = 6(\text{volume inside the cone}).$$

The volume of a right circular cone is $\frac{1}{3}\pi r^2 h$. Here, $r = 3$ and $h = 4$ since the apex is 4 units from the yz-plane. Thus, the flux is $6\left(\frac{1}{3}\pi(3)^2(4)\right) = 72\pi$.

◆ • ◆ • ◆ • ◆ • ◆

Example 54.5: Find the net flux of $F(x,y,z) = \langle 3,5,9 \rangle$ through a sphere of radius 1, centered at the origin.

Solution: The divergence of F is $\nabla \cdot F = 0 + 0 + 0 = 0$. Therefore, there is zero net flux through the sphere, or any other closed surface for that matter.

This does not mean that there is zero flux on all faces or sides of the closed surface. It means that equal amounts of matter are entering and leaving through the closed surface per unit time. For example, a four-sided object may have flux figures of 3, 5, 2 and –10 among its four sides. Its net flux is 0.

All constant vector fields F have div $F = 0$. Constant vector fields are incompressible. However, not all incompressible vector fields (those with div $F = 0$) are constant vector fields. Consider the case when $F(x,y,z) = \langle z, x, y \rangle$.

◆ • ◆ • ◆ • ◆ • ◆

55. Stokes Theorem

Recall that Green's Theorem allows us to find the work (as a line integral) performed on a particle around a simple closed loop path C by evaluating a double integral over the interior R that is bounded by the loop:

$$\text{Green's Theorem:} \quad \int_C \mathbf{F} \cdot d\mathbf{r} = \iint_R (N_x - M_y) \, dA.$$

Green's Theorem is restricted to closed loop paths in R^2. What about a closed loop path in R^3? For such paths, we use **Stokes Theorem**, which extends Green's Theorem into R^3.

If $\mathbf{F}(x, y, z) = \langle M(x, y, z), N(x, y, z), P(x, y, z) \rangle$ is a vector field and S is a simple oriented surface in R^3 with a boundary C, then Stokes Theorem is given by

$$\int_C \mathbf{F} \cdot d\mathbf{r} = \iint_S (\text{curl } \mathbf{F}) \cdot \mathbf{n} \, dS.$$

Recall that curl $\mathbf{F}$ is defined by

$$\text{curl } \mathbf{F} = \begin{vmatrix} \mathbf{i} & \mathbf{j} & \mathbf{k} \\ \partial_x & \partial_y & \partial_z \\ M & N & P \end{vmatrix}$$

$$= (P_y - N_z)\mathbf{i} - (P_x - M_z)\mathbf{j} + (N_x - M_y)\mathbf{k}$$

$$= \langle P_y - N_z, M_z - P_x, N_x - M_y \rangle.$$

A positive orientation of the surface is stated such that the path C is traversed counterclockwise. However, in R^3, the notion of counterclockwise can be less intuitive. Thus, a positively oriented surface is one where someone standing "up" would walk the loop with their left arm hanging over the surface.

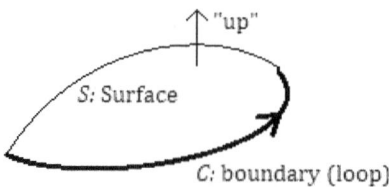

In Green's Theorem, the surface S is the region R in the xy-plane, and "up" is in the positive z direction. Since $\mathbf{F} = \langle M, N, 0 \rangle$ in R^2, then curl $\mathbf{F} = \langle 0, 0, N_x - M_y \rangle$ so that Green's Theorem is a special case of Stokes Theorem in R^2.

The integral $\iint_S (\text{curl } \mathbf{F}) \cdot \mathbf{n} \, dS$ needs to be expanded so that it can be useful. Suppose for now that the surface S is defined by $z = f(x,y)$. From this, we have that $\mathbf{n}$ is a normal vector to f by

$$\mathbf{n} = \frac{\langle f_x, f_y, -1 \rangle}{\sqrt{f_x^2 + f_y^2 + 1}} \quad \text{or} \quad \frac{\langle -f_x, -f_y, 1 \rangle}{\sqrt{f_x^2 + f_y^2 + 1}}.$$

Also, recall that $dS = \sqrt{f_x^2 + f_y^2 + 1} \, dA$. Thus, making substitutions, we have

$$\iint_S (\text{curl } \mathbf{F}) \cdot \mathbf{n} \, dS = \iint_S (\text{curl } \mathbf{F}) \cdot \frac{\langle -f_x, -f_y, 1 \rangle}{\sqrt{f_x^2 + f_y^2 + 1}} \sqrt{f_x^2 + f_y^2 + 1} \, dA.$$

This simplifies to

$$\iint_R (\text{curl } \mathbf{F}) \cdot \langle -f_x, -f_y, 1 \rangle \, dA.$$

The vector $\langle -f_x, -f_y, 1 \rangle$ is chosen depending on what direction "up" is stated. We may choose $\langle f_x, f_y, -1 \rangle$ in certain cases. A useful tactic is to note that if a path C is stated first, we can choose *any* surface f that is bounded by that path. Obviously, we choose "easy" surfaces in such a case.

The usual routine is:

You will be asked to find the value of a line integral $\int_C \mathbf{F} \cdot d\mathbf{r}$ around a simple loop path C in R^3. Path C may be stated explicitly or may be implied by some surface S given by $z = f(x,y)$. You will also be given the vector field $\mathbf{F} = \langle M, N, P \rangle$.

1. Find curl $\mathbf{F}$. For now, it will be in terms of x, y and z.
2. Determine $\langle -f_x, -f_y, 1 \rangle$ or $\langle f_x, f_y, -1 \rangle$, depending on the context. Usually, the first version is used because we can always declare that positive z is "up".
3. Find $(\text{curl } \mathbf{F}) \cdot \langle -f_x, -f_y, 1 \rangle$. If variable z remains, substitute with $z = f(x,y)$. You now have an expression in terms of x and y.
4. Determine the region of integration R, which will be the footprint cast by S onto the xy-plane.
5. Integrate the result in step (3) over region R.

Normal adjustments would be made, *e.g.* if the surface was stated as $x = f(y,z)$.

Example 55.1: Find $\int_C \mathbf{F} \cdot d\mathbf{r}$, where $\mathbf{F}(x,y,z) = \langle xy, x+y+z, x^2 \rangle$ and C is a circle of radius 1, centered at the origin, in the xy-plane, traverse counterclockwise where "up" is the positive z direction.

Solution: No surface S is specified, just a boundary path C. So let's try a couple different surfaces that have C as its boundary. First, we will let S be the interior of the circle in the xy-plane. That is, $z = f(x,y) = 0$. Thus, $\mathbf{n} = \langle 0,0,1 \rangle$.

Next, we find curl $\mathbf{F}$:

$$\text{curl } \mathbf{F} = \langle P_y - N_z, M_z - P_x, N_x - M_y \rangle = \langle -1, -2x, 1-x \rangle.$$

Thus, (curl $\mathbf{F}$) $\cdot \mathbf{n} = 1 - x$. This is integrated over the region inside the circle of radius 1, centered at the origin. We use polar coordinates, where $x = r \cos \theta$:

$$\iint_S (\text{curl } \mathbf{F}) \cdot \mathbf{n} \, dS = \iint_S (1-x) \, dA$$
$$= \int_0^{2\pi} \int_0^1 (1 - r \cos \theta) \, r \, dr \, d\theta$$
$$= \int_0^{2\pi} \int_0^1 (r - r^2 \cos \theta) \, dr \, d\theta.$$

We have

$$\int_0^1 (r - r^2 \cos \theta) \, dr = \left[\frac{1}{2} r^2 - \frac{1}{3} r^3 \cos \theta \right]_0^1 = \frac{1}{2} - \frac{1}{3} \cos \theta.$$

Then, we have

$$\int_0^{2\pi} \left(\frac{1}{2} - \frac{1}{3} \cos \theta \right) d\theta = \left[\frac{1}{2} \theta - \frac{1}{3} \sin \theta \right]_0^{2\pi} = \pi.$$

Therefore, with S as the portion of the xy-plane inside the circle of radius 1 centered at the origin, we have

$$\int_C \mathbf{F} \cdot d\mathbf{r} = \pi.$$

Let's try a different surface: Let S be the paraboloid $z = f(x,y) = 1 - x^2 - y^2$ that lies above the xy-plane. Note that C is the same bounding curve. We find $\mathbf{n}$:

$$\mathbf{n} = \langle -f_x, -f_y, 1 \rangle = \langle -(-2x), -(-2y), 1 \rangle = \langle 2x, 2y, 1 \rangle.$$

The curl **F** has not changed. Thus,

$$(\text{curl } \mathbf{F}) \cdot \mathbf{n} = \langle -1, -2x, 1 - x \rangle \cdot \langle 2x, 2y, 1 \rangle = -3x - 4xy + 1.$$

The region of integration is the same—the interior of the circle of radius 1, centered at the origin. Once again, we use polar coordinates:

$$\iint_S (\text{curl } \mathbf{F}) \cdot \mathbf{n} \, dS = \iint_S (-3x - 4xy + 1) \, dA$$

$$= \int_0^{2\pi} \int_0^1 (-3r \cos \theta - 4(r \cos \theta)(r \sin \theta) + 1) \, r \, dr \, d\theta$$

$$= \int_0^{2\pi} \int_0^1 (-3r^2 \cos \theta - 4r^3 \cos \theta \sin \theta + r) \, dr \, d\theta.$$

The inside integral, evaluated with respect to r, is

$$\int_0^1 (-3r^2 \cos \theta - 4r^3 \cos \theta \sin \theta + r) \, dr \, d\theta$$

$$= \left[-r^3 \cos \theta - r^4 \cos \theta \sin \theta + \frac{1}{2} r^2 \right]_0^1$$

$$= -\cos \theta - \cos \theta \sin \theta + \frac{1}{2}.$$

Then this is integrated with respect to θ:

$$\int_0^{2\pi} \left(-\cos \theta - \cos \theta \sin \theta + \frac{1}{2} \right) d\theta = \left[-\sin \theta - \frac{1}{2} \sin^2 \theta + \frac{1}{2} \theta \right]_0^{2\pi} = \pi.$$

Note that at $\theta = 0$ and 2π, the sine terms vanish. Thus, we get the same result, $\int_C \mathbf{F} \cdot d\mathbf{r} = \pi$.

Try this with another surface, for example, the hemisphere of radius 1, $z = \sqrt{1 - x^2 - y^2}$.

Example 55.2: Find $\int_C \mathbf{F} \cdot d\mathbf{r}$, where $\mathbf{F}(x,y,z) = \langle x+y, zy, 3x \rangle$ and C is the triangle traversed from $(4,0,0)$ to $(0,6,0)$ to $(0,0,12)$, back to $(4,0,0)$. Assume "up" is in the direction of positive z.

Solution: Since no surface is specified, let's use a plane passing through the vertices of the triangle. Below is an image of the path C and the eventual region of integration R:

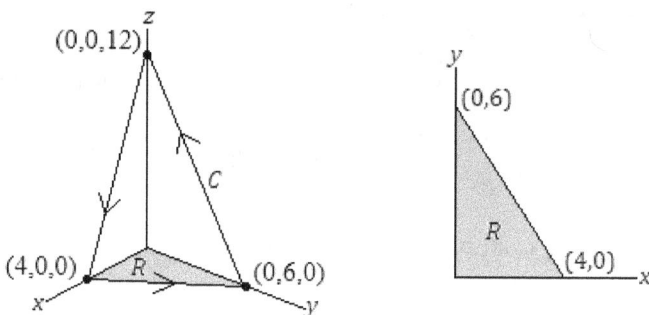

The plane is $\frac{x}{4} + \frac{y}{6} + \frac{z}{12} = 1$, or $3x + 2y + z = 12$ when fractions are cleared. We can read off a normal vector from the plane's equation: $\mathbf{n} = \langle 3,2,1 \rangle$. This is a useful vector since it has a 1 in the z position, agreeing with the upward direction. We find curl $\mathbf{F}$, which is $\langle -y, -3, -1 \rangle$. Thus,

$$\iint_S (\text{curl } \mathbf{F}) \cdot \mathbf{n} \, dS = \iint_R (-3y - 7) \, dA = \int_0^4 \int_0^{6-(3/2)x} (-3y - 7) \, dy \, dx.$$

The inside integral is

$$\int_0^{6-(3/2)x} (-3y - 7) \, dy = \left[-\frac{3}{2} y^2 - 7y \right]_0^{6-(3/2)x}$$

$$= -\frac{3}{2}\left(6 - \frac{3}{2}x\right)^2 - 7\left(6 - \frac{3}{2}x\right)$$

$$= -\frac{27}{8}x^2 + \frac{75}{2}x - 96.$$

The outside integral is

$$\int_0^4 \left(-\frac{27}{8}x^2 + \frac{75}{2}x - 96\right) dx = \left[-\frac{9}{8}x^3 + \frac{75}{4}x^2 - 96x\right]_0^4 = -156.$$

Therefore, $\int_C \mathbf{F} \cdot d\mathbf{r} = -156$. Let's verify this by finding the line integral along each segment of the triangle.

From (4,0,0) to (0,6,0), we have $\mathbf{r}(t) = \langle 4 - 4t, 6t, 0 \rangle$ for $0 \leq t \leq 1$, so that $d\mathbf{r} = \langle -4, 6, 0 \rangle$. Meanwhile,

$$\mathbf{F}(t) = \langle x + y, zy, 3x \rangle$$
$$= \langle (4-4t) + (6t), (0)(6t), 3(4-4t) \rangle \quad \begin{cases} x = 4 - 4t \\ y = 6t \\ z = 0 \end{cases}$$

or after simplification, $\mathbf{F}(t) = \langle 4 + 2t, 0, 12 - 12t \rangle$. Thus, the dot product is $\mathbf{F} \cdot d\mathbf{r} = -4(4 + 2t) = -16 - 8t$, and the line integral is

$$\int_0^1 (-16 - 8t)\, dt = [-16t - 4t^2]_0^1 = -20.$$

From (0,6,0) to (0,0,12), we have $\mathbf{r}(t) = \langle 0, 6 - 6t, 12t \rangle$ for $0 \leq t \leq 1$, so that $d\mathbf{r} = \langle 0, -6, 12 \rangle$. Also, $\mathbf{F}(t) = \langle 6 - 6t, 72t - 72t^2, 0 \rangle$ after simplification.

Thus, $\mathbf{F} \cdot d\mathbf{r} = -6(72t - 72t^2) = -432(t - t^2)$, and the line integral is

$$\int_0^1 -432(t - t^2)\, dt = -432 \left[\frac{1}{2}t^2 - \frac{1}{3}t^3 \right]_0^1 = -72$$

From (0,0,12) to (4,0,0), we have $\mathbf{r}(t) = \langle 4t, 0, 12 - 12t \rangle$ for $0 \leq t \leq 1$. This gives $d\mathbf{r} = \langle 4, 0, -12 \rangle$. Also, $\mathbf{F}(t) = \langle 4t, 0, 12t \rangle$ after simplification. Therefore,

$$\mathbf{F} \cdot d\mathbf{r} = 4(4t) - 12(12t) = 16t - 144t = -128t.$$

Finally, the line integral is

$$\int_0^1 -128t\, dt = -128 \left[\frac{1}{2}t^2 \right]_0^1 = -64.$$

The sum of these three line integrals is $-20 - 72 - 64 = -156$, agreeing with the result found by Stokes Theorem.

◆ • ◆ • ◆ • ◆ • ◆

See an error? Have a suggestion?
Please see www.surgent.net/vcbook

Test Yourself

These are problems you might see in a vector calculus course. They are general questions and are meant for practice. The key follows, but only with the answers. Can you "fill in the blanks" between the question and answer?

Let vectors $\mathbf{u} = \langle 2,1,-3 \rangle$, $\mathbf{v} = \langle 5,4,2 \rangle$ and $\mathbf{w} = \langle -4,1,6 \rangle$. For Exercises 1-20, find each of the following. If the answer does not exist, explain why.

1. $\mathbf{u} \cdot \mathbf{v}$

2. $\mathbf{v} \cdot \mathbf{w}$

3. $(\mathbf{u} \cdot \mathbf{w})\mathbf{v}$

4. $\mathbf{v} \times \mathbf{w}$

5. $\mathbf{u} + \mathbf{v} \cdot \mathbf{w}$

6. $\mathbf{u} \cdot \mathbf{v} \cdot \mathbf{w}$

7. $(\mathbf{u} \times \mathbf{v}) \times \mathbf{w}$

8. $(\mathbf{v} \cdot \mathbf{u}) \times \mathbf{w}$

9. $2\mathbf{u} + 3\mathbf{v} - \mathbf{w}$

10. $|\mathbf{u}|$

11. The angle in degrees between $\mathbf{u}$ and $\mathbf{w}$.

12. A vector parallel to $\mathbf{v}$, but of length 2.

13. A vector parallel to $\mathbf{w}$, pointed in the opposite direction, of length 3.

14. The projection of $\mathbf{u}$ onto $\mathbf{v}$.

15. The projection of $\mathbf{v}$ onto $\mathbf{w}$.

16. The equation of the plane that contains $\mathbf{u}$ and $\mathbf{v}$ and includes the point (2,0,1).

17. The acute angle that vector $\mathbf{w}$ makes with the plane from Exercise 16.

18. The area of the parallelogram formed by $\mathbf{u}$ and $\mathbf{v}$.

19. The area of the triangle formed by **v** and **w**.

20. Are **u** and **v** acute or obtuse?

21. A sphere has points (2, 5, 3) and (7, 9, 10) directly opposite one another. Find the equation of this sphere.

22. A sphere has center (4,7,2). Find the equation of the sphere with the largest possible radius such that it is fully contained within the first octant.

23. Find the radius of the circle that the sphere $(x-1)^2 + (y+2)^2 + (z-4)^2 = 25$ makes when it intersects the xy plane.

24. Find the equation of the line in R^3 that passes through point (4, 2, –6) and is parallel to $\mathbf{u} = \langle 2, -1, 5 \rangle$.

25. The points $A = (1,0,2)$, $B = (4,1,1)$, $C = (6,3,1)$ and $D = (10,5,4)$. Find the distance from point D to the plane formed by A, B and C.

26. Find the equation of the plane made by points $A = (1,0,2)$, $B = (4,1,1)$ and $C = (6,3,1)$.

27. Find the equation of the plane passing through (4,0,0), (0,5,0) and (0,0,–8).

28. If vector **u** points north and **v** points southeast, then **u** × **v** points which way?

29. Find the acute angle that the planes $x - 2y + 3z = 6$ and $2x + y - 7z = 0$ meet.

30. Let $\mathbf{r}(t) = \langle t^2, 2t - 3, 4t - t^3 \rangle$ be a curve in space traced out by an object. Find $\mathbf{v}(3)$ and $\mathbf{a}(3)$, then find the object's speed at $t = 5$.

31. Find the equation of the tangent line to **r** in the previous exercise, in parametric form, when $t = 1$.

32. Find the arc length of $\mathbf{r}(t) = \langle t, 3 \sin t, 3 \cos t \rangle$ for $0 \le t \le 2\pi$.

33. Find the arc length of $\mathbf{r}(t) = \langle t^3, t \rangle$ for $-1 \le t \le 2$.

34. Find the curvature of $y = x^3$ at $x = 1$.

35. Let $\mathbf{a}(t) = \langle 0, 2, t \rangle$, $\mathbf{v}(0) = \langle 1,2,1 \rangle$ and $\mathbf{r}(0) = \langle -3,0,2 \rangle$. Find $\mathbf{r}(1)$.

36. You start at (5,1) and start walking toward (1,6), intending to visit your friend's home at (3,7). If you are allowed one right-angle turn, find the point

at which you should make this turn so as to arrive at your friend's place in time for cartoons.

37. Find the coordinate of A in the diagram below.

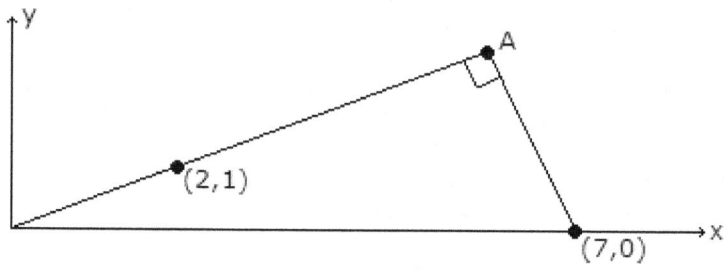

38. A force of 100 N hangs in the center of a cable between two anchor points that are the same height off the ground and 30 m apart horizontally. The object creates a sag of 2 m in the cable. Find the force exerted by each end of the cable on its anchor point.

39. A rock is propelled off a pedestal that is 10 meters off the level ground. The rock leaves the pedestal with a speed of 18 meters per second at an angle above the horizontal of 20 degrees. How high does the rock get, and how far downrange from the pedestal does the rock land?

40. Find the domain of $f(x,y) = \frac{1}{\sqrt{2x-y}}$.

41. Find the domain of $g(x,y) = \ln(x^2 + 4y)$.

42. Find $\lim_{(x,y) \to (0,0)} \left(\frac{2x+y}{x+2y}\right)$.

43. Find $\lim_{(x,y) \to (0,0)} \left(\frac{x+y+1}{x^2+y^2+1}\right)$.

44. Find all first and second-order partial derivatives of $f(x,y,z) = x^2 y^3 z^5$.

45. Find the slope of the tangent line to $f(x,y) = x^3 y - y^2$ when $x = 1$ and $y = 2$, in the direction of $x = 4$ and $y = 3$.

46. Find the direction of the steepest slope of $g(x,y) = \frac{x}{y^2}$ at $x = 3$ and $y = -2$. Then find the slope.

47. Find a vector normal to the surface $h(x,y) = 3xy + y^2 - x^3$ at $x = 4$ and $y = 2$.

48. Find the equation of the tangent plane to $h(x,y) = 3xy + y^2 - x^3$ at $x = 4$ and $y = 2$.

49. Find the equation of the tangent plane to $f(x,y) = 2x^2y + xy^2$ at $(2,1)$ and use it to estimate the value of $f(2.1, 0.9)$.

50. Larry is measuring a circular cylinder. He measures the height to be 8 m, with a tolerance of ±2 cm, and the radius as 3 m, with a tolerance of ±3 cm. Using differentials, find the approximate range of tolerance in the volume of this solid.

51. Let $g(x,y) = x^3 + y^3 - 3x - 12y + 1$. Find all critical points and classify them as min, max or saddle points.

52. Find all absolute minimum and maximum points on the surface $g(x,y) = xy + y^2 - x$ over the region R in the xy-plane bounded by the triangle with vertices (4,4), (−4,−4) and (−4,4).

53. Find the point on the surface $3x + 2y + z = 10$ closest to the origin.

54. Find the largest box by volume such that one vertex lies on the origin, its sides are parallel (or on) the x-, y- and z-axes, and one corner lies on the plane $x + 2y + 5z = 20$.

Use the contour map below of $z = f(x,y)$ to answer Questions 55-57. Assume C is a saddle, D is a minimum and A is a maximum, and that the surface is everywhere differentiable.

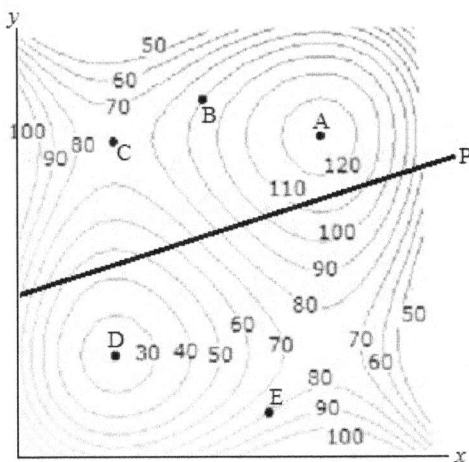

55. Write in the sign (+, −, 0) for the following partial derivatives.

$$f_x(A) = \underline{} \quad f_x(B) = \underline{} \quad f_x(E) = \underline{}$$
$$f_y(A) = \underline{} \quad f_y(B) = \underline{} \quad f_y(E) = \underline{}$$

56. Approximate the minimum and maximum values of f constrained to path P.

57. On the map, draw in the gradient vector at E.

58. Let $f(x,y) = 3\sin(x^2 - 4y)$. Find ∇f.

59. Let $y = f(x(s,t), y(s,t))$, and suppose that $\frac{\partial f}{\partial t} = 10, \frac{\partial f}{\partial x} = 2, \frac{\partial f}{\partial y} = 1, \frac{\partial x}{\partial s} =$
$-3, \frac{\partial x}{\partial t} = 4$ and $\frac{\partial y}{\partial s} = 5$. Find the value of $\frac{\partial y}{\partial t}$.

60. Find $\frac{dx}{dy}$ where $x^2 y^3 - 2xy^2 = 5x^3 - 2y$.

61. Find the volume between $z = 2e^{x^2+y^2}$ and the quarter-circle region in the xy-plane centered at the origin with radius 1.

62. Let $z = f(x,y) = 2x^2 + xy - \frac{1}{4}y^4 + 4x - 2y + 1$. Find the equation of the tangent plane at $x_0 = -2$ and $y_0 = 3$, then use it to estimate $f(-2.05, 3.1)$.

63. Find $\int_{-1}^{2} \int_{3}^{4} x(1-y)\, dy\, dx$.

64. Rewrite as a single double-integral: $\int_0^2 \int_0^{2x} xy\, dy\, dx + \int_2^6 \int_0^{6-x} xy\, dy\, dx$.

65. Find the volume below $f(x,y) = \sqrt{1 - x^2 - y^2}$ over the quarter-circle region in the xy-plane centered at the origin with radius 1.

66. Find the volume contained below the paraboloid $z = 4 - x^2 - y^2$ and above the xy-plane.

67. Reverse the order of integration: $\int_0^9 \int_{-\sqrt{9-y}}^{\sqrt{9-y}} dx\, dy$.

68. Evaluate $\int_0^9 \int_{-\sqrt{9-y}}^{\sqrt{9-y}} dx\, dy$.

69. Reverse the order of integration of $\int_0^4 \int_{-\sqrt{x}}^{\sqrt{x}} f(x,y)\, dy\, dx + \int_4^9 \int_{x-6}^{\sqrt{x}} f(x,y)\, dy\, dx$.

70. A transformation is given by $u = 2x + y$ and $v = -3x - 2y$. Find $J(u, v)$.

71. Evaluate $\iint_R (x - y) \, dA$, where R is the parallelogram with vertices $(0,0)$, $(5,2)$, $(7,5)$ and $(2,3)$.

72. Evaluate $\iint_R 2xy \, dA$, where R is the region in the xy-plane such that $x + 2y \leq 4$, $x \geq 0$ and $y \geq 0$.

73. Evaluate $\int_0^{\sqrt{2}/2} \int_{-x}^{x} (x^2 + y^2) \, dy \, dx + \int_{\sqrt{2}/2}^{1} \int_{-\sqrt{1-x^2}}^{\sqrt{1-x^2}} (x^2 + y^2) \, dy \, dx$

74. Evaluate $\int_{-4}^{-3} \int_0^{\sqrt{16-x^2}} f(x,y) \, dy \, dx + \int_{-3}^{3} \int_{\sqrt{9-x^2}}^{\sqrt{16-x^2}} f(x,y) \, dy \, dx + \int_3^4 \int_0^{\sqrt{16-x^2}} f(x,y) \, dy \, dx$, where $f(x,y) = x^2 + y^2$.

75. Suppose region E is between two hemispheres of radius 2 and radius 5 above the xy-plane, centered at the origin. Set us and evaluate $\iiint_E x^2 + y^2 + z^2 \, dV$.

76. Set up an integral and find the volume contained in the solid bounded by the xy-plane, the plane $z = x$, the paraboloid $x = 9 - y^2$ such that x is positive.

77. Find the volume within the region bounded by $z = x^2 + y^2$ and $z = 32 - x^2 - y^2$.

78. Find the volume of the solid bounded by $x = 0$, $y = 0$, $z = 0$, the cylinder $y^2 + z^2 = 9$ and the plane $x + y = 3$.

79. Set up and evaluate $\iiint_E dV$ where E is the tetrahedron with vertices $(0,0,0)$, $(2,0,0)$, $(0,3,0)$ and $(0,0,6)$.

80. Convert the rectangular coordinate $(2, -2, 5)$ into spherical coordinates (ρ, θ, φ).

81. A solid is bounded below by a circular cone (vertex at the origin) and above by a sphere (center at the origin) such that $(2,1,5)$ lies on the rim where the cone and sphere intersect. Find its volume.

82. Find $\int_C f(x,y) \, ds$ where $f(x,y) = x^2 + 3y$ and C is the straight line from $(1,2)$ to $(3,1)$.

83. Find $\int_C f(x,y) \, ds$ where $f(x,y) = xy^2$ and C is the path along $y = x^3$ from $(0,0)$ to $(2,8)$.

84. Find $\int_C f(x,y)\, ds$ where $f(x,y) = 2x + y^2$ and C is the portion of a circle of radius 1, centered at the origin, starting at (1,0) and ending at (0,1) in the first quadrant.

85. A particle follows a straight-line path from (1,2) to (5,7) within the vector field $F(x,y) = \langle xy, y^2 \rangle$. Find the work. (That is, find $\int_C \mathbf{F} \cdot d\mathbf{r}$ where C is the path of the particle.)

86. Show that $\mathbf{F}(x,y) = \langle 6x + 5y, 5x + 4 \rangle$ is conservative, then find $f(x,y)$ such that $\nabla f = \mathbf{F}$.

87. Find $\int_C \mathbf{F} \cdot d\mathbf{r}$, where $\mathbf{F}(x,y) = \langle 4xy^3, 6x^2y^2 \rangle$ and C is a sequence of straight lines from (0,0) to (1,3) to (4,7) to (9,5) to (2,1).

88. Find $\int_C \mathbf{F} \cdot d\mathbf{r}$, where $\mathbf{F}(x,y) = \langle 3y, -2x \rangle$ and C is a path starting at (0,0) to (4,0) to (4,4) back to (0,0).

89. Find $\int_C \mathbf{F} \cdot d\mathbf{r}$, where $\mathbf{F}(x,y) = \langle 10y, 12x \rangle$ and C is a circle of radius 4 centered at the origin traced clockwise.

90. Find $\int_C \mathbf{F} \cdot d\mathbf{r}$, where $\mathbf{F}(x,y) = \langle \sin y, x \cos y \rangle$ and C is an ellipse centered at (5,4) with minor axis 7 and major axis 4.

91. Find $\int_C \mathbf{F} \cdot \mathbf{n}\, ds$, where $\mathbf{F}(x,y) = \langle 3x, x - 2y \rangle$ and C is a straight line from (3,0) to (0,5).

92. Find $\int_C \mathbf{F} \cdot \mathbf{n}\, ds$, where $\mathbf{F}(x,y) = \langle x^2, 6xy \rangle$ and C is the path along $y = x^2$ from (1,1) to (4,16).

93. Find $\int_C \mathbf{F} \cdot \mathbf{n}\, ds$, where $\mathbf{F}(x,y) = \langle 2x, x + y^2 \rangle$ and C is a sequence of straight-line segments from (0,0) to (5,0) to (5,3) to (0,3) to (0,0).

94. Find the surface area of the paraboloid $z = x^2 + y^2$ bounded above by the plane $z = 10$.

95. Find the surface area of the cone $z = 2\sqrt{x^2 + y^2}$ for $1 \le z \le 8$.

96. Find $\iint_S x\, dS$, where S is the plane in the first octant with axis-intercepts (0,0,2), (0,5,0) and (10,0,0).

97. Find the work done by $\mathbf{F}(x,y) = \langle z, y, x+y \rangle$ along a triangle from $(1,0,0)$ to $(0,3,0)$ to $(0,0,5)$ back to $(1,0,0)$. Assume positive direction is in the positive z-direction.

98. Find the flux of $\mathbf{F}(x,y) = \langle z, y, x+y \rangle$ through the simply-closed tetrahedron bounded by the points $(0,0,0)$, $(1,0,0)$, $(0,3,0)$ and $(0,0,5)$.

99. Find the flux of $\mathbf{F}(x,y) = \langle x^2, y+2z, z \rangle$ through the cube with opposite corners $(0,0,0)$ and $(2,2,2)$.

100. Vector field $\mathbf{F}(x,y) = \langle z, x^2, e^y \rangle$ flows through a simply-closed cylindrical solid. If the flux through one end is 10, and the flux through the side is 15, find the flux through the other end.

Answers to "Test Yourself"

1. 8
2. −4
3. $\langle -125, -100, -50 \rangle$
4. $\langle 22, -38, 21 \rangle$
5. Impossible, $\mathbf{v} \cdot \mathbf{w}$ is a scalar, $\mathbf{u}$ a vector, vector-to-scalar addition is not defined.
6. Impossible, $\mathbf{u} \cdot \mathbf{v}$ is a scalar, and the dot product of a scalar to a vector is not defined.
7. $\langle -117, -96, -62 \rangle$
8. Impossible, $\mathbf{v} \cdot \mathbf{u}$ is a scalar, and the cross product of a scalar to a vector is not defined.
9. $\langle 23, 13, -6 \rangle$
10. $\sqrt{14}$
11. 156.6 degrees
12. $\left\langle \frac{10}{\sqrt{45}}, \frac{8}{\sqrt{45}}, \frac{4}{\sqrt{45}} \right\rangle$
13. $\left\langle \frac{12}{\sqrt{53}}, -\frac{3}{\sqrt{53}}, -\frac{18}{\sqrt{53}} \right\rangle$
14. $\left\langle \frac{8}{9}, \frac{32}{45}, \frac{16}{45} \right\rangle$
15. $\left\langle \frac{16}{53}, -\frac{4}{53}, -\frac{24}{53} \right\rangle$
16. $14x - 19y + 3z = 31$
17. 19.214 degrees
18. $\sqrt{566} \approx 23.79$ square units
19. $\frac{1}{2}\sqrt{2369} \approx 24.34$ square units
20. Acute, since the dot product is positive.

21. $\left(x-\frac{9}{2}\right)^2 + (y-7)^2 + \left(z-\frac{13}{2}\right)^2 = \frac{45}{2}$
22. $(x-4)^2 + (y-7)^2 + (z-2)^2 = 4$
23. $r = 3$
24. $x(t) = 4 + 2t$, $y(t) = 2 - t$, $z(t) = -6 + 5t$
25. $\frac{4}{3}\sqrt{6} \approx 3.266$ units
26. $x - y + 2z = 5$
27. $\frac{x}{4} + \frac{y}{5} - \frac{z}{8} = 1$ or $10x + 8y - 5z = 40$
28. Into the page.
29. 40.2 degrees.
30. $\mathbf{v}(3) = \langle 6,2,-23 \rangle$; $\mathbf{a}(3) = \langle 2,0,-18 \rangle$, $|\mathbf{v}(5)| = |\langle 10,2,-71 \rangle| = \sqrt{5145} \approx 71.73$ units/time.
31. When $t = 1$, the position is $\mathbf{r}(1) = \langle 1,-1,3 \rangle$, which can be treated as a point $(1, -1, 3)$, and the velocity (tangent) vector is $\mathbf{v}(1) = \langle 2,2,1 \rangle$. The tangent line is $\langle 1 + 2t, -1 + 2t, 3 + t \rangle$.
32. $\sqrt{10} \cdot 2\pi$ units.
33. About 10.178 units.
34. $6/10^{3/2} \approx 0.19$.
35. $r(1) = \left\langle -2, 3, \frac{19}{6} \right\rangle$
36. $x = 5 - \frac{152}{41} \approx 1.293$, $y = 1 + \frac{190}{41} \approx 5.634$
37. $(5.6, 2.8)$
38. About 378 N.
39. The rock reaches its highest point at $t = 0.628$ seconds, with a height of 11.93 meters, and the rock lands $t = 2.189$ seconds after being released, with a horizontal distance of 37.02 meters.
40. $\{(x,y)|y < 2x\}$
41. $\left\{(x,y)\big|y > -\frac{1}{4}x^2\right\}$
42. Does not exist
43. 1
44. $f_x = 2xy^3z^5$, $f_y = 3x^2y^2z^5$, $f_z = 5x^2y^3z^4$, $f_{xx} = 2y^3z^5$, $f_{xy} = 6xy^2z^5$, $f_{xz} = 10xy^3z^4$, $f_{yx} = 6xy^2z^5$, $f_{yy} = 6x^2yz^5$, $f_{yz} = 15x^2y^2z^4$, $f_{zx} = 10xy^3z^4$, $f_{zy} = 15x^2y^2z^4$, $f_{zz} = 20x^2y^3z^3$
45. $\frac{15}{\sqrt{10}} \approx 4.743$
46. Direction: $\left(\frac{1}{4}, \frac{3}{4}\right)$, slope: $\frac{\sqrt{10}}{4} \approx 0.791$
47. $\langle -42, 16, -1 \rangle$ or any non-zero multiple.
48. $42x - 16y + z = 100$
49. $9x + 12y - z = 20$ or $z = 9x + 12y - 20$; $f(2.1, 0.9) \approx 9.7$.

50. $dV = \pm 5.089$ cubic meters
51. $(1,2,-17)$ min, $(-1,2,-13)$ saddle, $(1,-2,15)$ saddle, $(-1,-2,19)$ max
52. Absolute maximum: $(-4,-4,36)$, absolute minimum: $\left(\frac{1}{4},\frac{1}{4},-\frac{1}{8}\right)$
53. $\left(\frac{15}{7},\frac{10}{7},\frac{5}{7}\right)$
54. $800/27 \approx 29.629$ cubic units.
55. $f_x(A) = 0 \quad f_x(B) = + \quad f_x(E) = +$
    $f_y(A) = 0 \quad f_y(B) = - \quad f_y(E) = -$
56. Minimum about 45, maximum about 115

57.
    (Orthogonal to the contour, positive slope.)
58. $\nabla f = \langle 6x\cos(x^2 - 4y), -12\cos(x^2 - 4y)\rangle$
59. Starting with $\frac{\partial f}{\partial t} = \frac{\partial f}{\partial x}\frac{\partial x}{\partial t} + \frac{\partial f}{\partial y}\frac{\partial y}{\partial t}$ we have $10 = (2)(4) + (1)\frac{\partial y}{\partial t}$ which gives $\frac{\partial y}{\partial t} = 10 - 8 = 2$.
60. $\frac{dx}{dy} = -\frac{F_y}{F_x} = \frac{4xy - 3x^2y^2 - 2}{2xy^3 - 2y^2 - 15x^2}$
61. $\frac{\pi}{2}(e - 1) \approx 2.7$ square units
62. The equation of the tangent plane is given by $z + 31.25 = -(x + 2) - 31(y - 3)$ or by $x + 31y + z = 59.75$, and $f(-2.05, 3.1) \approx -34.3$
63. $-15/4$
64. $\int_0^4 \int_{y/2}^{6-y} xy\, dx\, dy$
65. $\pi/6$
66. 8π
67. $\int_{-3}^{3} \int_0^{9-x^2} dy\, dx$
68. 36
69. $\int_{-2}^{3} \int_{y^2}^{y+6} f(x,y)\, dx\, dy$
70. 1
71. 11
72. $16/3$
73. $\pi/8$
74. $175\pi/4$
75. $6186\pi/5$
76. $648/5$
77. 256π
78. $27\pi/4 - 9$
79. $\int_0^2 \int_0^{3-\frac{3}{2}x} \int_0^{6-3x-2y} dz\, dy\, dx = 6$.

80. $(\sqrt{33}, 7\pi/4, 0.5148)$
81. $20\pi(\sqrt{30} - 5)$
82. $53\sqrt{5}/6$
83. Approximately 308.97
84. $2 + \pi/4$
85. $517/3$
86. $M_y = 5 = N_x$, and $f(x, y) = 3x^2 + 5xy + 4y$
87. 8
88. -40
89. -32π
90. 0
91. 12
92. -255
93. 75
94. $\frac{\pi}{6}(41^{3/2} - 1) \approx 136.94$ square units
95. $63\sqrt{5}\pi/4 \approx 110.64$ square units
96. $250\sqrt{30}/3$
97. $15/2$
98. $5/2$
99. 32
100. -25

If you see an error, please visit www.surgent.net/vcbook

Thank you!

Appendix

Proof of the Dot Product, $\mathbf{u} \cdot \mathbf{v} = |\mathbf{u}||\mathbf{v}| \cos \theta$

Let $\mathbf{u} = \langle u_1, u_2 \rangle$ and $\mathbf{v} = \langle v_1, v_2 \rangle$ be two vectors in R^2. The dot product of $\mathbf{u}$ and $\mathbf{v}$ is denoted $\mathbf{u} \cdot \mathbf{v}$ and is defined by the formulas

$$\mathbf{u} \cdot \mathbf{v} = u_1 v_1 + u_2 v_2 \quad \text{and} \quad \mathbf{u} \cdot \mathbf{v} = |\mathbf{u}||\mathbf{v}| \cos \theta,$$

where θ is the angle between $\mathbf{u}$ and $\mathbf{v}$ when their feet are placed together.

There is also the property relating a vector's magnitude to the dot product, $|\mathbf{u}|^2 = \mathbf{u} \cdot \mathbf{u}$.

The proof that $\mathbf{u} \cdot \mathbf{v} = |\mathbf{u}||\mathbf{v}| \cos \theta$ uses the Law of Cosines. Given any triangle with side lengths a, b and c, and let θ be the angle between sides a and b, opposite side c, as shown below.

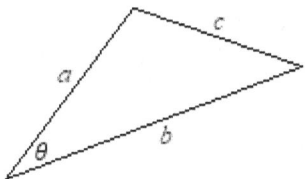

The Law of Cosines relates the three side lengths and the angle θ by the formula

$$c^2 = a^2 + b^2 - 2ab \cos \theta.$$

Replace a with vector $\mathbf{u}$, and b with vector $\mathbf{v}$. Thus, the length of vector $\mathbf{u}$ is $|\mathbf{u}|$ and the length of vector $\mathbf{v}$ is $|\mathbf{v}|$. Side c is then the difference of $\mathbf{u}$ and $\mathbf{v}$, so that the length of c is given by $|\mathbf{u} - \mathbf{v}|$. Making replacements into the Law of Cosines, we have

$$|\mathbf{u} - \mathbf{v}|^2 = |\mathbf{u}|^2 + |\mathbf{v}|^2 - 2|\mathbf{u}||\mathbf{v}| \cos \theta.$$

Using the property $|\mathbf{u}|^2 = \mathbf{u} \cdot \mathbf{u}$, we have

$$(\mathbf{u} - \mathbf{v}) \cdot (\mathbf{u} - \mathbf{v}) = \mathbf{u} \cdot \mathbf{u} + \mathbf{v} \cdot \mathbf{v} - 2|\mathbf{u}||\mathbf{v}| \cos \theta.$$

Distribute on the left side:

$$\mathbf{u} \cdot \mathbf{u} - 2\mathbf{u} \cdot \mathbf{v} + \mathbf{v} \cdot \mathbf{v} = \mathbf{u} \cdot \mathbf{u} + \mathbf{v} \cdot \mathbf{v} - 2|\mathbf{u}||\mathbf{v}| \cos \theta.$$

Terms cancel, leaving

$$-2\mathbf{u} \cdot \mathbf{v} = -2|\mathbf{u}||\mathbf{v}| \cos \theta.$$

Dividing by –2 gives the desired result:

$$\mathbf{u} \cdot \mathbf{v} = |\mathbf{u}||\mathbf{v}| \cos \theta.$$

◆ • ◆ • ◆ • ◆ • ◆ •

Proof of the Magnitude of the Cross Product, $|\mathbf{u} \times \mathbf{v}| = |\mathbf{u}||\mathbf{v}| \sin \theta$

Let $\mathbf{u} = \langle u_1, u_2, u_3 \rangle$ and $\mathbf{v} = \langle v_1, v_2, v_3 \rangle$ be two vectors in R^3. The cross product of $\mathbf{u}$ and $\mathbf{v}$ is denoted $\mathbf{u} \times \mathbf{v}$ and is defined by the formula

$$\mathbf{u} \times \mathbf{v} = \begin{vmatrix} \mathbf{i} & \mathbf{j} & \mathbf{k} \\ u_1 & u_2 & u_3 \\ v_1 & v_2 & v_3 \end{vmatrix}$$

$$= \begin{vmatrix} u_2 & u_3 \\ v_2 & v_3 \end{vmatrix} \mathbf{i} - \begin{vmatrix} u_1 & u_3 \\ v_1 & v_3 \end{vmatrix} \mathbf{j} + \begin{vmatrix} u_1 & u_2 \\ v_1 & v_2 \end{vmatrix} \mathbf{k}$$

$$= \langle u_2 v_3 - u_3 v_2, \; u_3 v_1 - u_1 v_3, \; u_1 v_2 - u_2 v_1 \rangle.$$

The dot product is defined two ways:

Geometrically: $\mathbf{u} \cdot \mathbf{v} = |\mathbf{u}||\mathbf{v}| \cos \theta$, where θ is the angle formed when the feet of $\mathbf{u}$ and $\mathbf{v}$ are placed together.

Formula: $\mathbf{u} \cdot \mathbf{v} = u_1 v_1 + u_2 v_2 + u_3 v_3$.

Also, recall the formula that relates magnitude of a vector with the dot product:

$$\mathbf{u} \cdot \mathbf{u} = |\mathbf{u}|^2.$$

Let expression A be the magnitude-squared of the cross product. This is

$$|\mathbf{u} \times \mathbf{v}|^2 = (\mathbf{u} \times \mathbf{v}) \cdot (\mathbf{u} \times \mathbf{v})$$
$$= (u_2 v_3 - u_3 v_2)^2 + (u_3 v_1 - u_1 v_3)^2 + (u_1 v_2 - u_2 v_1)^2.$$

Fully expanded, this is:

$$(u_1 v_2)^2 + (u_2 v_1)^2 + (u_1 v_3)^2 + (u_3 v_1)^2 + (u_2 v_3)^2 + (u_3 v_2)^2$$
$$- 2 u_1 u_2 v_1 v_2 - 2 u_1 u_3 v_1 v_3 - 2 u_2 u_3 v_2 v_3.$$

Let expression B be the dot product squared of **u** and **v**. This is

$$(\mathbf{u} \cdot \mathbf{v})^2 = (u_1 v_1 + u_2 v_2 + u_3 v_3)^2$$

Fully expanded, this is

$$(u_1 v_1)^2 + (u_2 v_2)^2 + (u_3 v_3)^2 + 2 u_1 u_2 v_1 v_2 + 2 u_1 u_3 v_1 v_3 + 2 u_2 u_3 v_2 v_3.$$

Let expression C be the product of the magnitudes of **u** and **v** squared. This is

$$(|\mathbf{u}||\mathbf{v}|)^2 = |\mathbf{u}|^2 |\mathbf{v}|^2 = (u_1^2 + u_2^2 + u_3^2)(v_1^2 + v_2^2 + v_3^2).$$

Fully expanded, this is

$$(u_1 v_1)^2 + (u_1 v_2)^2 + (u_1 v_3)^2 + (u_2 v_1)^2 + (u_2 v_2)^2 + (u_2 v_3)^2 \\ + (u_3 v_1)^2 + (u_3 v_2)^2 + (u_3 v_3)^2.$$

Looking carefully at expressions A, B and C, note that A + B = C.

Thus,

$$|\mathbf{u} \times \mathbf{v}|^2 + (\mathbf{u} \cdot \mathbf{v})^2 = (|\mathbf{u}||\mathbf{v}|)^2.$$

Isolate $|\mathbf{u} \times \mathbf{v}|^2$:

$$|\mathbf{u} \times \mathbf{v}|^2 = (|\mathbf{u}||\mathbf{v}|)^2 - (\mathbf{u} \cdot \mathbf{v})^2.$$

Use the geometric formula for the dot product and make the replacement:

$$|\mathbf{u} \times \mathbf{v}|^2 = (|\mathbf{u}||\mathbf{v}|)^2 - (|\mathbf{u}||\mathbf{v}| \cos \theta)^2.$$
$$= (|\mathbf{u}||\mathbf{v}|)^2 - (|\mathbf{u}||\mathbf{v}|)^2 \cos^2 \theta.$$

Factor the $(|\mathbf{u}||\mathbf{v}|)^2$ to the front:

$$|\mathbf{u} \times \mathbf{v}|^2 = (|\mathbf{u}||\mathbf{v}|)^2 (1 - \cos^2 \theta).$$

Recall that $\sin^2 \theta = 1 - \cos^2 \theta$:

$$|\mathbf{u} \times \mathbf{v}|^2 = (|\mathbf{u}||\mathbf{v}|)^2 \sin^2 \theta.$$

Since two vectors never open wider that 180 degrees (π radians), $\sin \theta$ is always non-negative. Thus, taking square roots, we have

$$|\mathbf{u} \times \mathbf{v}| = |\mathbf{u}||\mathbf{v}| \sin \theta.$$

More Discussion on Parametrically-Defined Surfaces in R^3

If a surface is defined **explicitly** in R^3, where one of its variables is isolated (and thus, is dependent), then the surface is, in a sense, already defined parametrically in terms of its other two independent variables. The bounds of the independent (parameter) variables must be stated.

◆ ● ◆ ● ◆ ● ● ◆ ● ◆

Example: Write $z = x^2 + xy^3$ as a parametrically-defined surface in vector form.

Solution: Here, x and y are independent, and z is dependent, and the surface can be written in vector form as $\mathbf{r}(x, y) = \langle x, y, x^2 + xy^3 \rangle$. It is common to use different letters representing parameter variables. Thus, we could let $x = u$ and $y = v$, so that the surface is written in vector form as $\mathbf{r}(u, v) = \langle u, v, u^2 + uv^3 \rangle$, with $-\infty < u < \infty$ and $-\infty < v < \infty$.

◆ ● ◆ ● ◆ ● ● ◆ ● ◆

Example: Write $y = x^2 \sin z$ as a parametrically-defined surface in vector form.

Solution: We have $\mathbf{r}(x, z) = \langle x, x^2 \sin z, z \rangle$ with $-\infty < x < \infty$ and $-\infty < z < \infty$. Variables x and z are independent, y is dependent.

◆ ● ◆ ● ◆ ● ● ◆ ● ◆

Example: Write $x + 2z = 6$ as a parametrically-defined surface in vector form.

Solution: In R^3, the equation $x + 2z = 6$ is a plane. It can be visualized as a line confined to the xz-plane, then allowed to extend into the positive and negative y directions. It may help to rewrite this plane as $x + 0y + 2z = 6$.

If we solve for z, we have $z = 3 - \frac{1}{2}x$. Then, with x and y as the parameter variables, the plane can be written parametrically in vector form as $\mathbf{r}(x, y) = \langle x, y, 3 - \frac{1}{2}x \rangle$ with $-\infty < x < \infty$ and $-\infty < y < \infty$. Although the value of z does not depend on y, we still need y as a parameter variable to account for the portion of the plane that extends in the positive and negative y directions.

Solving for x, we have $x = 6 - 2z$, and with y and z as the parameter variables, the plane can be written as $\mathbf{r}(y, z) = \langle 6 - 2z, y, z \rangle$ with $-\infty < y < \infty$ and $-\infty < z < \infty$.

Example: Write the plane that passes through the points (3,0,0), (0,4,0) and (0,0,8) as a parametrically-defined surface in vector form.

Solution: Recall that a plane whose x, y and z intercepts are $(a,0,0)$, $(0,b,0)$ and $(0,0,c)$ respectively can be written as $\frac{x}{a} + \frac{y}{b} + \frac{z}{c} = 1$. Thus, in this example, we have

$$\frac{x}{3} + \frac{y}{4} + \frac{z}{8} = 1, \quad \text{or} \quad 8x + 6y + 3z = 24.$$

Letting x and y be the parameter variables and solving for z, we can write this plane parametrically as $\mathbf{r}(x,y) = \langle x, y, 8 - \frac{8}{3}x - 2y \rangle$, with $-\infty < x < \infty$ and $-\infty < y < \infty$.

Similarly, letting x and z be the parameter variables and solving for y, we can write this plane parametrically as $\mathbf{r}(x,z) = \langle x, 4 - \frac{4}{3}x - \frac{1}{2}z, z \rangle$ with $-\infty < x < \infty$ and $-\infty < z < \infty$. Lastly, if y and z are the parameter variables, and we solve for x, we have $\mathbf{r}(y,z) = \langle 3 - \frac{3}{4}y - \frac{3}{8}z, y, z \rangle$, with $-\infty < y < \infty$ and $-\infty < z < \infty$. All three vector-value functions parametrically describe the same plane

♦ ♦ ♦ ♦ • • ♦ • ♦

A surface is defined **implicitly** as $g(x,y,z) = 0$, and is usually best suited for cases where isolating any of the variables is difficult. Note that all explicit surfaces can be written implicitly by moving all terms to one side of the equal sign. For example, the surface $z = x^2 + xy^3$ in the first example can be written implicitly as $x^2 + xy^3 - z = 0$.

♦ ♦ ♦ ♦ • • ♦ • ♦

Example: A circular cylinder is defined by a circle or radius 2 centered at the origin on the xy-plane and extending both ways into the z direction. Write this as a parametrically-defined surface in vector form.

Solution: We use polar variables to define the circle, so that $x = 2\cos\theta$ and $y = 2\sin\theta$. Meanwhile, variable z can be any real number. Thus, the cylinder is defined parametrically in terms of θ and z by

$$\mathbf{r}(\theta, z) = \langle 2\cos\theta, 2\sin\theta, z \rangle, \text{ with } 0 \leq \theta \leq 2\pi \text{ and } -\infty < z < \infty.$$

Note that in this setting, the radius of the cylinder is fixed as $r = 2$, so that r as a variable is not needed in this parametrization.

Example: A circular cone has its vertex at the origin and opens in the positive z direction such that $0 \leq z \leq 10$, and that when $z = 10$, the radius of the circular cross section is also 10 units. Write this cone as a parametrically-defined surface in vector form.

Solution: Cones can be defined explicitly. In this case, we have $z = \sqrt{x^2 + y^2}$, and by assuming x and y to be the parameter variables, we can define the cone parametrically as

$$\mathbf{r}(x, y) = \langle x, y, \sqrt{x^2 + y^2} \rangle, \quad \text{where } x^2 + y^2 \leq 100.$$

Alternatively, note that as z increases, so does the radius of the circular cross section of the cone at any given z. For example, when $z = 0$ (the origin), the cone here is just a single point, so that the radius is $r = 0$, and when $z = 10$, the radius of the circular cross section is $r = 10$. This suggests that $z = r$. The circular cross sections can then be defined using polar coordinates, with $x = r \cos \theta$ and $y = r \sin \theta$.

It may be awkward (but not incorrect) to use r as a parameter when we use $\mathbf{r}$ to represent the vector-valued function. To avoid the awkward notation, we will let $r = u$ and $\theta = v$, and the cone can be defined parametrically as

$$\mathbf{r}(u, v) = \langle u \cos v, u \sin v, u \rangle, \quad \text{where } 0 \leq u \leq 10 \text{ and } 0 \leq v \leq 2\pi.$$

This parameterization of the cone takes advantage of the properties of circles and allows for constant bounds of the parameter variables.

◆ • ◆ • ◆ • ◆ • ◆

Example: A circular cone has its vertex at the origin and opens in the positive x direction, where $0 \leq x \leq 8$. Assume that the angle at the vertex is $\frac{\pi}{3}$ radians. Write this cone as a parametrically-defined surface in vector form.

Solution: We need a sense of how fast the radii of the circular cross sections increase as x increases. In the sketch at below, y is 0, so we see a triangular region on the xz-plane, a "shadow" cast by the cone onto this plane.

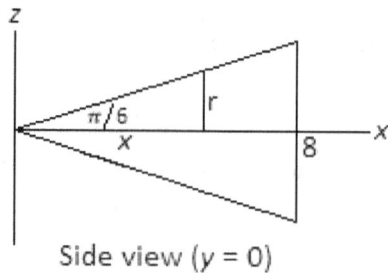

Side view ($y = 0$)

In general, a cone with x as the dependent variable would be written as $x = a\sqrt{y^2 + z^2}$. When $x = 8$ (and assuming $y = 0$), then $\frac{z}{8} = \tan\frac{\pi}{6}$, or $z = 8\left(\frac{\sqrt{3}}{3}\right)$. These values are then substituted into $x = a\sqrt{y^2 + z^2}$, and we find that $a = \sqrt{3}$.

This cone, written in explicit form, is $x = \sqrt{3}\sqrt{y^2 + z^2} = \sqrt{3(y^2 + z^2)}$, and with y and z as parameter variables, the cone is written in vector form as

$$\mathbf{r}(y, z) = \langle \sqrt{3(y^2 + z^2)}, y, z \rangle, \text{ where } y^2 + z^2 \leq 3.$$

Note that the region $y^2 + z^2 \leq 3$ is a circular disk lying in the yz-plane with radius $\sqrt{3}$. This is the "shadow" cast by the cone onto the yz-plane, and we determined the radius when we solved for a.

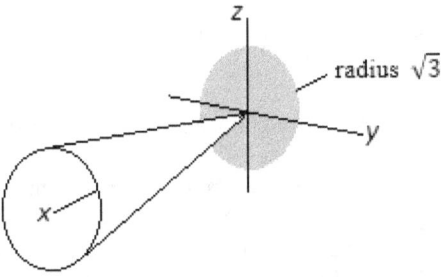

The above representation of the cone is correct, but states the restrictions on y and z implicitly, rather than as separate inequalities with constant bounds. Below, we use an alternative way to describe the cone in which the bounds on the parameter variables are constants.

Since the angle at the vertex is $\frac{\pi}{3}$ radians, then half that, $\frac{\pi}{6}$ radians, is the angle between the x-axis and one side of the cone. From this, we can find the radius of the circular cross-section for any given x:

$$\tan\left(\frac{\pi}{6}\right) = \frac{r}{x}, \quad \text{thus,} \quad r = x \tan\left(\frac{\pi}{6}\right) = \frac{\sqrt{3}}{3}x.$$

To parameterize this cone, we note that the circular cross sections occur in rectangular coordinates y and z, so we convert them to polar coordinates, $y = r\cos\theta$ and $z = r\sin\theta$. Since $r = \frac{\sqrt{3}x}{3}$, we have $y = \frac{\sqrt{3}x}{3}\cos\theta$ and $z = \frac{\sqrt{3}x}{3}\sin\theta$. Thus, the cone is parametrically defined as

$$\mathbf{r}(x, \theta) = \langle x, \frac{\sqrt{3}x}{3}\cos\theta, \frac{\sqrt{3}x}{3}\sin\theta \rangle, \text{ where } 0 \leq x \leq 8 \text{ and } 0 \leq \theta \leq 2\pi.$$

Example: A sphere of radius 7 is centered at the origin. Write this sphere as a parametrically-defined surface in vector form.

Solution: We could describe the sphere as $x^2 + y^2 + z^2 = 49$, so that $z = \pm\sqrt{49 - x^2 - y^2}$. Thus, one possible parametrically defined vector-valued function would be the union of

$$\mathbf{r}(x, y) = \langle x, y, \sqrt{49 - x^2 - y^2}\rangle \text{ and } \mathbf{r}(x, y) = \langle x, y, -\sqrt{49 - x^2 - y^2}\rangle, \text{ with}$$
$$x^2 + y^2 \leq 49.$$

However, this form is clumsy and unsatisfactory since it requires two separate vector-valued functions to describe one surface. Note that a full sphere cannot be written as an explicit function of two variables (roughly speaking, it fails the vertical line test).

To parametrically describe a sphere more elegantly, we use spherical coordinates. Recall that

$$x = \rho \sin \phi \cos \theta, \quad y = \rho \sin \phi \sin \phi, \text{ and } z = \rho \cos \phi.$$

In this example, the spherical radius is fixed at $\rho = 7$. Thus, we can use ϕ and θ as the parameter variables and write the sphere as

$$\mathbf{r}(\phi, \theta) = \langle 7 \sin \phi \cos \theta, 7 \sin \phi \sin \theta, 7 \cos \phi \rangle,$$
$$\text{with } 0 \leq \phi \leq \pi \text{ and } 0 \leq \theta \leq 2\pi.$$

This fully describes the sphere and uses two parameter variables both bound by constants.

◆ ● ◆ ● ◆ ● ◆ ● ◆

Example: Describe what $\mathbf{r}(\phi, \theta) = \langle 4 \sin \phi \cos \theta, 4 \sin \phi \sin \theta, 4 \cos \phi \rangle$, where $0 \leq \phi \leq \frac{\pi}{2}$ and $0 \leq \theta \leq \pi$, represents.

Solution: This describes one quarter of a sphere. If we rethink this in terms of x, y and z, this is the portion of the sphere of radius 4 bounded by the xy-plane (where $z \geq 0$), and by the xz-plane (where $y \geq 0$).

The images are on the next page.

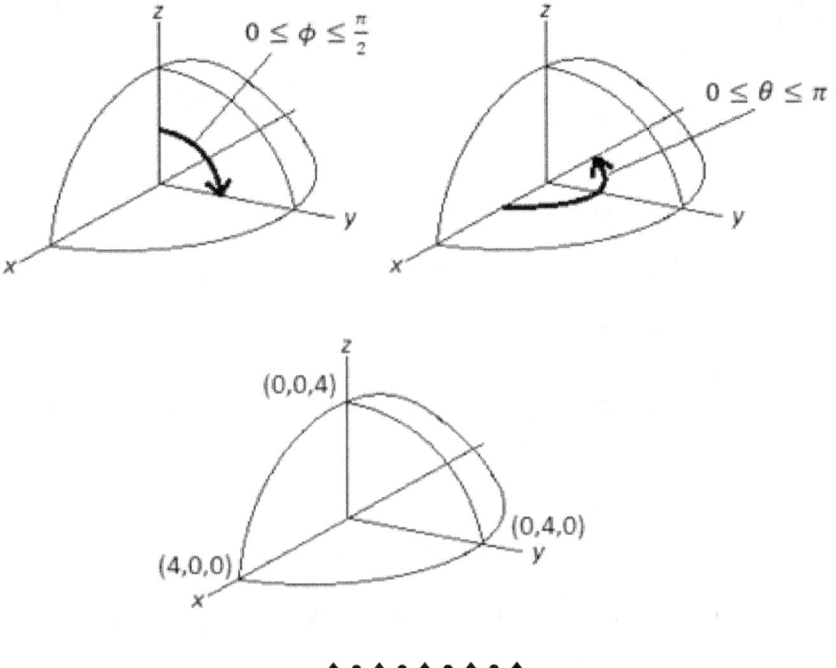

Example: Describe the surface given implicitly by $\mathbf{r}(u,v) = \langle 2u+v+1, u-3v-2, 4u+v \rangle$, where $-\infty \leq u \leq \infty$ and $-\infty \leq v \leq \infty$, then rewrite this same surface in explicit form.

Solution: Since the parameter variables are written as separate terms and raised to the first power, this surface is a plane written in implicit form. To rewrite this plane in explicit form, we can generate three points on this plane by choosing values for u and v. For example,

$$\mathbf{r}(0,0) = \langle 1,-2,0 \rangle, \quad \mathbf{r}(0,1) = \langle 2,-5,1 \rangle \text{ and } \mathbf{r}(1,0) = \langle 3,-1,4 \rangle.$$

These are actually vectors, but their feet are at the origin, and their heads are at the ordered triple representing a point on the plane. Thus, vector $\langle 1,-2,0 \rangle$ corresponds to the point $(1,-2,0)$ on the plane, and so on. From this, two vectors *within* the plane can be generated:

$$\mathbf{v}_1 = \mathbf{r}(1,0) - \mathbf{r}(0,0) = \langle 2,1,4 \rangle \quad \text{and} \quad \mathbf{v}_2 = \mathbf{r}(0,1) - \mathbf{r}(0,0) = \langle 1,-3,1 \rangle.$$

Their cross product is a vector that is normal to the plane:

$$\mathbf{n} = \mathbf{v}_1 \times \mathbf{v}_2 = \langle 13,2,-7 \rangle.$$

(Note: this same normal vector can also be generated by finding $\mathbf{r}_u \times \mathbf{r}_v$. We have $\mathbf{r}_u = \langle 2,1,4 \rangle$ and $\mathbf{r}_v = \langle 1,-3,1 \rangle$, and their cross product is $\mathbf{r}_u \times \mathbf{r}_v = \langle 13,2,-7 \rangle$, as we found above.)

Recall that the general form for a plane passing through (x_0, y_0, z_0) with normal $\mathbf{n} = \langle a, b, c \rangle$ is

$$a(x - x_0) + b(y - y_0) + c(z - z_0) = 0.$$

Thus, the plane passing through $(1, -2, 0)$ and normal to $\langle 13, 2, -7 \rangle$ is

$$13(x - 1) + 2(y + 2) - 7z = 0, \quad \text{or} \quad 13x + 2y - 7z = 9.$$

Solving for z, we have $z = -\frac{9}{7} + \frac{13}{7}x + \frac{2}{7}y$, and this plane can then be described parametrically in explicit form as

$$\mathbf{r}(x, y) = \langle x, y, -\frac{9}{7} + \frac{13}{7}x + \frac{2}{7}y \rangle,$$
$$\text{where } -\infty < x < \infty \text{ and } -\infty < y < \infty.$$

There are other equivalent forms. For example, we could have solved for y and let x and z be the parameter variables.

♦ • • ♦ • • ♦ • • ♦

Example: Rewrite $\mathbf{r}(u, v) = \langle u + v, 2u - v, u^2 + v^2 \rangle$ in explicit form using parameter variables x and y. Assume $-\infty \le u \le \infty$ and $-\infty \le v \le \infty$.

Solution: Let $x = u + v$ and $y = 2u - v$. We then solve for u and v in terms of x and y. Note that we can stack the equations as a system:

$$u + v = x$$
$$2u - v = y.$$

Adding the two equations, the variable v vanishes, and we have $3u = x + y$, so that $u = \frac{1}{3}x + \frac{1}{3}y$. Back substituting, we find that $v = \frac{2}{3}x - \frac{1}{3}y$. Thus,

$$u^2 + v^2 = \left(\frac{1}{3}x + \frac{1}{3}y\right)^2 + \left(\frac{2}{3}x - \frac{1}{3}y\right)^2$$
$$= \frac{1}{9}x^2 + \frac{2}{9}xy + \frac{1}{9}y^2 + \frac{4}{9}x^2 - \frac{4}{9}xy + \frac{1}{9}y^2$$
$$= \frac{5}{9}x^2 - \frac{2}{9}xy + \frac{2}{9}y^2.$$

Thus, the same surface written parametrically in explicit form is

$$r(x,y) = \langle x, y, \frac{5}{9}x^2 - \frac{2}{9}xy + \frac{2}{9}y^2 \rangle,$$

where $-\infty < x < \infty$ and $-\infty < y < \infty$. This is an elliptical paraboloid that opens in the positive z direction.

◆ • ◆ • ◆ • ◆ • ◆ •

Geometrical Proofs of the Jacobians: Polar and Spherical Cases

Polar case: Consider a point in R^2 that is r units from the origin at an angle of θ relative to the positive x-axis:

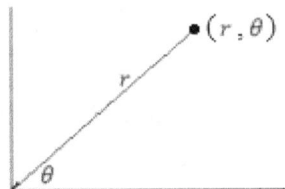

Now, extend the radius by a small amount, Δr, and the angle by a small amount, $\Delta\theta$. This creates an area element called a "polar rectangle", the shaded region shown below:

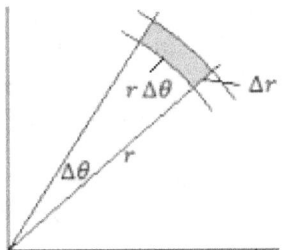

The formula for length of a circular arc of radius r subtended by an angle θ is $S = r\theta$. Thus, the length of the small arc of radius r and angle θ is $r\Delta\theta$.

When both Δr and $\Delta\theta$ are small, the region is approximated well by a rectangle. Using the length times width area formula for a rectangle, we have that the area of this polar rectangle is approximately $\Delta A \approx (r\Delta\theta)(\Delta r) = r\, \Delta r\, \Delta\theta$.

Allowing both differentials to trend to zero as a limit results in the Jacobian $dA = r\, dr\, d\theta$.

Spherical case: A point in R^3 is given by (ρ, θ, ϕ). Each coordinate is extended by small amounts, $\Delta\rho$, $\Delta\theta$ and $\Delta\phi$, respectively.

In the image below, small changes in ρ and ϕ are shown. This allows us to determine the length of two of the three sides of a "spherical rectangular solid". The formula $S = r\theta$ was used to find the length $\rho\Delta\phi$.

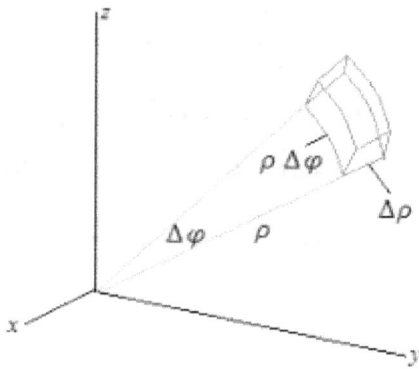

Now, we study the effect of a small change in θ:

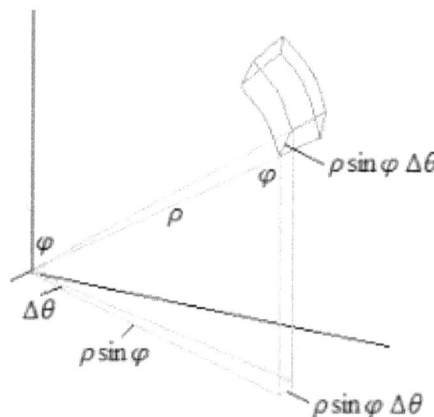

We see a right triangle with ρ as the hypotenuse and by transversals, ϕ is the angle shown nearest the cube figure. Projecting to the xy-plane, the opposite side is $\rho \sin \phi$. Then, with $\Delta\theta$ being the small change in θ, we obtain a small arc length on the xy-plane, $\rho \sin \phi \, \Delta\theta$. This is the third side of the spherical rectangular solid.

The volume of this spherical rectangular solid is approximately $dV = (\Delta\rho)(\rho\Delta\phi)(\rho \sin \phi \, \Delta\theta)$. Allowing the differentials to trend to zero as a limit results in the Jacobian $dV = \rho^2 \sin \phi \, d\rho \, d\theta \, d\phi$.

www.ingramcontent.com/pod-product-compliance
Lightning Source LLC
Chambersburg PA
CBHW071348210526
45465CB00001B/15